Ne

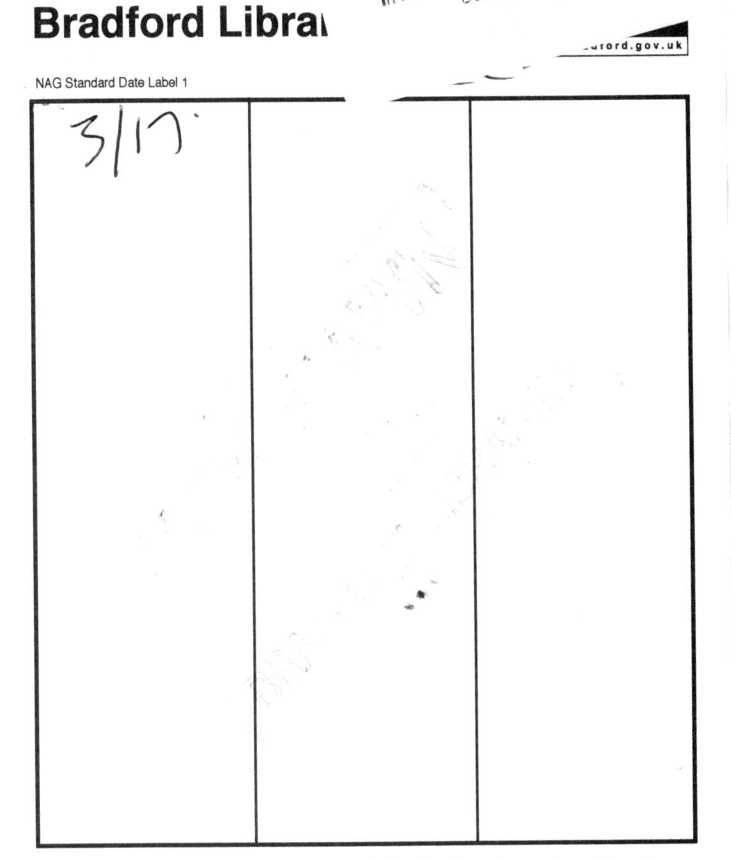

Manchester University Press

Network neutrality

From policy to law to regulation

Christopher T. Marsden

Manchester University Press

Published by Manchester University Press
Altrincham Street, Manchester M1 7JA
www.manchesteruniversitypress.co.uk

British Library Cataloguing-in-Publication Data
A catalogue record for this book is available from the British Library

Library of Congress Cataloging-in-Publication Data applied for

ISBN 978 1 5261 0727 5 hardback
ISBN 978 1 5261 0548 6 paperback
ISBN 978 1 5261 0547 9 open access

First published 2017

Typeset by Out of House Publishing

Printed in Great Britain
by CPI Group (UK) Ltd,
Croydon CR0 4YY

Table of contents

Figures

Tables

European Court and competition cases

C(2007) 5617 State aid C-55/2007 (ex NN 63/2007 (ex CP 106/2006)) – United Kingdom Crown guarantee to BT Pension Scheme, Brussels, 28 November 2007

C-58/08 *Vodafone and Others v. Secretary of State for Business, Enterprise and Regulatory Reform*, 3 CMLR 44

C-360/10, Reference for a preliminary ruling under Article 267 TFEU from the rechtbank van eerste aanleg te Brussel (Belgium), made by decision of 28 June 2010, received at the Court on 19 July 2010, in the proceedings *Belgische Vereniging van Auteurs, Componisten en Uitgevers CVBA (SABAM) v. Netlog NV*, OJ C288/18

C-461/10 *Bonnier Audio AB and others v. Perfect Communication Sweden AB* [2010] OJ C317/24; final judgment 19 April 2012

C(2011) 7279, Case No COMP/M.6281 Microsoft-Skype, Commission decision pursuant to Article 6(1)(b) of Council Regulation No 139/2004, Brussels, 7 October 2011

C-131/12, *Google Spain SL and Google Inc. v. Agencia Española de Protección de Datos (AEPD) and Mario Costeja González*, of 13 May 2014

Joined cases C-293/12, *Digital Rights Ireland v. Minister for Communications, Marine and Natural Resources and Others* and C-594/12 *Kärntner Landesregierung and Others*, ECLI:EU:C:2014:238, judgment of 8 April 2014

C(2012) 8223 final State aid SA.33671 (2012/N) United Kingdom National Broadband scheme for the UK – Broadband Delivery UK

C-620/13 P – *British Telecommunications v. Commission*, Judgment of the Court (Second Chamber) of 22 October 2014

C-362/14, *Maximillian Schrems v. Data Protection Commissioner*, Judgment of Grand Chamber 6 October 2015, ECLI:EU:C:2015:650

COMP/M.7612 *Hutchison 3G UK/Telefonica UK*, OJ C310/05 of 19 September 2015

European Commission Communications

COM(2004) 28 Communication on unsolicited commercial communications or 'spam', 22 January 2004

COM(2006) 334 Communication on the review of the EU Regulatory Framework for electronic communications networks and services, 29 June 2006

COM(2007) 698 Proposal for a Directive of the European Parliament and of the Council amending Directive 2002/22/EC on universal service and users' rights relating to electronic communications networks, Directive 2002/58/EC concerning the processing of personal data and the protection of privacy in the electronic communications sector and Regulation (EC) No 2006/2004 on consumer protection cooperation

COM(2010) 245 A Digital Agenda for Europe, 19 May 2010

COM(2010) 253 (final) Progress Report on the Single European Electronic Communications Market 2009 (15th Implementation Report), 25 May 2010

COM(2011) 206 Single Market Act, Twelve levers to boost growth and strengthen confidence, 'Working together to creat new growth'

COM(2011) 222 Communication on the open internet and net neutrality in Europe, 19 April at 2011

COM(2011) 942 A coherent framework for building trust in the Digital Single Market for ecommerce and online services

COM(2012) 09 Safeguarding Privacy in a Connected World: A European Data Protection Framework for the 21st Century, 25 January 2012

COM(2012) 11 Proposal for a Regulation on the protection of individuals with regard to the processing of personal data and on the free movement of such data (General Data Protection Regulation), 25 January 2012

COM(2012) 573 Single Market Act II, Together for new growth, 3 October 2012

COM(2013) 627 Proposal for a Regulation laying down measures concerning the European single market for electronic communications and to achieve a Connected Continent, and amending Directives 2002/20/EC, 2002/21/EC and 2002/22/EC and Regulations (EC) No 1211/2009 and (EU) No 531/2012, 11 September 2013

COM(2013) 846 final, Rebuilding Trust in EU–US Data Flows, 27 November 2013

COM(2013) 847 Communication on the functioning of the Safe Harbour from the Perspective of EU Citizens and Companies Established in the EU, 27 November 2013

COM(2015) 192 Proposal for a Regulation laying down measures concerning the European single market for electronic communications and to achieve a Connected Continent, and amending Directives 2002/20/EC, 2002/21/EC and 2002/22/EC and Regulations (EC) No 1211/2009 and (EU) No 531/2012 (first reading) – Adoption a) of the Council's position b) of the statement of the Council's reasons Brussels, 29 September 2015

European Commission Decisions

Decision 2000/520/EC of 26 July 2000 pursuant to Directive 95/46/EC of the European Parliament and of the Council on the adequacy of the protection provided by the safe harbour privacy principles and related frequently asked questions issued by the US Department of Commerce, OJ L215/7, 25 August 2000

Decision 2009/703/EC concerning the State aid C 55/07 (ex NN 63/07, CP 106/6) UK Crown guarantee to BT Pension Scheme, OJ L242/21, 11 February 2009

Decision of 10 August 2010 setting up the Expert Group on the Internet of Things, OJ C217/10, 11 August 2010

European Directives, Recommendations and Regulations

Directive 95/46/EC of 24 October 1995 on the protection of individuals with regard to the processing of personal data and on the free movement of such data, OJ L281/31, 11 November 1995

Directive 98/34/EC of 22 June laying down a procedure for the provision of information in the field of technical standards and regulations, OJ L204/37, 21 July 1998

Directive 98/48/EC of 20 July 1998 amending Directive 98/34/EC laying down a procedure for the provision of information in the field of technical standards and regulations, OJ L217/18, 5 August 1998

Directive 2000/31/EC of 8 June 2000 on certain legal aspects of information society services, in particular electronic commerce, in the Internal Market, OJ L178/1, 7 July 2001 (Electronic Commerce Directive)

Directive 2002/19/EC of 7 March 2002 on access to, and interconnection of, electronic communications networks and associated facilities (Access Directive), OJ L108/7, 24 April 2002

Directive 2002/20/EC of 7 March 2002 on the authorisation of electronic communications networks and services (Authorisation Directive), OJ L108/21, 24 April 2002

Directive 2002/21/EC of the European Parliament and of the Council of 7 March 2002, on a common regulatory framework for electronic communications networks and services (Framework Directive), OJ L108/33, 24 April 2002

Directive 2002/22/EC of 7 March 2002 on universal service and users' rights relating to electronic communications networks and services (Universal Service Directive), OJ L108/51, 24 April 2002

Directive 2002/58/EC of 12 July 2002 concerning the processing of personal data and the protection of privacy in the electronic communications sector (Directive on privacy and electronic communications), OJ L201/37, 31 July 2002 (E-Privacy Directive)

Directive 2006/24/EC of 15 March 2006 on the retention of data generated or processed in connection with the provision of publicly available electronic communications services or of public communications networks and amending Directive 2002/58/EC, OJ L105/54, 13 April 2006 (Data Retention Directive)

Directive 2009/136/EC of 25 November 2009 amending Directive 2002/22/EC on universal service and users' rights relating to electronic communications networks and services, Directive 2002/58/EC concerning the processing of personal data and the protection of privacy in the electronic communications sector and Regulation (EC) No 2006/2004 on cooperation between national authorities responsible for the enforcement of consumer protection laws (Citizens' Rights Directive) OJ L337/11, 18 December 2009

Directive 2009/140/EC of 25 November 2009 amending Directives 2002/21/EC on a common regulatory framework for electronic communications networks and services, 2002/19/EC on access to, and interconnection of, electronic communications networks and associated facilities, and 2002/20/EC on the authorisation of electronic communications networks and services (Better Regulation Directive) OJ L337/37, 18 December 2009

Directive 2010/13/EU, of 10 March 2010 on the coordination of certain provisions laid down by law, regulation or administrative action in Member States concerning the provision of audiovisual media services (Audiovisual Media Services Directive) (codified version), OJ L95/1, 15 April 2010

Recommendation of 17 December 2007 on relevant product and service markets within the electronic communications sector susceptible to *ex ante* regulation in accordance with Directive 2002/21/EC of the European Parliament and of the Council on a common regulatory framework for electronic communications networks and services, OJ L344/65, 28 December 2007

Recommendation C(2009) 3359 final of 7 May 2009 on the Regulatory Treatment of Fixed and Mobile Termination Rates in the EU

Recommendation CM/Rec(2010) 13 of 23 November 2010 on the protection of individuals with regard to automatic processing of personal data in the context of profiling

Recommendation C(2013) 5761 final of 11 September 2013 on consistent non-discrimination obligations and costing methodologies to promote competition and enhance the broadband investment environment

Recommendation 2014/710/EU of 9 October 2014 on relevant product and service markets within the electronic communications sector susceptible to *ex ante* regulation in accordance with Directive 2002/21/EC of the European Parliament and of the Council on a common regulatory framework for electronic communications networks and services, OJ L295/79, 11 October 2014

Regulation (EC) 717/2007 of 27 June 2007 on roaming on public mobile telephone networks within the Community and amending Directive 2002/21/EC, OJ L171/32, 29 June 2007

Regulation (EC) 1211/2009 of 25 November 2009 establishing the Body of European Regulators for Electronic Communications (BEREC) and the Office, OJ L337/1, 18 December 2009

Regulation (EU) 531/2012 of 13 June 2012 on roaming on public mobile communications networks within the Union, OJ L172/10, 30 June 2012

Regulation (EU) 2015/2120 of 25 November 2015 laying down measures concerning open internet access and amending Directive 2002/22/EC on universal service and users' rights relating to electronic communications networks and services and Regulation (EU) No 531/2012 on roaming on public mobile communications networks within the Union, OJ L301/1, 6 November 2015

Regulation (EU) 2016/679 of 27 April 2016 on the protection of natural persons with regard to the processing of personal data and on the free movement of such data, and repealing Directive 95/46/EC (General Data Protection Regulation), OJ L119/1, 4 May 2016

European Institution documents and public releases

European Commission documents

15th Implementation Report, Staff Working Document Part 1, 2010

Background for the public consultation on the regulatory environment for platforms, online intermediaries, data and cloud computing and the collaborative economy, 2015

Blueprint for next-generation quality-enabled network interconnection, Digital Agenda Newsroom, 3 June 2014

Community Guidelines for the application of State aid rules in relation to rapid deployment of broadband networks, OJ C235/7 30 September 2009

Conclusions from the Internet of Things public consultation, 2013.

Consultation on the Commission's comprehensive approach on personal data protection in the European Union, 4 November 2012

H2020-ICT-2015 Collective Awareness Platforms for Sustainability and Social Innovation, 2015

Internet Governance Forum 2015: Joint Declaration from European Commission Vice-President Andrus Ansip and Members of the European Parliament, 12 November

MEMO-15–5275 of 27 October 2015 Roaming charges and Open Internet: questions and answers

Public consultation on the regulatory environment for platforms, online intermediaries, data and cloud computing and the collaborative economy, 24 September 2015

Quality of Broadband Services in the EU: March 2012, contracted to SamKnows with Contract number: 30-CE-0392545/00-77; SMART 2010/0036. ISBN 978-92-79-30933-5. DOI: 10.2759/24341

SEC(2007) 1472 Commission Staff Working Document: Impact Assessment, 13 November 2007

SWD(2013) 153 final, Commission Staff Working Document, E-commerce Action Plan 2012–2015, State of Play 2013, 23 April 2013

SWD(2013) 331 final, Impact Assessment accompanying the document Proposal for a Regulation laying down measures concerning the European single market for electronic communications and to achieve a Connected Continent, and amending Directives 2002/20/EC, 2002/21/EC and 2002/22/EC and Regulations (EC) No 1211/2009 and (EU) No 531/2012, 11 September 2013

European Commission Press Releases

IP/09/570 Telecoms: Commission launches case against UK over privacy and personal data protection, 14 April 2009

IP/09/1626 Telecoms: Commission steps up UK legal action over privacy and personal data protection, 29 October 2009

IP/10/1215 Digital Agenda: Commission refers UK to Court over privacy and personal data protection, 30 September 2010

IP/10/1482 Digital Agenda: consultation reveals near consensus on importance of preserving open Internet, 9 November 2010

IP/10/860 Digital Agenda: Commission launches consultation on net neutrality, 30 June 2010

IP/11/905 European Commission, Digital Agenda: Commission starts legal action against 20 Member States on late implementation of telecoms rules, 19 July 2011

IP/12/46 Commission proposes a comprehensive reform of data protection rules to increase users' control of their data and to cut costs for businesses, 25 January 2012

IP/12/60 Digital Agenda: Commission closes infringement case after UK correctly implements EU rules on privacy in electronic communications, 26 January 2012

IP/12/1244 State aid: Commission clears UK umbrella support scheme for broadband investment 'BDUK', 21 November 2012

IP/14/1089 Antitrust: Commission closes investigation into internet connectivity services but will continue to monitor the sector, 3 October 2014

IP/15/5927 Bringing down barriers in the Digital Single Market: no roaming charges as of June 2017, 27 October 2015

Speech by Margrethe Vestager: Competition in telecom markets, 42nd Annual Conference on International Antitrust Law and Policy, Fordham University, 2 October 2015

SPEECH 13/498 Neelie Kroes: The EU, safeguarding the open internet for all, European Parliament: Guaranteeing Competition and the Open Internet in Europe, Brussels, 4 June 2013

SPEECH 11/285 The internet belongs to all of us, Neelie Kroes, European Commission Vice-President for the Digital Agenda, Press conference on Net Neutrality Communication, 19 April 2011

European Council documents

European Council, Resolution of 17 January 1995 on the lawful interception of telecommunications, OJ C329/1, 4 November 1996

European Parliament documents

Final report on the existence of a global system for the interception of private and commercial communications (ECHELON Interception System), Temporary Committee on the ECHELON Interception System, approved 5 September 2001

Parliamentary Questions, E-006146-13, Question for written answer to the Commission by Josef Weidenholzer (S&D), 31 May 2013

Parliamentary Questions, E-004461/2015, Answer given by Mr Oettinger on behalf of the Commission, 10 June

European regulatory reports

Article 29 Working Party

Letter addressed to Ms Le Bail to deliver input to the Commission on the current practices at national level, the problems encountered in implementing the Directive as well as some suggestions for improvements or changes in relation to special categories of data ('sensitive data'), notification and the practical implementation of the Article 28(6) of the Directive 95/46/EC, 20 April 2011

Opinion 2/2010 on online behavioural advertising WP 171, 22 June 2010

Opinion 16/2011 on EASA/IAB Best Practice Recommendation on Online Behavioural Advertising WP 188, 8 December 2011

Press Release Brussels, 15 December 2011

WP203, 00569/13/EN Opinion 03/2013 on purpose limitation

BEREC

BoR (10) 42 BEREC Response to the European Commission's consultation on the open Internet and net neutrality in Europe, 30 September 2010 at www.erg.eu.int/doc/berec/bor_10_42.pdf

BoR (11) 44 Draft BEREC Guidelines on Net Neutrality and Transparency, 3 October 2011

BoR (11) 53 A framework for Quality of Service in the scope of Net Neutrality, 8 December 2011

BoR (11) 67 Guidelines on transparency as a tool to achieve net neutrality, 14 December 2011

BoR (12) 30 A view of traffic management and other practices resulting in restrictions to the open Internet in Europe – Findings from BEREC's and the European Commission's joint investigation, 29 May 2012

BoR (12) 31 Differentiation practices and related competition issues in the scope of Net Neutrality – Draft report for public consultation, 29 May 2012

BoR (12) 32 BEREC Guidelines for Quality of Service in the scope of Net Neutrality – Draft for public consultation, 29 May 2012

BoR (12) 33 An assessment of IP-interconnection in the context of Net Neutrality – Draft report for public consultation, 29 May 2012

BoR (12) 34 BEREC public consultations on Net Neutrality Explanatory paper, 29 May 2012

BoR (12) 120 Statement with observations about net neutrality for ETNO's proposal to International Telecommunications Union (ITU) World Conference on International Telecommunications, 14 November 2012

BoR (12) 131 Guidelines for quality of service in the scope of net neutrality, 26 November 2012

BoR (12) 132 Differentiation practices and related competition issues in the scope of net neutrality: Final report, 26 November 2012

BoR (13) 117 Ecosystem Dynamics and Demand Side Forces in Net Neutrality: Progress Report and Decision on Next Steps, 17 September 2013

BoR (14) 50 BEREC views on the European Parliament first reading legislative resolution on the European Commission's proposal for a Connected Continent Regulation, 16 May 2014

BoR (14) 117 Monitoring quality of Internet access services in the context of net neutrality, 25 September 2014

BoR (15) 72 Fixed and mobile termination rates in the EU January 2015, 25 May 2015

BoR (15) 90 Report on how consumers value net neutrality, 8 June 2015

BoR (15) 190 Draft Agenda for the 25th meeting of the BEREC Board of Regulators 10 December 2015, London (United Kingdom), hosted by Ofcom, 1 December 2015

BoR (15) 213 Work Programme 2016: BEREC Board of Regulators, 10 December 2016

BoR (15) 226 Statement on BEREC's work to produce guidelines for the implementation of net neutrality provisions of the TSM regulation, 15 December 2015

BoR (16) 49 Update on BEREC work to produce guidelines for the implementation of net neutrality provisions of the TSM regulation, 24 February 2016

Outcomes of the BEREC–EMERG–EAPEREG–REGULATEL Summit 3 July 2015

Questions for BEREC stakeholder dialogue with representatives of end-users/ consumers and civil society, 24 November 2015

Statement on the publication of a European Commission proposal for a Regulation, 2013

Views on the European Parliament first reading legislative resolution on the proposal for a Regulation, 2014

European Data Protection Supervisor (EDPS)

Opinion on net neutrality, traffic management and protection of privacy and personal data, 2011

Opinion on the Proposal for a Regulation of the European Parliament and of the Council laying down measures concerning the European single market for electronic communications and to achieve a Connected Continent, and amending Directives 2002/20/EC, 2002/21/EC and 2002/22/EC and Regulations (EC) No 1211/2009 and (EU) No 531/2012, 2013

United Kingdom Regulator reports

Competition and Markets Authority

BT Group/EE merger inquiry, 2015
BT–EE Provisional findings report: Appendices, 30 October 2015
Online reviews and endorsements, Opened: 26 February 2015
Commercial use of consumer data, Opened: 27 January 2015
CAT refers superfast broadband price control appeals to CMA, 6 January 2016

Competition Commission

Report on the charges made by mobile operators for terminating calls – Reports on references under section 13 of the Telecommunications Act 1984 on the charges made by Vodafone, O2, Orange and T-Mobile for terminating calls from fixed and mobile networks, 18 February 2003

Ofcom

12th Annual Communications Market Report, 2015
Average UK broadband speed continues to rise, 7 August 2013
BBC new on-demand video proposals: Market Impact Assessment, 2006
BT Pensions Statement, 2010
ConnectedNations, 2015
Consumer research into the transparency of traffic management information provided by ISPs, 4 September 2013
CW/00946/02/07 Own-Initiative Enforcement Programme to Give Effect to General Condition 22 (Service Migrations), 2007
Draft Annual Plan 2012/13, 2012
Future regulation of on-demand programme services: Consultation published 18 December 2015

Guidance on individual General Conditions: links to Guidance on the General Conditions (undated)
Mobile call termination market review 2015–18: draft decision, 6 February 2015
Ofcom's approach to net neutrality, 2011
Response to the European Commission Consultation on Content Online in the Single Market, 20 October 2006
Traffic Management and net neutrality, a Discussion Document, 24 June 2010
UK home broadband performance, November 2015: the performance of fixed-line broadband delivered to UK residential customers, 24 March 2016
Voluntary Code of Practice: Broadband Speeds, 2008

Oftel

Beyond the telephone, the television and the PC, Culture, Media And Sport Committee Inquiry Into Audio-Visual Communications And The Regulation Of Broadcasting, March 1998
Consumers' use of mobile telephony, 2002
Press Statement Ref 02-01, 2002
Mobile Numbering Consultation, 1996, Appendix 1
Statement on mobile termniation rates, 26 September 2001

United Kingdom court cases

Coggs v. Bernard (1703) 2 Ld Raym 909, 13 William III

Colt v. Office of Communications [2013] CAT 29

Edward Darcy Esquire v. Thomas Allin of London Haberdasher (1599) 74 ER 1131

Everything Everywhere Limited v. Competition Commission, Office of Communications, Hutchison 3G (UK) Limited, British Telecommunications plc [2013] EWCA Civ 154 of 6 March

Everything Everywhere Limited v. Ofcom (Mobile Call Termination) [2013] EWCA Civ 154

Godfrey v. Demon Internet Service [2001] QB 201

Lane v. Cotton (1701) 1 Ld Raym 646

Peek v. The North Staffordshire Railway Company (1863) 10 HLC 473

R (T-Mobile, Vodafone, Orange) v. Competition Commission [2003] EWHC 1566 (Admin) [2003] EuLR 769

Shetland Times Ltd v. Jonathan Wills and Another, 1997 FSR (Ct Sess. OH), 24 October 1996

Southcote's Case (1601) 4 Co Rep 83b; Cro Eliz 815

Spillers and Bakers Ltd v. Great Western Railway Company (1911) 1 KB 386

Telefónica O2 UK Limited v. Ofcom [2012] EWCA Civ 1002

Trustees of the BT Pension Scheme v. HMRC FTC/91 & 92/2011 [2013] UKUT 105 (TCC) (28 February 2013)

Trustees of the BT Pension Scheme v. HMRC [2015] EWCA Civ 713

Vaughan v. Taff Vale Railway (1860), 5 H&N 679 157 ER 1351

United Kingdom statutes

Carriers Act 1830
Communications Act 2003
Data Protection Act 1998
Digital Economy Act 2010
Office of Communications Act 2002
Railway and Canal Traffic Act 1854
Railway Regulation Act 1844
Railways Act 1993
Regulation of Investigatory Powers Act 2000
Statute of Monopolies 1623 c. 3 (Regnal. 21_Ja_1)
Telecommunications Act 1984
Transport Act 1962

Statutory instruments

SI 2000/2665 Investigatory Powers Tribunal Rules
SI 2003/1372 Competition Appeal Tribunal Rules 2003
SI 2003/1900 Communications Act 2003 (Commencement No. 1) Order 2003
SI 2003/2426 Privacy and Electronic Communications (EC Directive) Regulations 2003
SI 2003/3142 Office of Communications Act 2002 (Commencement No. 3) and Communications Act 2003 (Commencement No. 2) Order 2003
SI 2004/2068 Competition Appeal Tribunal (Amendment and Communications Act Appeals) Rules 2004
SI 2011/1340 Regulation of Investigatory Powers (Monetary Penalty Notices and Consents for Interceptions) Regulations 2011
SI 2015/1648 Competition Appeal Tribunal Rules 2015

United States cases

American Civil Liberties Union v. Reno (1997) 21 US 844 of 27 June No. 96–511
Comcast v. FCC (2010) No. 08-1291, delivered 6 April
Cubby v. CompuServe (1991) 766 F. Supp 135
Dowling v. United States (1985) 473 US 207, 222
National Cable & Telecommunications Association et al. v. Brand X Internet Services et al. (2005) 545 U.S. 967
Playboy Enterprises, Inc. v. Frena (1993), 839 F. Supp. 1552 (M.D. Fla.)
Stratton Oakmont Inc v. Prodigy (1995) NY Misc. 23 Media L. Rep. 1794
United States v. Domain Names (defendants in rem) (31 January 2011) (No. 18 MAG 262) Affidavit in Support of Application for Seizure Warrant Pursuant to 18 USC §§2323, 981
Verizon v. Federal Communications Commission 740 F.3d 623 (D.C. Cir. 2014); 11–1355
Western Union Telegraph Co v. Call Publishing Co. 181 US 92 (1901)

United States legislation

American Recovery and Reinvestment Act 2009

Clayton Antitrust Act of 1914 Pub.L. 63–212, 38 Stat. 730, enacted 15 October 1914, codified at 15 U.S.C. §§ 12–27, 29 U.S.C. §§ 52–53

Communications Act of 1934 as amended by Communications (Deregulatory) Act of 1996, 47 USC

Communications Assistance to Law Enforcement Act of 1994 Pub. L. No. 103–414, 108 Stat. 4279

Copyright Act 1976, 17 USC

Digital Millennium Copyright Act 1998

Foreign Intelligence Surveillance Act of 1978

Foreign Intelligence Surveillance Act of 1978 Amendments Act of 2008, H.R. 6304, Stat. 2436, Public Law 110–261

Online Copyright Infringement Liability Limitation Act (OCILLA) 1998, Section 512 to Title 17 USC

Pacific Telegraph Act of 1860, 18 June, 36 Cong., 1 Sess., Chapter 137, Section 2

Sherman Act 1890, 26 Stat. 209, 15 U.S.C. §§ 1–7

Telecommunications Act 1934 47 USC

Federal Communications Commission (FCC) documents

Formal Complaint of Free Press and Public Knowledge Against Comcast Corporation for Secretly Degrading Peer-to-Peer Applications; Broadband Industry Practices; Petition of Free Press et al. for Declaratory Ruling that Degrading an Internet Application Violates the FCC's Internet Policy Statement and Does Not Meet an Exception for 'Reasonable Network Management', File No. EB-08-IH-1518, WC Docket No. 07-52, Memorandum Opinion and Order, 23 FCC Rcd 13028, 13054, 13057, 2008

In AT&T Inc and BellSouth Corp, Application for Transfer of Control, 22 FCC Rcd 5562, 2007

In AT&T Inc and BellSouth Corp Application for Transfer of Control, 22 FCC Rcd 5562, 2008

Internet Policy Statement 05–151, 2005

In the Matter of Applications of AT&T Inc. and DIRECTV For Consent to Assign or Transfer Control of Licenses and Authorizations MB Docket No. 14-90, FCC 15-94, 2015

In the Matter of Applications of Comcast Corporation, General Electric Company and NBC Universal, Inc. For Consent to Assign Licenses and Transfer Control of Licensees, MB Docket No. 10-56, FCC 11-4FCC (2011) MB Docket No. 10-56, FCC 11-4, 2011

In the Matter of Further Inquiry into Two Under-Developed Issues in the Open Internet Proceeding Preserving the Open Internet, Report and Order, GN Docket No. 09-191 Broadband Industry Practices WC Docket No. 07-52, , 25 FCC Rcd 17905, 17910, 2010

In the Matter of Protecting and Promoting the Open Internet GN Docket No. 14–28, 29 FCC Rcd 5561, 2014

In the Matter of Protecting and Promoting the Open Internet, GN Docket No. 14–28 Report and Order on Remand, Declaratory Ruling, and Order

Adopted: 26 February 2015, Released 12 March. FCC 15-24 (Open Internet Order)

Madison River Communications, LLC, Order, DA 05–543, 20 FCC Rcd 4295, 2005

Memorandum Opinion and Order, 23 FCC Rcd 13028, 2008

Notice of proposed rulemaking, Adopted: 15 May 2014, Comment Date: 15 July 2014, Reply Comment Date: 10 September 2014

Open Internet Advisory Committee Annual Report, Released 20 August 2013

Open Internet Advisory Committee, Policy Issues in Data Caps and Usage-Based Pricing, Economic Impacts of Open Internet Frameworks Working Group, 2013

Open Internet Advisory Opinion Procedures, Protecting and Promoting the Open Internet, GN Docket No. 14-28, 2 July 2015

Preserving the Open Internet Broadband Industry Practices, GN Docket No. 09-191, WC Docket No. 07-52, FCC10-201, 2010

Report and Order, In the Matter of Preserving the Open Internet; Broadband Industry Practices; GN Docket No. 09-191, WC Docket No. 07-52, FCC 09-93, 21 December 2009 (Released 23 December)

Report and Order Preserving the Open Internet, 25 FCC Rcd 17905, 2010

Report on a Rural Broadband Strategy, 22 May 2009

Other national laws, regulatory decisions and guidelines

Brazil

Law No. 12.965, 23 April 2014 by the Presidency of the Republic, Civil House Legal Affairs Subsection

Decreto No. 8.771, de 11 de maio de 2016 de Regulamenta a Lei no. 12.965, de 23 de abril de 2014, para tratar das hipóteses admitidas de discriminação de pacotes de dados na internet e de degradação de tráfego, indicar procedimentos para guarda e proteção de dados por provedores de conexão e de aplicações, apontar medidas de transparência na requisição de dados cadastrais pela administração pública e estabelecer parâmetros para fiscalização e apuração de infrações

Canada

Railways Act 1906 R. S. C. (1906)

Telecommunications Act S.C. 1993, c. 38 Assented to 1993-06-23

Canadian Radio-television and Telecommunications Commission (CRTC)

CRTC 2008–108 Telecom Decision: The Canadian Association of Internet Providers' application regarding Bell Canada's traffic shaping of its wholesale Gateway Access Service, File no. 8622-C51-200805153, Ottawa, 20 November 2008

CRTC 2009–657 Review of the Internet traffic management practices of Internet service providers, Telecom Regulatory Policy, File No. 8646-C12-200815400, Ottawa, 21 October 2009

CRTC 2011–703, Withdrawal of ITMP Letter: Billing practices for wholesale residential high-speed access services, leading to abandonment of explicit P2P traffic throttling

Statement from Jean-Pierre Blais, Chairman of the CRTC, on monetary penalties and paper bill fees, 17 December 2014

Unlimited Music Service 8661-P8-201510199 Telecom Commission Letter addressed to David Watt (Rogers Communications Inc) and Dennis Béland (Québécor Média), 2016

Chile

Decree 368 of 15 December 2010

Ley 20.453 de 18 de agosto que consagra el principio de neutralidad en la red para los consumidores y usuarios de Internet 2010

Costa Rica

Andrés Oviedo Guzmán, Fabio Isaac Masís Fallas y Juan Manuel Campos Ávila, v. Ministerio De Ambiente, Energía y Telecomunicaciones, Ministerio De La Presidencia Sala Constitucional De La Corte Suprema De Justicia. San José, 13 July 2010

Finland

Information Society Code (917/2014) Part I, Ministry of Transport and Communications, 2014

France

Conseil d'Etat Decision No. 360397/360398 of 10 July 2013

India

Prohibition of Discriminatory Tariffs for Data Services Regulations No. 2 of 2016

Telegraphy Act 1885

Israel

Ministry of Communications, Internet (over-the-top) services and challenges to regulation, 2015

The Netherlands

Aanhangsel Handelingen II (Appendix Official Report), 2008/09.

Netherlands Department of Economic Affairs, Net Neutrality Guidelines, 15 May 2015,, for the Authority for Consumers and Markets (ACM) for the enforcement by ACM of Article 7.4a of the Netherlands Telecommunications Act 2012 (official translation)

Telecommunications Act 2012, official translation by the Dutch government available at www.government.nl/documents/policy-notes/2012/06/07/dutch-telecommunications-act (Accessed 15 September 2016)

Norway

Act No. 54 of of 14 June 2013 amending Act No. 83 of 4 July 2003 relating to electronic communications (The Electronic Communications Act)

Norwegian Communications Authority, BEREC and net neutrality, 2012

Norwegian Communications Authority, Net neutrality guidelines, 2013

Singapore

Info-communications Development Authority of Singapore, 'IDA's Decision and Explanatory Memorandum for the public consultation on Net Neutrality', 2011, available at www.ida.gov.sg/Policies-and-Regulations/Consultation-Papers-and-Decisions/Store/Consultation-on-Policy-Framework-for-Net-Neutrality (Accessed 15 September 2016)

Slovenia

Law on Electronic Communications (ZEKOM), No. 003-02-10/2012-32, 20 December 2012

Abbreviations

3G	Third-generation mobile networks, providing voice and data capacity at speeds above 128 kilobits per second.
3GPP	3rd Generation Partnership Project: a collaboration between telecommunications associations to make a globally applicable 3G mobile phone system specification within the scope of the International Telecommunication Union (ITU).
ADSL	Asymmetric Digital Subscriber Line: technology for sending data over copper telephone wires, using asymmetrical speeds: higher download and slow uploading speed.
ADSL2+	Asymmetric Digital Subscriber Line 2+: a later higher speed variant of ADSL.
AKOS	Agencija za komunikacijska omrežja in storitve Republike Slovenije/Communications Networks and Services Agency of the Republic of Slovenia, Slovenian regulator.
ARCEP	Autorité de régulation des communications electroniques et des postes, the French regulator.
ASQ	Assured Service Quality.
BBC	British Broadcasting Corporation: a publicly owned and publicly financed broadcaster (see PSB).
BEREC	Body of European Regulators of Electronic Communications: regulatory body set up to help implement 2009 European telecoms laws.
BITAG	Broadband Internet Technical Advisory Group.
BSG	Broadband Stakeholders Group.
BT	British Telecom: UK incumbent.
CAS	content, applications or services.
CAT	Competition Appeal Tribunal (UK).

CDN	Content Delivery Network: a means of caching content closer to the end user's IAP.
CoE	Council of Europe: socio-cultural organisation established in 1948, currently with 47 members. See also ECHR.
CPP	Calling Party Pays.
CRTC	Canadian Radio-television and Telecommunications Commission: the converged federal regulator of broadcasting and telecoms for federal Canada.
DiffServ	differentiated services.
DMCA	US Digital Millennium Copyright Act 1998: a statute which obliges ISPs to take down material whenever they are notified of copyright infringement, under the Notice and Take Down (NTD) procedure.
DOCSIS3.0	Data Over Cable Service Interface Specification: the third generation of these cable broadband data standards.
DoS	Denial of Service.
DPI	Deep Packet Inspection: means by which IAPs can read into the packets of data they carry to analyse the contents as well as the header, in order to prioritise, deprioritise or even block the packets.
DSL	Digital Subscriber Line.
E2E	End to End: design principle governing Internet architecture.
EC	European Commission: executive body of the EU, responsible for developing and implementing the *acquis communautaire*, the body of EU law.
ECD	Electronic Commerce Directive, 2000/31/EC, which limits ISPs liability for packets they host or carry over their networks without knowledge of the content.
ECHR	European Convention on Human Rights, more formally the Convention for the Protection of Human Rights and Fundamental Freedoms, signed in 1950 by Member States of the Council of Europe.
ECPS	Electronic Communications Service Provider.
EDPS	European Data Protection Supervisor.
EDRi	European Digital Rights Initiative, a non-profit lobbying group on behalf of national privacy and Internet rights groups across Europe.
EEA	European Economic Area.
EFTA	European Free Trade Association.
EINS	European Internet Science.
ERG	European Regulators Group: advisory body set up by the 2002 regulatory framework for European telecoms, the grouping of the Member State NRAs.

ETNO	European Telecommunications Network Operators: association of predominantly incumbent network owners.
EU	European Union, as established in the Treaty of Maastricht 1992 and formerly the European Economic Community (EEC).
European Council	Council of Ministers of EU Member States, responsible for proposing legislation to the European Parliament.
Exabyte	1000 Petabytes (1 million Terabytes or 1 billion Gigabytes).
FCC	Federal Communications Commission: the converged broadcast and telecoms regulator for the US at federal level.
FRAND	fair, reasonable and non-discriminatory terms, where a monopoly provider of facilities (whether patents and other intellectual property or physical goods) provides access to its competitors.
FT	France Telecom: domestic incumbent in France, also owner of Orange mobile networks and formerly branded as Wanadoo IAP internationally.
FTTx	fibre to the home: high-speed Ethernet-ready transmission wire offered as FTTH (home), FTTP (premises) and FTTC (cabinet – street furniture for telecoms normally available to each neighbourhood, therefore more local than the exchange), FTTrN (to remote nodes), FTTB (to building or basement) varieties.
GB	Gigabyte (1024 megabytes).
Gbps	Gigabit per second (1/8th of a Gigabyte per second, or 128 Mbps).
GCHQ	Government Communications Headquarters.
GSM	Global System for Mobile Communication, also known as 2G: second-generation mobile telephony.
HADOPI	Haute Autorité pour la Diffusion des Oeuvres et la Protection des Droits sur Internet (High Authority for the Diffusion of Works and Protection of (Copy)Rights on the Internet): an agency established under the 2009 French Law against copyright infringement, more formerly known as the loi favorisant la diffusion et la protection de la création sur Internet.
HDTV	high definition television.
IAP	Internet Access Provider: company providing access to the Internet for consumers and businesses. The largest in most Member States is the incumbent telecommunications provider. Mobile networks are also IAPs.

IAS	Internet Access Service.
ICC	Interception of Communications Commissioner.
ICO	Information Commissioner's Office.
ICT	information communication technology.
IETF	Internet Engineering Task Force: a self-regulating technical standards body.
IGF	Internet Governance Forum: United Nations multi-stakeholder discussion forum initially held in Athens 2006, and to be held annually for at least four years thereafter.
IM	Instant Messenger.
IoT	Internet of Things.
IP	Internet Protocol.
IPTV	Internet Protocol Television: video programming delivered over IP networks rather than broadcast (cable, terrestrial and satellite) networks.
IRG	Independent Regulators Group
ISP	Internet Service Provider
ITRE	Committee on Industry, Research and Energy
ITU	International Telecommunications Union: United Nations body established to coordinate global telecommunications, successor to International Telegraph Union founded in 1865.
IWF	Internet Watch Foundation: UK 'hotline' for illegal content reporting, established 1996.
KB	Kilobyte.
Kbps	Kilobits per second.
LLU	Local Loop Unbundling: the regulated process whereby competitors can access the incumbent telecommunications provider's connections from telephone exchanges to the customer premises, using regulated access prices and conditions.
LTE	Long Term Evolution.
MAU	monthly active user.
MB	Megabyte (1024 kilobytes).
Mbps	Megabits per second.
Member	State Member State of the EU: 28 in total as at 2016.
MEP	Member of the European Parliament.
MNO	mobile network operator.
MPLS	Multiprotocol Label Switching: a standard set for NGNs.
MSG	multi-stakeholder governance: the process by which civil society groups are included in regulatory discussions with governments and corporate interests.
NGA	Next Generation Access: the use of new technologies (such as FTTx) to offer high-speed connections between the subscriber's premises and the main NGN.

NGNs	Next Generation Networks: all-Internet Protocol (IP) networks.
NGO	non-governmental organisation.
NRA	National Regulatory Authority: in reference to independent national bodies established under national law in the Member States of the European Union, which implement the European communications framework. NRA can also be used to refer generically to any national authority, such as the Canadian CRTC or US FCC.
NTD	Notice and Take Down: regime by which ISPs can avoid liability for potentially damaging content by removing such content on receipt of notice from a third party.
OECD	Organisation for Ecomic Cooperation and Development: a 'think-tank' for developed nations, with 30 national members. Membership is limited by commitment to a market economy and a pluralistic democracy. Formed in 1961 and grew out of the Organisation for European Economic Co-operation (OEEC), established in 1947.
Ofcom	Office of Communications Regulation: UK converged regulator of broadcasting and telecoms, established in 2002 and operational from December 2003.
OIAC	The FCC Open Internet Advisory Committee.
P2P	peer-to-peer: usually used in reference to file sharing amongst many peers, an efficient form of many-to-many information sharing as compared to a broadcast model using a central server. P2P is the method of distribution used by Skype, BitTorrent and many other information-sharing programmes.
PECP	provider of electronic communications to the public.
Petabyte	1000 Terabytes (1 million Gigabytes).
PIAS	Providers of Internet Access Service.
PSB	Public service broadcaster, granted special licensing conditions ostensibly in exchange for educational and news programming. Members of European Broadcasting Union (EBU).
QoS	Quality of Service: protocols and standards designed to offer guaranteed QoS have been mooted for many years, but none has yet been successfully marketed on the public Internet.
RFID	Radio Frequency Identification.
RIO	Reference Interconnection Offer.
RPP	Receiving Party Pays.
RTNDP	transparent, non-discriminatory and proportionate.
SDN	Software Defined Network.
SMP	significant market power: measure of dominance in European competition law, with a specific application to telecoms law.
SpS	Specialised Services.

TCP Transmission Control Protocol.
telco telecommunications provider: the term normally used for incumbent former national monopoly providers. There are also 'competitive telcos' – all other providers of switched telecommunications services except the national incumbent.
Terabyte 1000 Gigabtyes (1 million Megabytes).
TMP Traffic Management Practices.
UDP User Datagram Protocol.
UHDTV ultra-high definition video streaming and downloading.
USO Universal Service Obligation: for European consumers the right to a 33 Kbps telephone line. The USO will be upgraded as broadband network speeds increase.
VDSL very high bit rate digital subscriber line.
VoIP Voice over Internet Protocol: technology to digitise sound in packets sent over the Internet. Its primary advantage is that distance does not affect the cost of the call between two VoIP-enabled phones (or PCs attached to the phone or a data system).
VPN Virtual Private Network.
W3C WWW Consortium, a self-regulatory organisation.
Web2.0 social networking applications using blogs, podcasts, wikis, social networking websites, search engines, auction websites, games, VoIP and P2P services. These services, which are based in part on the Ajax mark-up language, makes user-generated and distributed content central to consumers' internet experiences.
Wifi standard for WLAN designed to Institute of Electrical and Electronic Engineers (IEEE) 802.11a/b/g specification.
WIK Wissenschaftliches Institut für Kommunikationsdienste GmbH: a telecoms economics research institute based in Bonn, well known for its work on behalf of the EC, German regulators and DT and its subsidiaries, and many other clients.
WiMAX Worldwide Interoperability for Microwave Access: a broadband wireless technology.
WLAN Wireless Local Area Networks, which often use the Wifi technology standards.
WWW The World Wide Web: a set of standards including those for graphical user interfaces using hypertext mark-up languages for displaying Internet information, invented by Tim Berners-Lee, now standardised by the World Wide Web Consortium.
Zettabyte 1000 Exabytes: ameasure of annual network traffic.

Acknowledgements

This book has been seven years in the writing and researching – beginning with sending the manuscript for the prequel to Bloomsbury in May 2009. I repeat my thanks to those who were acknowledged in that preface. This acknowledgements section also functions as a partial methodology as there are so many collaborators, inspirations and interviewees to acknowledge. The book is methodologically grounded in empirical interviews and analysis of primary and secondary materials, the latter often accompanied by interviews with the authors in person, by telephone, social media or by email. In spring 2009, no-one had written a book about net neutrality, though there have been many since, including several very impressive PhD theses in European policy and law, notably those by my sometime co-bloggers Katerina Maniadaki (2015) and Jasper Sluijs, those I examined by Alissa Cooper (2013) and Angela Daly, and somewhat tangentally the doctorates I examined by Andres Guadamuz (2011) and T. J. McIntyre. My own previous book on net neutrality appeared online in January 2010, and I was awarded a PhD at the University of Essex in summer 2010 for that work. I became a professor there, then was appointed to Sussex University in spring 2013. I have continued to work on net neutrality since 2009, writing about developments in mobile/wireless (2010), privacy (2011/13), Internet engineering (2012/13), European law (2012), human rights and developing countries (2013), and censorship and privacy (2014).[1] But this book is far from a rehash of 2009 with added sections on developments since then.

This book was written in part using social media, a new development since the prequel was finished in spring 2009. Many multimedia connections have supplied material, inspiration and networking opportunities. I joined Facebook in 2007, Twitter in May 2009 (@ChrisTMarsden), Slideshare in 2010,[2] and while the former is only used for actual 'friends' (as I do not wish to dilute my

[1] Latest papers at http://ssrn.com/author=220925.

[2] Latest presentations at www.slideshare.net/EXCCELessex.

Dunbar number further than modern life forces), the latter pair have provided a very economical form of bibliographical alerts and other interactions, and my most recent public presentations. Similarly, my blog on net neutrality, with well over 1,200 entries since completing the prequel,[3] has had almost half a million views, by bots and occasionally real people. To view updates on my thoughts after April 2016, do follow my work on Slideshare and @ChrisTMarsden.

I begin with the projects which formed a background to this work, either directly or tangentially.

The first thank you is to Professor Ian Brown, my collaborator on the interim book *Regulating Code* (MIT Press, 2013), in which the chapter on 'Smart Pipes' forms the net neutrality case study, and where my views on prosumer law (which dominate this book) were first sharpened. The genesis of that book was our work together at the Cambridge–MIT Institute in 2005/06, and our initial presentation on social media dominance to Gikii (the finest European Internet law symposium) in 2008. The work was mainly written in 2011/12 once I finished working on a book entirely about *Internet Co-regulation* (Cambridge University Press, 2011), finished in Melbourne in February 2012. Some of the ideas were further sharpened in the book tour of spring 2013. Putting prosumer law at the centre of our regulatory theory, shaped by both consumer, innovator and human rights perspectives, is key to the understanding of Internet regulation generally, and net neutrality in particular. It is why most telecoms regulators just do not 'get it' on net neutrality – they don't care about less than entirely dumb citizens using the pipes they regulate. Ian has been a continued inspiration to my work for over a decade and it is to him as much as anyone that I intellectually dedicate this work.

The second thank you is to the academics and especially the engineers and Internet scientists, the people who made sure I understood the artefact that I discuss in this book. That begins with Cambridge, Professor Jon Crowcroft in particular, and MIT, Professor David Clark in particular, as well as Narseo Vallina Rodriguez and Hamed Haddadi. It also includes the various collaborators in Internet Science[4] and Openlaws.eu, notably Thanassis Tiropanis, Kave Salmatian at Savoie, Juan Carlos de Martin, Elena Pavan and the fellows at NEXA Torino,[5] Ziga Turk, Francesca Musiani, Meryem Marzouki, colleagues at IvIR at the University of Amsterdam, notably Nico van Eijk, Alison Powell, Damian Tambini, Sally Broughton-Micova and Monica Horten at LSE, and my 'brilliant mind' economist mentor Jonathan Cave. In particular, my thanks to my sometime research assistant and guide to Brussels and Den Haag, Ben Zevenbergen. I also must acknowledge the brilliant academic technologists at

[3] Latest blog posts at http://chrismarsden.blogspot.co.uk/2009/05/book-book-book.html.

[4] See www.internet-science.eu/groups/governance-regulation-and-standards.

[5] See http://nexa.polito.it/fellows.

the FCC, notably Scott Jordan and Jon Peha, as well as the US net neutrality law pioneers Barbara Cherry, Barbara van Schewick and Rob Frieden, as well as Kevin Werbach and Andrea Matwyshyn. This area of research would not have been possible without Mark Lemley, Lawrence Lessig and Tim Wu, of course. I thank Harold Feld, Bill Lehr, Jesse Sowell, Julie Brill, Jonathan Sallet, Lawrence Spiwak, Christopher Yoo, Rene Arnold, George Ford, Sandra Bramann, Tom Hazlett, Roslyn Layton, Milton Mueller and the many other Beltway insider and outsiders who offered advice and critique at TPRC'15, IAMCR'15 (or earlier).[6] Everyone at Gikii over the years has helped to develop my thinking, especially Daithi McSithigh, @technollama and Lilian Edwards.

At Sussex, my new home where I finished the book's writing, I have been supported by Ed Steinmuller, Ian Wakeman and the Information Law group (@pillrabbit, @MMFrabboni and the great @technollama). Note that '2nd editions' are not 'REF-able',[7] but this is a 'sequel with added law'. My thanks to the Law Department for giving me the space on sabbatical in autumn 2015 to complete the manuscript and to make it distinctive from the prequel, with Andres Guadamuz manfully directing our brand-new LLM in IT & IP. Essex, especially Geoff Gilbert and Sabine Michalowski, had been very supportive of the prequel in 2009 and the 'Regulating Code' period of 2012.

I must also thank former collaborators who have since moved on to Ofcom, especially where they have since disagreed with me, or rather followed the 'company line' that the self-regulatory solution of a Code, combined with switching and transparency, can work. Ofcom conducted some fantastic research in the period, notably in 2015 with four blockbuster reports, two of which were published in December as I finished this book. While I did not agree with the speed of Ofcom's progress towards implementing net neutrality, I acknowledge that many current and former Ofcom experts provided both critical friendly advice and evidence when the corporate speed of progress was not always what some may have desired. Similarly, though many may think the European Commission a barrier to net neutrality implementation, there are many colleagues there who have conducted preliminary extraordinary work to make the new Regulation 2015/2120 a reality. They include Herbert Ungerer, the godfather of telecoms law in Europe, Kevin Coates, Anna Buchta, Constantijn van Oranje, Robert Madelin, Bettina Klein, Anna Herold, Kamila Kloc, Loretta Anania, Richard Cawley, Tony Shortall, Christian d'Cunha, Nicole Dewandre and Gordon Lennox. Members of the European Parliament whom I must acknowledge with great thanks include Amelia Andersdottir, Julia Reda (@senficon), Sabine Verheyen, Michael Reimon and Marietje Schaake, and their staffers past and present. I also must acknowledge those experts on European legislative affairs and

[6] See www.tprcweb.com/tprc43-program-overview.

[7] See www.ref.ac.uk/about/guidance/faq/all.

the consumer at BEUC, notably Guillermo Beltra amd Thomas Myhr, and the European Digital Rights Initiative (EDRi), notably Joe MacNamee, together with La Quadrature du Net, BoF, ORG, Digital Rights Ireland, Statewatch, Article 19 and other EDRi members.[8] For co-regulatory expertise and friendship, I also acknowledge my debt to Linda 'Soft Law' Senden, Yves Blondeel and Winston Maxwell. At the EBU, I thank Michael Wagner and Jenny Metzdorf, and the Open Forum Europe, of which I am a fellow. At ICANN Europe, I thank Jean Jacques Sahel, Adam Peake and Frederic Donck. At OECD past and present, I thank Sam Paltridge, Rudolf van der Berg, Taylor Reynolds, Verena Weber.

In chapters on their laws, I also thank experts from Norway, Netherlands, Slovenia, Council of Europe, OSCE, UN CEPAL, Canada, Brazil, South Korea, India and Japan. It would be remiss of me not to mention specifically Frode Sørensen, Carolina Botero, Pranesh Prakash, Sunil Abraham, Michael Joyce, Michael Geist, Craig McTaggart, Greg Taylor, Kevin Martin, Martin Husovec, and corporate experts at Vodafone, BT, Sky, Telefónica, Telecom Italia, Deutsche Telekom, Google, Microsoft, Facebook, BBC, Verizon, Comcast and many other companies, as well as a huge variety of start-ups and shut-downs, venture capitalists and others in the ecology affected by net neutrality.

I enjoyed research fellowships and academic support during this research at Center for Technology & Society at Fundação Getulio Vargas (Rio), Seoul National University, GLOCOM International University of Japan, University of Melbourne School of Law, Foundación Telefónica, Comitê Gestor da Internet no Brasil (CGI), the South Korean Prime Minister's office, Irish regulator ComReg and the Internet Science Network of Excellence.[9] To all, my thanks. No-one else is responsible for any errors or omissions in this book.

The book's heart was written in Brazil during three visits in 2015, especially as Fellow at FGV in October–November. The colleagues and friends there I must thank include Eduardo Magrani, Konstantinos Styliano, Nico Zingales, Nathalia Foditsch, Konstantina Bania, Louise Marie Hurel, Jamila Venturini, Pedro Mizukami, Marilia Maciel, Luiz Fernando Marrey Moncau. Most of all, for fellowship, friendship, intellectual discussion and co-authorship, Machiavellian strategic discourse, and companionship over the years at the Council of Europe, Internet Governance Forum, and FGV, I owe a huge debt of thanks to Luca Belli (and Marion). *Abraços* to all my friends in Rio, and the CGI in Sao Paolo, especially Flávia Lefèvre, Vinicius Santos, Diego Canabarro (and Pedro Ramos at InterLab). IGF2015 was also an extraordinary experiment in

[8] See http://history.edri.org/about/members.

[9] Internet Science Network of Excellence funded under the European Commission's Seventh Framework Programme: Information and Communication Technologies Grant Agreement Number 288021.

net neutrality discussion, and I thank all the many discussants there, notably Vint Cerf, Joe Cannataci, David Kaye, Hernan Galperin, Stefaan Verhulst and the Association for Progressive Communications participants. Here's to another 25 years of APC, 20 years of CGI and a second decade of IGF!

If you helped but I forgot to thank you here or in the prequel, email. I will thank you in the blog 'roll of honour'.

Finally, I can announce this is my last full-length monograph of this generation. I have authored five monographs in a decade, from *Codifying Cyberspace* (Routledge, 2007, with @damiantambini) to the prequel in 2009, to *Internet Co-regulation* (Cambridge University Press, 2011), *Regulating Code* (MIT Press, 2013, with @IanBrownOII) to this book. I am writing from the sunset window of our new-old house in Raynes Park, and while the future may hold many exciting open access contributions to the literature on Internet law, regulation and the digital socio-economic environment, they will not be sole-authored monographs. I dedicate that thought and this book with all my love to Kenza, without whom this book could never have been written.

Chris Marsden
Raynes Park, Surrey

Introduction: neutrality, discrimination and common carriage

> We cannot allow Internet service providers to restrict the best access or to pick winners and losers in the online marketplace for services and ideas. That is why today, I am asking the Federal Communications Commission [FCC] to answer the call of almost 4 million public comments, and implement the strongest possible rules to protect net neutrality. When I was a candidate for this office, I made clear my commitment to a free and open Internet, and my commitment remains as strong as ever ... The FCC is an independent agency, and ultimately this decision is theirs alone.
>
> President Barack H. Obama[1]

Net neutrality is a zombie[2] that has sprung to life recently. It is a policy of Internet[3] non-discrimination based on innovation, free speech, privacy and content provider commercial self-interest, imposed on the technocratic economic regulation of telecommunications (telco) local access networks. The regulators, telcos and governments don't like it one bit. The laws and regulations are formally 'Open Internet' not 'network neutrality', as I will explain. It is Net Neutrality 2.0, to use a cliché in explaining the second wave of a technology-led innovation.[4] Net neutrality is the principle that Internet Access Providers (IAPs) do not censor or otherwise manage content which individual users are attempting to access. That means that telcos should not block or 'throttle' Voice over Internet Protocol (VoIP, e.g. Skype, WhatsApp) or video (e.g. YouTube, BBC iPlayer or NetFlix) except

[1] Obama (2014).

[2] I have previously referred to net neutrality for a decade as an 'undead' debate which telecoms lawyers and economists have been unable to kill, for instance Marsden (2009), stating: 'No matter how many economists plant a stake in its heart, or come to bury it not praise it, net neutrality will not die.'

[3] The 'Internet' is a network of Autonomous Systems, of which about 40,000 are of a scale that is relevant. See Haddadi *et al.* (2009).

[4] van Eijk (2011b).

under narrowly defined conditions. Net neutrality regulation is critical to the future of Internet access for businesses and micro-enterprises, as well as students, citizens and all domestic users – and therefore to the future mass adoption of the Internet of Things (IoT), cloud computing and Big Data. In this book I claim, as Zhou Enlai stated of the French revolutionary movement in 1968 (not 1789),[5] that it is too early for advocates to claim success: legal prohibition on discrimination by IAPs has not yet resulted in effective net neutrality regulation.

Net neutrality directly regulates the relationship between IAPs and content providers, providing rules about how IAPs may contract with and treat the traffic of those content providers, specifically that they may not discriminate against certain providers (either blocking their content or favouring commercial rivals such as IAP affiliates). It does not regulate those content providers directly. Internet access is a very special communications service, recognised and reinstated in the US in 2015 as common carriage after a strange 11-year experiment with deregulation. Common carriage, or variants thereof, is a status it has always enjoyed in most other countries. Without Internet access, there is no content, application or service to enjoy: it is the *sine qua non* of Internet use. For that reason, IAPs have a Faustian bargain with government: in exchange for the privilege of providing such a unique public service and the attendant rights to dig up the streets and conduct other works (for instance, enforcing a public Right of Way), IAPs accept that their service is subject to special rules. If this was not previously clear to the general public and politicians, the Snowden revelations of 2013 made it crystal clear. IAPs provide you with access to the Internet and thus can track your every click of a mouse and every bit that is transferred. Their cooperation with law enforcement is vital to mass or individual surveillance, and their long-term cooperation with law enforcement is part of that special Faustian bargain with the state.

This book marks the shift from net neutrality policy towards legal effect in Europe, and hence more intensive engagement in national law and regulation. In this comparative analysis of law and regulation towards net neutrality policy (with some reference to the Americas, East Asia and India), I examine how human rights, behavioural science and innovation economics have been brought to bear on the typically neo-classical economic models used to regulate telco networks through which we connect to the Internet. The telcos and their regulators' corporatist modes of regulation, co- and self-regulation (state–firm bargaining with former state monopolies like British Telecom (BT) and foreign investors such as T-Mobile) are now exposed to the clamorous demand of civil society organisations for multi-stakeholder governance (MSG).[6] No

[5] McGregor (2011).

[6] Powell (2015).

one can ignore it, from the most vaunted technocrat or thrusting entrepreneur developing 5G mobile, cloud and IoT policies, to the Dutch grandmother using a video Instant Messenger (IM) service with her grandchildren, asking 'Why is my Internet so slow now?'[7]

Ever since the broadband Internet was brought to users in the late 1990s, this has been the messiest issue in telecoms regulation narrowly, MSG for the Internet generally, and affects anybody who previously, currently or in future uses the Internet: all of us. It is fiendishly difficult to identify abuses of net neutrality, and many corporate economists have claimed it is a solution in search of a problem or abuse: Schrödinger's net neutrality? It is and will be a nightmare to regulate; it is a joy for an academic case study in law, technological regulation more generally, economics and regulatory policy. In a major study of democratising regulatory decision making, Faris *et al.* 'conclude that a diverse set of actors working in conjunction through the networked public sphere played a central, arguably decisive, role in turning around the FCC policy on net neutrality'.[8] Civil society roles in Internet governance generally and lobbying of government and regulators has been critical to the emergence of net neutrality.

I advance an argument that the Internet itself was a radical departure from 'normal business' for telecoms regulation, which was largely ignored for as long as possible given the collapse of broadband competitors to incumbents after 2001. Web2.0 prosumers – web users who tweet and express their views to politicians – are not business as usual. This has been made abundantly clear in copyright as well as net neutrality policy over the period since Lessig predicted its arrival in 1999, with corporate response well explored by Wu and Zittrain, and in the European context Horten and Kron.[9] Telecom lawyers and economists have largely overlooked this clash of cultures. Sutherland argues that:

> Telecommunications policy research appears fundamentally different from [media and Internet studies], lacking the constant, fractious disputes between different schools of thought (e.g., realist v. interpretivist) ... telecommunications policy research seems to be an instance of Kuhnian 'normal science', a discipline which has adopted a research paradigm that is unchallenged. Taking the hegemonic research paradigm for granted, rigor has been assumed ... since other perspectives are excluded.[10]

In this book I argue that it is not business as usual for telecoms regulators, and that net neutrality cannot be dismissed as a technical or competition challenge

[7] The most popular blog post by far on my net neutrality blog is that referencing former European Commission Vice President Neelie Kroes' decision to adopt only net neutrality 'lite' in 2013. See Marsden, Chris (2014a).

[8] Faris *et al.* (2015).

[9] See Lessig (1999), Wu (2003a), Zittrain (2008), and in the European context Horten (2011). See also Kron (2012).

[10] Sutherland (2014).

without serious implications for consumers and civil society interaction. I deliberately broaden the debate to consider the structure of the radical changes which Internet Protocol (IP) networks are wreaking on mass communications systems, for voice, data and audiovisual networks. At the time of writing, it is not at all clear how regulators will react to the mass mobilisations of users in the US (2014), India (2015) and potentially Europe (2016). Political scientists will find this a challenging and illuminating area to study, but for regulatory lawyers it is enough to remark that competition law will not entirely meet this challenge, as I explore in Chapter 2.

A word on how you, the reader, should approach the book. For those who recall my 2010 book,[11] this is not a second edition; it is a new book about legislation and regulation, how these have developed from policy, and why they have not solved the problems, which themselves have evolved. It is written with lawyers, social scientists generally and the well-informed general reader in mind. In 2014–15, over 6 million citizens replied to various net neutrality consultations, including over 2 million in India and 4 million in the US. Law and regulation have developed significantly since 2010, and the audience for this book will be well aware of many net neutrality issues. It is not an 'idiot's guide' introduction to what the Internet is and how it works (though footnotes provide you with background reading). In this book I do not repeat myself by discussing the pre-regulatory theoretical background in depth.[12] While I am well informed by technical insights from computer scientists, I make no claims to innovative technical insight, and technologists should bear this in mind when reading especially this Introduction and Chapter 3, which outlines some technical issues. Books can and will be written about the technologies and policies of interception and intermediary liability, which I examine in Chapter 5, zero rating and mobile net neutrality examined in Chapter 7, and UK communications regulation in Chapter 6, and books have already been written about US net neutrality, competition law and net neutrality,[13] and human rights and net neutrality.[14] I refer to all of those.

What this book does offer to lawyers, other social scientists, policymakers and citizens is a state of the art analysis of how European net neutrality law and regulation has arrived at a liminal moment in 2016, when laws and regulations

[11] Marsden (2010).

[12] The two opposing law and economics camps on these issues are described in Marsden and Cave (2007). Hahn and Wallstein (2006), Yoo (2006), Speta (2004) and many economists were against regulation. Lemley and Lessig (1999, 2000), Wu (2003b, 2007), Frieden (2006), Cherry (2006), Economides and Tåg (2007), Weiser (2009), Frischman and van Schewick (2007) and van Schewick (2010) were in favour.

[13] Maniadaki (2015).

[14] See Belli and De Filippi (2015) and Nunziato (2009). For the opposing view, see Zelnick (2013).

have been drafted, passed and taken effect, but little actual enforcement or implementation has taken place.

As I explained in the Acknowledgements, this is the sum of many comparative research projects, transdisciplinary conversations, expert conversations and interviews over two decades. The book presents the results of fieldwork in South America, North America and Europe over an extended period (2003–15), the latter part of which focused on implementation.[15] This book is based on empirical interviews conducted in-field[16] with regulators, government officials, IAPs (known in European law formally as 'Electronic Communications Service Providers', as examined in Chapters 1 to 4), content providers, academic experts, non-governmental organisations (NGOs) and other stakeholders from Chile, Brazil, the United States, Canada, the United Kingdom, Netherlands, Slovenia and Norway. Collaboration between socio-legal scholars, senior computer scientists and economists is essential to serious investigation of network neutrality. Such intensive collaboration enhances much well-meaning law and humanities work in examining Internet law and telecommunications regulation, in such areas as behavioural advertising regulation and the implications of widespread commercial deployment of Deep Packet Inspection (DPI), which is explored in Chapter 5. The project builds on both my previous network neutrality research funded through the European Commission (EC), national governments in Europe and East Asia, the United Nations and Council of Europe, RAND Corporation, and the 37-partner European Internet Science (EINS) Network of Excellence, specifically its Joint Research Activity on Regulation and Governance.[17]

State of the law on net neutrality 2017

We begin at the end, or as I suggest in Chapter 8, the 'end of the beginning'. 'Open Internet' legislation has been passed in the European Economic Area (EEA) via European Union (EU) Regulation 2015/2120 of 25 November 2015, and has been regulated in the United States via the Open Internet Orders of 2010 (transparency provisions) and 2015 (anti-discrimination).[18] While regulatory and legislative logjams and litigation have resulted in delayed

[15] The final four years of research (2011–15) was funded by the European Commission EU FP7 EINS grant agreement No. 288021 and internal funding from both Sussex and Essex Universities. No ISP or content provider has provided funding to the project since 2008, though several of each funded earlier stages. I am grateful to have been appointed Research Fellow at Melbourne University School of Law in 2012, and FGV Centre for Technology and Society in 2015.

[16] With the exception of Chile, where the UN CEPAL in 2013 and Brazilian CGI in 2015 provided a forum for Chilean stakeholders to travel to workshops on comparative implementation.

[17] EINS (2015).

[18] FCC, Internet Policy Statement 05–151, 2005.

implementation of regulation in the US[19] and European Union in the period since their respective initial intentions to regulate were announced in 2009,[20] several countries passed legislation and/or implemented regulation of net neutrality.[21] Notable examples of laws with the date of regulation/legislation are: Chile (2010–11),[22] Brazil (2014–16),[23] Singapore (2011),[24] Israel (2011)[25] and Costa Rica (2010).[26] Other nations have introduced forms of self- and co-regulation for net neutrality with varying degrees of regulatory commitment, including Norway (2009),[27] the UK,[28] South Korea (2013) and Japan (2009).[29] Canada has used existing law to regulate.[30] European nations with pre-existing laws include Netherlands (2012),[31] Slovenia (2012)[32] and Finland (2014).[33] EU Member States Germany and France[34] issued policies which have not been translated into specific net neutrality regulatory action. In Belgium, Italy and Luxembourg, proposals have been put forward for legislation, but no law was passed, in view of European Regulation negotiations from 2013[35] until the end of 2015.[36]

In 2016 net neutrality policy was paused on the edge of enforcement, on both sides of the Atlantic. On the first anniversary of the entry into force of its Open Internet Order, the US FCC was in pre-election mode, implementing privacy regulation for its newly reinstated Title II common carrier IAPs but ignoring the obvious infringements of net neutrality by IAP zero rating services. This led 53 consumer organisations to write to the FCC Chair to

[19] Crawford (2011).

[20] Marsden, Chris (2012).

[21] Marsden (2013a, 2013b); Shin and Han (2012); Jitsuzumi (2012); Candeub and McCartney (2012).

[22] Chile, Ley 20.453 de 18 de agosto 2010.

[23] Brazil, Law No. 12.965, 23 April 2014.

[24] Info-communications Development Authority of Singapore, 'IDA's Decision and Explanatory Memorandum for the public consultation on Net Neutrality', 2011.

[25] Amending Article 51C(b) Telecommunications Law 1982, Summarized by the Ministry of Communications, Israel, Internet (over-the-top) services and challenges to regulation, 2015. See also Greenbaum (2014).

[26] Costa Rica, *Guzmán et al. v. Ministerio De Ambiente, Energía y Telecomunicaciones* (2010).

[27] See Norwegian Communications Authority, Net neutrality guidelines, 2013.

[28] Marsden (2014a).

[29] Jitsuzumi (2015).

[30] CRTC (2009). See also Marsden, Chris (2010).

[31] Netherlands Telecommunications Act 2012.

[32] Slovenia, Law on Electronic Communications, No. 003-02-10/2012–32, 20 December 2012, Article 203(4), available in Slovenian at www.uradni-list.si/1/content?id=111442 (Accessed 24 September 2016).

[33] Finland, Information Society Code (917/2014), 2014. I am grateful to Professor Päivi Korpisaari, of Helsinki University for pointing me to the Finnish regulation.

[34] Jasserand (2013).

[35] COM(2013) 627.

[36] Olsen, T. (2015).

urge enforcement to begin.[37] In Europe, the eighth year of austerity economics meant starvation of public funds to deploy high speed networks or to reinforce the staffing of the national telecoms regulators. While Regulation 2015/2120 was published in November 2015, the Body of European Regulators of Electronic Communications (BEREC) was drafting Guidelines to be published in August 2016, with significant doubts that all – or many – of the 28 national regulators would take their enforcement responsibilities seriously. That remains to be seen in 2017 and thereafter. The UK was itself embroiled in a non-binding #Brexit referendum held on 23 June on whether to leave or stay in the European Union, which may eventually impact substantially on its telecoms policies. Telecoms is largely a Digital Single Market responsibility, liberalisation and competition primarily negotiated since the mid-1980s in Brussels rather than London. 2016 promised to be the year after the regulations were passed, but not the decisive year for net neutrality enforcement.

Chapters 1 to 2 explain the beginnings of net neutrality regulation in the US and Europe, before Chapter 3 explains some of the current debate over access to Specialised Services: fast lanes with higher Quality of Service (QoS). Chapter 4 then examines the new European law of 2015, with Chapter 5 examining the interaction between that law and interception/privacy. Chapter 6 takes a deep dive into UK self- and co-regulation of net neutrality. In each of the national case studies, initial confusion at lack of clarity in net neutrality laws[38] gave way to significant cases, particularly since 2014, which have given regulators the opportunity to clarify their legislation or regulation. The majority of such cases relate to mobile (or in US parlance 'wireless') net neutrality, and in particular so-called 'zero rating' practices, which I explore in Chapter 7. Finally, Chapter 8 offers a toolkit for regulation and conclusions. The conclusion notes the limited political and administrative commitment to effective regulation thus far, and draws on that critical analysis to propose reasons for failure to implement effective regulation. It compares results and proposes a regulatory toolkit for those jurisdictions that intend effective practical partial or complete implementation of net neutrality. It sets out a future research agenda for exploring implementation of regulation. This book marks the real start of net neutrality law after many false dawns.

I offer a solution that I term the co-regulatory common carriage answer, but I do not expect it to be implemented: regulators and lobbyists have spent two decades trying to ignore user demands for net neutrality and they will find co-existence with the zombie extraordinarily difficult. I argue against social or economic justifications for either barring any proprietary high-speed traffic at all, or for strict versions of net neutrality that would not allow any traffic

[37] Marsden, Chris (2016).

[38] Marsden, Chris (2013c).

prioritisation. There is too much at stake either to expect government to supplant the market in providing higher speed connections, or the market to continue to deliver openness without basic policy and regulatory backstops to ensure some growth.[39]

I argue that higher QoS for higher prices on Specialised Services should be offered on fair, reasonable and non-discriminatory (FRAND) terms to all-comers, a modern equivalent of common carriage. As common carriage dictates terms but not the specific market conditions,[40] transparency and non-discrimination would not automatically result in a plurality of services. The type of service which may be entitled to FRAND treatment could result in short-term exclusivity in itself, for instance as wireless/mobile cell towers may only be able to carry a single high-definition video stream at any one point in time and therefore a monopoly may result. My argument is also that a minimum level of service should be provided which offers Open Internet access without blocking or degrading specific applications or protocols – an updated form of the Universal Service Obligation (USO),[41] proposed by Ofcom and the UK government at 10Mbps by 2020.[42] That provides a basic level of service which all subscribers should eventually receive, though already insufficient for the new generation of ultra-high definition TV (UHDTV). This minimum speed promise is controversial in many countries, because it interferes with private deals between IAPs and multinational content providers to discriminate in providing what is now called 'zero rating'. This practice of 'sponsored data plans' provides free content within the 'walled garden'[43] of the IAP's special offer, paid for either by the IAP itself or the content provider. The former case is much cheaper than regular Internet traffic, because the content should be cached in the IAP's own network and otherwise made more efficient to deliver, reducing the cost of hauling the data over the Open Internet.[44] There is nothing unusual in such deals to provide accelerated tailored content over the fixed Internet, and it has become the most popular means to receive BBC video content online in the UK, for instance.[45] It is, however, controversial in mobile IAPs, especially where there is no Internet access included in the subscriber's bundle, yet it remains described as accessing the 'Internet' when it is actually only that content pre-negotiated. The Organisation for Economic Cooperation and Development (OECD) explains: 'zero rating can clearly be pro-competitive … [and] becomes less of an issue with … higher or unlimited data allowances.

39 Meisel (2010), p. 20.
40 Cherry (2006).
41 Mueller (1998).
42 Department for Culture, Media and Sport (2015a).
43 The concept of a 'walled garden' is discussed further in Chapter 5.
44 See Farman (2015a, 2015b).
45 Sweney (2016). See also Parnwell (2014).

Regulators need to be vigilant'.[46] This impacts most severely in poorer developing countries for new Internet users, and was the most controversial issue discussed at the United Nations Internet Governance Forum in November 2015.[47] I explore the issue in Chapter 7.

Net neutrality definition and policy

'Net neutrality' is a difficult term to define accurately, as it is a principle not a regulatory process.[48] I help to define it in this Introduction, but for those confused by the many claims by politicians and IAPs to support net neutrality, here is a clarification. Hahn and Wallsten explain that net neutrality:

> usually means that broadband service providers charge consumers only once for Internet access, don't favor one content provider over another, and don't charge content providers for sending information over broadband lines to end users.[49]

Note that all major consumer IAPs are vertically integrated to some extent, with proprietary video, voice, portal and other services. Conventional US economic arguments have been broadly negative towards the concept of net neutrality, preferring the introduction of tariff-based congestion pricing.[50] The lack of trust and security on the Internet, combined with a lack of innovation in the QoS offered in the core network over the entire commercial period of the Internet since NSFNet was privatised in 1995 meant that development was focused almost entirely in the application layer. Peer-to-peer (P2P) programmes, such as low-grade VoIP and file-sharing as well as the World Wide Web (WWW), were designed during this period.[51] 'Carrier-grade' VoIP, data and video transmission was restricted to commercial Virtual Private Networks (VPNs) that could guarantee security, with premium content attempting to replicate the same using Content Delivery Networks (CDNs) such as Akamai (examined in Chapter 3), or IAPs' own offerings deployed within their network.

Network congestion and lack of bandwidth at peak times is a feature of the Internet: it has always existed. That is why video over the Internet was, until the late 1990s, simply unfeasible. It is why VoIP often has patchy quality, and why engineers have been trying to create better QoS. Prior to commercialisation in 1995, the Internet had never been subject to regulation beyond that needed for interoperability and competition, building on the Computer I and II inquiries

[46] OECD (2015), pp. 187–192.
[47] See Belli and De Filippi (2015).
[48] See Marsden (2013a).
[49] Hahn and Wallsten (2006).
[50] See David (2001).
[51] Brown and Marsden (2013a), pp. 36–39; Cannon (2003) and references therein.

by the FCC in the United States, and the design principle of End to End (E2E). That principle itself was bypassed by the need for greater trust and reliability in the emerging broadband network by the late 1990s, particularly as spam email led to viruses, botnets and other risks. The E2E principle governing Internet architecture is a two-edged sword, with advantages of openness and a dumb network, and disadvantages of congestion, jitter and ultimately a slowing rate of progress for high-end applications such as high definition television (HDTV).[52] E2E may have its disadvantages as compared with QoS. Steinberg, in a classic 1996 article, explained the cultural and engineering gulf between telecoms and Internet traffic designers. 'BellHeads':

> are the engineers and managers who grew up under the watchful eye of Ma Bell and who continue to abide by Bell System practices out of respect for Her legacy. They believe in solving problems with dependable hardware techniques and in rigorous quality control ... Opposed to the Bellheads are the Netheads, the young Turks who connected the world's computers to form the Internet. These engineers see the telecom industry as one more relic that will be overturned by the march of digital computing.[53]

This argument continues unabated to this day, though the 'relic' continues to dominate local access for consumers. Bubley recently stated:

> A lot of proposed SDN [Software Defined Network] models have Net Neutrality implications. For example, I heard many discussions about 'app-aware service chains; and ways to 'slice' the network so it behaves differently for particular services or DPI-detected flows.[54]

It is worth noting European regulator group BEREC's remarks: 'Over the Internet, a guaranteed end-to-end QoS offer is ... neither commercially nor technically realistic. Differentiated services (DiffServ), which fall just short of guaranteed end-to-end QoS, exist but continue to be exceptional'.[55] They add that: 'where end-to-end QoS arrangements are currently in use, they almost always consist of specialised services (e.g. IPTV [Internet Protocol Television]), provided not over the Internet but within a closed network within the Internet Access Provider's own network',[56] but that:

> mechanisms other than end-to-end QoS traffic classes have been developed over time for improving [E2E] network performance, including end-point based congestion control for reduction of the traffic load, Internet Exchange Points and the increased use of peering. CDNs are also used to improve the user's experience

[52] Saltzer, Reed and Clark (1984), p. 288.
[53] Steinberg (1996).
[54] Bubley (2015).
[55] BEREC, BoR (12) 120.
[56] *Ibid.*

> of an application's quality (QoE). All of these mechanisms have evolved through commercial innovation, without the need for regulatory intervention. Furthermore, they do not threaten the system of decentralised efficient routing of Internet traffic, since they are applied at endpoints.[57]

In December 2012 BEREC appeared determined to prevent attempts by former monopoly IAPs to assert control over Internet traffic, stating 'Put simply, ETNO [European Telecommunications Network Operators] is trying to extract additional revenues from its existing network assets, in a bid to reassert control over a changing communications ecosystem'.[58] Chapter 4 explains that this determination did not survive negotiations to create the 'horse designed by a committee' that is the dromedarian European law: Regulation 2015/2120.

Net neutrality lite and heavy

Dividing net neutrality into its forward-looking positive and backward-degrading negative elements is the first step in unpacking the term, in comprehending that there are two types of problem: charging more for more, and charging the same for less. IAPs can discriminate against all content or against the particular content that they compete with when they are vertically integrated.

Backward-looking 'net neutrality lite' claims that Internet users should not be disadvantaged due to opaque and invidious practices by their current IAP. That means no throttling, blocking of rival content (e.g. Skype, BitTorrent, NetFlix or WhatsApp), and ensuring the 'Four Freedoms' for Internet users: their own choice of content, applications, services and devices to connect to the Internet. In the US, regulators have a long history of fighting such discrimination, and in 2016 it is fairly clear that they would take action against such private censorship.

Forward-looking 'positive net neutrality' is the new focus of the problem: network owners with vertical integration into content or alliances have enhanced incentives to require content owners (who may also be consumers) to pay a toll to use the higher speed networks that they offer to end users. These 'fast lanes' are typically upgraded fibre to the customer's neighbourhood (or even road, or even house) rather than traditional copper networks. Positive 'net neutrality' or 'heavy' neutrality is argued to place a burden on investors in upgrading to faster IAP lines, though the costs have fallen dramatically and in 2016 most European and US Internet users can access a 30Mbps line. Infamously, the head of AT&T Ed Whitacre argued in 2005 that: 'The Internet can't be free in that sense, because we and the cable companies have made an

[57] *Ibid.*

[58] *Ibid.*

investment and for a Google or Yahoo! or Vonage or anybody to expect to use these pipes [for] free is nuts![59]

In 2016 the 'net neutrality lite' element is becoming regulated, and is less controversial than previously. Governments and regulators that fiercely fought net neutrality law have finally conceded that it may not be as disastrous as they had claimed, as I detail in the UK case in Chapter 6. They are only conceding net neutrality 'lite' because they have secured approval for their IAPs to discriminate on their fast lane services, which are now called 'Specialised Services' (examined in Chapter 3). For example, in November 2014 the European mobile IAPs finally agreed to stop fighting for the right to throttle their users' Internet use when it became inevitable that a new European law would be passed in some shape. They declared that: 'We are committed to maintaining an open Internet and to treating providers of similar content and services in a non-discriminatory manner, provided that they are legally and fairly offered according to Europe's laws.'[60] But they argued they should discriminate on 'fast lane' services:

> providing a range of services at different levels of quality and price, in order that all sectors of European industry can maximise their commercial opportunities from advanced services, and to providing affordable Internet services for consumers to help eliminate the digital divide.[61]

US and EU regulation of net neutrality 2017

The US regulator FCC has acted on several network neutrality complaints (notably those against Madison River in 2005 and Comcast in 2008), as well as introducing the principle in part through several merger conditions placed on dominant IAPs.[62] The 26 February 2015 Open Internet Order applied from 12 June 2015 and promised to enforce net neutrality.[63] The FCC announced in July 2015 how to receive case-by-case advice about future plans, for instance zero rating schemes or Specialised Services, that may risk breaching net neutrality: 'new process involves requesting and receiving an advisory opinion on specific, prospective business practices.'[64] At paragraphs 30–31 it explains that 'Although advisory opinions are not binding on any party, a requesting party may rely on an opinion if the request fully and accurately contains all the material facts and

[59] Business Week International Online Extra (2005).

[60] Make the Net Work (2014).

[61] *Ibid.*

[62] *Comcast v. FCC* (2010) No. 08-1291.

[63] FCC, Internet Policy Statement 05–151, 2005.

[64] FCC, Open Internet Advisory Opinion Procedures, Protecting and Promoting the Open Internet, GN Docket No. 14-28, 2015

representations necessary for the opinion and the situation conforms to the situation described in the request for opinion.' The FCC 'may later rescind an advisory opinion, but any such rescission would apply only to future conduct and would not be retroactive.' The FCC claimed in 2015 that the Order offered 'Bright Line Rules':

- No Blocking: broadband providers may not block access to legal content, applications, services, or non-harmful devices.
- No Throttling: broadband providers may not impair or degrade lawful Internet traffic on the basis of content, applications, services, or non-harmful devices.
- No Paid Prioritization: broadband providers may not favor some lawful Internet traffic over other lawful traffic in exchange for consideration of any kind – in other words, no 'fast lanes.' This rule also bans IAPs from prioritizing content and services of their affiliates.[65]

That final provision should eliminate zero rating, but it does continue. Zero rating is a common practice in the US; for instance T-Mobile has offered 33 zero-rated music services since 2014.[66]

As seen previously in the mergers of Bell Atlantic into Verizon and the formation of AT&T in 2005/06 and Comcast/NBC Universal in 2011, the US government has found itself most able to enforce net neutrality with decisions inserted into merger approvals. The 2015 merger of DirecTV into AT&T imposed such conditions on zero rating.[67] In its AT&T/DirecTV approval of 27 July 2015, the FCC stated at paragraph 395: 'we require the combined entity to refrain from discriminatory usage-based allowance practices for its fixed broadband Internet access service.[68] In response to accusations that AT&T ignored previous commitments in mergers, the FCC at paragraph 398 'require that AT&T retain both an internal company compliance officer and an independent, external compliance officer'. This regulation has some teeth. Comcast's attempted takeover of Time Warner Cable abandoned in 2015 would also have been likely to see such conditions imposed alongside interoperability/neutrality in its dealing with third party

[65] FCC, In the Matter of Protecting and Promoting the Open Internet, GN Docket No. 14–28, 2015 (Open Internet Order).

[66] Northrup (2015).

[67] Telecom Paper (2015): 'If approved by the commissioners, 12.5 million customer locations will have access to a competitive fibre connection from AT&T. The additional roll-out is around ten times the size of AT&T's current FttP deployment and increases the national residential fibre build by over 40 percent … AT&T will not be permitted to exclude affiliated video services and content from data caps on its fixed broadband connections. It will also be required to submit all completed interconnection agreements with the FCC.'

[68] FCC, In the Matter of Applications of AT&T Inc. and DIRECTV For Consent to Assign or Transfer Control of Licenses and Authorizations MB Docket No. 14-90, 2015.

device authentication – which concerns the freedom to attach devices to the network.[69]

In Europe, more complete confusion over zero rating and Specialised Services existed amongst governments, European institutions and regulators in 2016. This will be the focus of the BEREC consultations that conclude in August 2016, explored at the end of Chapter 4. The European Parliament had negotiated 'net neutrality lite' rules on blocking/throttling in 2009 – with emphasis on the 'lite' – to be implemented via regulatory action and reporting from 2011 under the amended Electronic Communications package.[70] It essentially permitted discrimination (under certain conditions) on speed and price for new network capacity, but insists that existing networks do not discriminate 'backwards' – that is, do not reduce the existing levels of service or block content without clear and transparent notice to users, and demonstrable reasonableness of those actions. This had to be adopted by national parliaments in June 2011 – though many delayed.

The European Parliament, European Commission and newly formed BEREC, on behalf of the 28 national telecoms regulators, all announced investigations into the implementation of net neutrality carried out in the second half of 2010, at the end of which the European Commission presented to the European Parliament its first annual findings in the area. Development of European legal implementation of the network neutrality principles has been slow, with the European Commission referring much of the detailed work to BEREC, which undertook an extensive work programme on net neutrality from 2011, leading slowly towards European legislative activity in 2013–15, and BEREC Guidelines to enforce the 2015 regulation by August 2016.

In 2012/14 three Member States implemented laws that were much stricter than the 2009 rules. The most famous case was the Netherlands, where questions were asked in parliament about Skype blocking in 2009,[71] and by spring

[69] Brodkin (2014): 'Roku is pleased to inform the Commission that effective November 25, 2014, Roku and Comcast entered into an agreement pursuant to which Comcast has, among other things, agreed to authenticate the HBO GO and Showtime Anytime apps on Roku video streaming devices for Comcast's subscribers whose subscriptions entitle them to access the content and services made available through such apps' – filed in the decison on the Comcast–Time Warner Cable merger, stopping selective blocking of HBO and AShowtime apps on Roku and Playstation 3 but not Apple TV or Comcast-affiliated devices. Yet Roku in August 2014 had argued that 'Rather than prioritizing platform support by customer interest or software compatibility, MVPDs [cable distributors] can use their power of authentication to favor one streaming platform over another. A large and powerful MVPD may use this leverage in negotiations with content providers or operators of streaming platforms, ultimately favoring parties that can either afford to pay for the privilege of authentication, or have other business leverage that can be used as a counterweight to discriminatory authentication. Additionally, MVPDs with affiliated ISPs can abuse their power over authentication by choosing to authenticate only their own or affiliated offerings.'

[70] Directive 2009/136/EC (Citizens' Rights Directive) and the Declaration appended to Directive 2009/140/EC (Better Regulation Directive).

[71] Aanhangsel Handelingen II (Appendix Official Report), 2008/09, nr 2765 and 2766.

Table 1 BEREC papers on net neutrality 2010–15

BoR (10) 42	BEREC Response to the European Commission's consultation on the open Internet and net neutrality in Europe
BoR (11) 44	Draft BEREC Guidelines on Net Neutrality and Transparency
BoR (11) 67	Guidelines on transparency as a tool to achieve net neutrality
BoR (12) 30	A view of traffic management and other practices resulting in restrictions to the open Internet in Europe – Findings from BEREC's and the European Commission's joint investigation
BoR (12) 31	Differentiation practices and related competition issues in the scope of Net Neutrality
BoR (12) 32	BEREC Guidelines for Quality of Service in the scope of Net Neutrality
BoR (12) 33	An assessment of IP-interconnection in the context of Net Neutrality
BoR (12) 34	BEREC public consultations on Net Neutrality Explanatory paper
BoR (12) 120	Statement with observations about net neutrality for ETNO's proposal to International Telecommunications Union (ITU) World Conference on International Telecommunications
BoR (13) 117	Ecosystem Dynamics and Demand Side Forces in Net Neutrality: Progress Report and Decision on Next Steps
BoR (14) 117	Monitoring quality of Internet access services in the context of net neutrality
BoR (15) 90	Report on how consumers value net neutrality

2011 the largest IAP, KPN, was boasting on investor calls about spying on user behaviours so as to block the new messaging app, WhatsApp, that was leading users to stop texting.[72] As a result, when the Netherlands voted on adoption of the 2009 package on 22 June 2011, it strengthened the powers of its regulator substantially to outlaw discrimination and the use of the spying technology that KPN had used. The law was later confirmed by the Dutch Senate in April 2012, though the regulations to make detailed rules were not passed until mid-2013. Nevertheless, by June 2011 the 2009 laws had unravelled before they had even been implemented in most countries. This, together with Slovenia adopting even more stringent rules than the Netherlands in its telecoms law of December 2012, meant that the rules would inevitably have to be revisited.

At national level, other EU Member States have been slow to recognise net neutrality problems, despite strong anecdotal evidence arising, which I analyse in the UK case in Chapter 6. The UK government opposes net neutrality, and

[72] Sterling (2011).

Ofcom's role has been both restricted to encouraging self-regulation and since 2009 funding research by SamKnows into detection of Traffic Management Practices (TMP).[73] and its effect on consumers. I explore the opaque practices of co-regulatory forums where governments or regulators have decided on partial private rather than public diplomacy with IAPs. Empirical analysis of UK IAP practices has showed that net neutrality violations have been far more frequent than in the US.[74] The government itself has been inert, even erroneously reporting to the European Commission in its 15th Annual Implementation Report on telecoms liberalisation that no problems were occurring in 2009–10.[75] The UK regulatory regime has focused on behavioural 'nudge' responses to net neutrality violations, though it has also conducted technical measurement of both broadband speeds and traffic measurement, as well as a recent study into types of monitoring,[76] so that 'regulators keep a close watch on the operations of the market, using frequent detailed traffic measurement reports'.[77] It has a fatal flaw in its light touch co-regulatory regime: the consumer's major obsession when switching provider is the headline speed that IAPs promise in their advertising, not any unannounced and typically concealed blocking. The UK situation is further considered in detail in Chapter 6, as well as in Chapter 4, given it is subject to European law, at least as is the case at the time of writing.

European net neutrality was intended to sink slowly below the waves of symbolic public safety legislation and self-interested pro-industry regulation after 2009. As the ink was drying on the national laws implementing the 2009 rules, the European Commission had to write new rules, which it announced in June 2013 and presented the Connected Continent proposal on 11 September 2013.[78] Malcolm Harbour MEP's Opinion on implementation of the 2009 Directives was amended that month by his committee to add these pro-regulatory comments:

> there is a potential for anti-competitive and discriminative behaviour in traffic management and calls, therefore, on the Member States to prevent any violation of net neutrality;
>
> 5. Underlines that end to end quality of service prioritisation alongside best effort delivery could undermine the principle of net neutrality; calls on the Commission and regulators to monitor these trends and, if appropriate, to deploy

[73] It was proven by SamKnows in 2008 that British Telecom throttled all P2P traffic aggressively during the evening peak: see Collins (2008).

[74] Cooper and Brown (2015).

[75] COM(2010) 253.

[76] Ofcom, 12th Annual Communications Market Report, 2015. See also the Ofcom-commissioned study by Predictable Network Solutions Limited (2015).

[77] Crowcroft (2015).

[78] COM(2013) 627.

> the quality of service obligation tools set out in Article 22 of Directive 2002/22/EC on universal service and users' rights relating to electronic communications networks and services, and if necessary to consider additional EU legislative measures.[79]

The Member State governments and European Parliament returned to negotiation of these rules for over two years to autumn 2015, culminating in a vote in the European Parliament to adopt the rules on 27 October 2015. This is the focus of Chapters 4 and 5 of the book, followed by consideration of comparative case studies in implementation in Chapters 6 and 7.

Net neutrality 'lite' law is 'on the books' in 2016, but regulating the fast lane Specialised Services and zero rating has been prevented in Europe, and is undecided in the United States. These problems will be the focus of the book. Before we can explore the future and present, however, we need to look to how we got here: the history of net neutrality and its predecessor, common carriage.

History of common carriage: forerunner to net neutrality

Network neutrality is the latest phase of an eternal argument over control of communications media. The Internet was held out by early legal and technical analysts to be special, due to its decentred construction, separating it from earlier 'technologies of freedom' including radio and the telegraph.[80] Net neutrality has been variously defined, most prominently by regard to its forerunner: common carriage. Common carriers who claim on the one hand the benefits of rights of way and other privileges, yet on the other claim traffic management for profit rather than network integrity, are trying both to have their cake and to eat it.[81] Common carriage is defined by the duties imposed on public networks in exchange for their right to use public property as a right of way, and other privileges. The telecommunications network is a common carrier, as is the public road. Noam explains that:

> When historically they [infrastructure services] were provided in the past by private firms, English common law courts often imposed some quasi-public obligations, one of which one was common carriage. It mandated the provision of service to willing customers, bringing common carriage close to a service obligation to all once it was offered to some.[82]

We need to explore this history in order to explain that the policy impetus behind net neutrality is not 'a problem in search of a solution' but a return to

[79] Harbour (2013).
[80] De Sola Pool (1983).
[81] See Frieden (2010) and Werbach (2010).
[82] Noam (1994).

classical ideas of how communications networks serve the public interest. In this history, I draw on eminent US communications scholars, mathematician Odlyzko[83] and lawyer Cherry,[84] as well as UK-based legal academics Atiyah, Otto-Freund, and the great Victorian legislator William Ewart Gladstone, author of the Railways Act 1844, the model of modern communications legislation.[85] This section may appear to consist of nineteenth-century 'train spotting' at first glance, but it is essential to realise that Internet access in the twenty-first century is comparable to railways in the nineteenth: it transforms economy, society and ecology in extraordinary ways. Internet access, like train access, is not a 'widget' problem of competition law, but a communications network with deep roots in the body politic and citizens' daily lives. That does not mean it needs no competition, but that it requires far more innovative practices that must not be closed off by the company that provides local access. This was recognised by Gladstone in 1844, and should be recognised by legislators today.[86]

Common carriers in mediaeval times included farriers and public houses (every horse to be shoed and person to be allowed shelter without discrimination between travellers). In *Lane v. Cotton* (1701), Sir John Holt CJ stated: 'If a man takes upon him a public employment, he is bound to serve the public as far as the employment extends; and for refusal an action lies, as … Against a carrier refusing to carry goods when he has convenience, his wagon not being full.'[87] Holt CJ limited liability in the landmark case of *Coggs v. Bernard* in 1703,[88] which led to an expansion of common carriers taking advantage of the special legal status. In refusing to impose strict liability, he relied on his interpretation of what was known of Roman law, overturning the 1601 precedent in *Southcote's Case*.[89] Citizens have ancient rights of way and of service 'by the custom of the realm', inherited by the American colonies far before the original Tea Party.[90]

The UK Carriers Act of 1830 was the first legislation for carriage of goods by land, codifying the common law and replacing the traditional tort of bailment as condition for carriage of goods.[91] The Act applied to all common carriers by

[83] The application of his modelling to Internet traffic has been groundbreaking, see Odlyzko (1998, 2004, 2014a) and Table 4.

[84] Cherry (2006, 2008). This culminated in Cherry and Peha (2014).

[85] See Railway Regulation Act 1844, s.6, and Kahn-Freund (1963).

[86] Marsden (2015).

[87] *Lane v. Cotton* (1701) 1 Ld Raym 646, at 654.

[88] *Coggs v. Bernard* (1703) 2 Ld Raym 909, 13 William III. See Holmes (1881), p. 132.

[89] *Southcote's Case* (1601) 4 Co Rep 83b; Cro Eliz 815, discussed by Jones and Theobald (1833), extensively in footnote 13 at pp. 38–39, and in the text at pp. 41, 58–62 and xli.

[90] Jones and Theobold (1833) Appendix: Common Carriers, p. v. They devote the entire Appendix to describing the definitions and duties of common carriers.

[91] Carriers Act 1830, Chapter 68 11 Geo 4 and 1 Will 4, s.1, for 'more effectual Protection of Mail Contractors, Stage Coach Proprietors, and other Common Carriers'.

land, and defined a common carrier as any individual, firm or company (other than the government) who or which transports goods as a business, for money, from place to place, over land or inland waterways, for all persons (consignors) without any discrimination between them.[92] Sir William Jones argued that the 1830 Act left open the possibility that carriers are 'still at liberty to make a special contract in the usual form'.[93] The UK Railways Act 1844 included common carriage provisions for common carriage and 'Parliamentary trains':

> all Passenger Railway Companies ... shall, by means of One Train at least to travel along their Railway from one End to the other of each Trunk, Branch, or Junction Line ... once at least each Way on every Week Day ... provide for the Conveyance of Third Class Passengers ... The Fare or Charge for each Third Class Passenger by such Train shall not exceed One Penny for each Mile travelled.[94]

Common carriage should not be confused with charging tolls for higher speed networks, though the Turnpike Riots of eighteenth-century England were associated with turning the King's Highway into a private road, and UK opposition to road charging continues. Common carriage is not a flat rate for all packets, or necessarily a flat rate for all packets of a certain size. It is a *non-discrimination bargain*: for the privileges of classification as a common carrier, those private actors are granted the rights and benefits that an ordinary private carrier would not have. It should be noted that telegraph lines ran alongside railway lines, which led to a provision in the Railways Act 1844 that government could take over railways in time of war, a power reproduced in the US Pacific Telegraph Act 1860, and that modern telecommunications run in part alongside railway lines, with the original alternative infrastructure to British Telecom in the UK being that of British Rail (later Racal) Telecommunications, which contracted with BT's competitors to offer them backhaul – not least because, as a common carrier, railways provide secure rights of way for their services.[95]

The monopoly of railways with the historic UK 'mania' booms in investment in the 1830s and 1840s (to be repeated in the US in the 1860s and 1870s),[96] and their clear superiority over canals and all other forms of conveyance of goods, led to calls to extend the law beyond the 1844 Act. The common carriage requirement was fully brought in by the Railway and Canal Traffic Act 1854 s.7,

[92] Note that a carrier must carry goods of the consignor for hire and not free of charge in order to be called a common carrier. Further, he must be engaged in the business of carrying goods for others for money from one place to another. A person who carries goods occasionally or free of charge is not a common carrier.

[93] Jones and Theobold (1833), Appendix: Common Carriers, p. xii.

[94] Later extended by the Railway and Canal Traffic Act 1854, s.7, abolished in the Transport Act 1962, s.46(3), but maintained in the Standard Conditions of Carriage of the British Railways Board, and for carriage by road, canal and carriage of goods by sea.

[95] Kessell (2015).

[96] See Odlyzko (2014b).

which imposed liability on the railway company for 'neglect or default'. Where the operator attempted to limit or exclude his liability for such loss in the contract for carriage, the exclusion clause would only operate where it was 'just and reasonable'. The 1854 Act led in the 1860s to a perceived need to further tighten regulation of charges, and in 1865 a Royal Commission and Select Committee inquiry led by Lord Carlingford, president of the Board of Trade. This concluded that his own Board of Trade was not 'sufficiently judicial', that the courts were insufficiently expert on railways, and that Parliament was ill-equipped for such a permanent role. The Railway and Canal Traffic Act 1873 then created the specialist Court of the Railway and Canal Commission to enforce the 1854 Act.

In court practice, the common carriage requirement led the 'fanatical adherent to freedom of contract'.[97] Bramwell B held in *Vaughan v. Taff Vale* (1860) that where railways offered insurance to cover liability for customers, then declining that insurance would shift liability entirely onto the customer.[98] This watering down of Parliament's 1854 intention was continued by Blackburn J in *Peek v. The North Staffordshire Railway Company* (1863), who stated that:

> a condition exempting the carriers wholly from liability for the neglect or default of their servants is prima facie unreasonable. I do not go as far as to say that it is necessarily in every case unreasonable and void, if [a carrier] offers in the alternative to carry on terms that he shall have no liability at all and holds forth as an inducement a reduction of the price below that which would be reasonable remuneration for carrying at owner's risk, [...] I think that a condition thus offered may be reasonable enough.[99]

The definition of traffic was later clarified and limited in *Spillers and Bakers Ltd v. Great Western Railway Company* (1911).[100] Hodges argued that rail transport was too great a public common carriage to be left to contracts adjudged by fanatical laissez-faire Victorian judges, approving of Cardwell's Railway and Canal Traffic Act of 1854:

> [t]he necessity of a supervision of some kind over the traffic on our railways has long been acknowledged and it was felt that it would be an intolerable abuse if the Queen's subjects were deprived (by the railways) of the protection which the crown formerly afforded them when travelling over the ancient highways. Moreover, it may be assumed that the need for rigorous control and supervision is even more necessary than formerly when before the railways there could be no monopoly of the means of conveyance.[101]

[97] Atiyah (1980).
[98] *Vaughan v. Taff Vale Railway* (1860), 5 H&N 679 157 ER 1351.
[99] *Peek v. The North Staffordshire Railway Company* (1863) 10 HLC 473, at 557.
[100] *Spillers and Bakers Ltd v. Great Western Railway Company* (1911) 1 KB 386.
[101] Hodges and Manley Smith (1876).

The final nail in the coffin of railways common carriage came with the Transport Act 1962 ss.43, 46(3), which removed all common carriage liabilities from the now-nationalised railways. Kahn-Freund stated: 'the Act goes much further in giving effect to laissez-faire in the law of transport than English law has ever done at any time since the seventeenth century.'[102] Common carriage thus has a somewhat unhappy judicial history in mainland UK, with the nationalised nature of telecommunications somewhat obscuring the picture in that industry. While I do not caution that UK judges are as fanatically free market and anti-consumerist as their early Victorian forebears, it is worth contemplating how a common law approach to the tortious liabilities of access providers may arrive at very different conclusions than European consumer law.

In mass communications in the US, 'the issues comprising Net Neutrality have been around since the *Pacific Telegraph Act of 1860* and they are here to stay whether the 2015 *Open Internet Order* survives judicial review or not.'[103] Article 2 of that Act states: 'messages received from any individual, company, or corporation, or from any telegraph lines connecting with this line at either of its termini, shall be impartially transmitted in the order of their reception, excepting that the dispatches of the government shall have priority.'[104] The US Supreme Court in 1901 confirmed that a public telegraph company (and more especially the largest) has a duty of non-discrimination towards the public.[105] Telecoms networks were established to be common carriers as they achieved maturity, following telegraphs, railways, canals and other networks. Noam explained in 1994: 'it is not the failure of common carriage but rather its very success that undermines the institution. By making communications ubiquitous and essential, it spawned new types of carriers and delivery systems.'[106] He forewarned that net neutrality would have to be the argument employed by those arguing for non-discriminatory access, as well as accurately predicting the death of common carriage ten years later.

Common carriers are under a duty to carry goods lawfully delivered to them for carriage. The duty does not prevent carriers from restricting the

[102] Kahn-Freund (1963). However, until the Railways Act 1993 (which not only privatised the railways but substituted freedom of contract), the 1962 Transport Act was observed more in the breach as contract replaced statute, maintained in the Standard Conditions of Carriage of the British Railways Board, and for carriage by road, canal and (at least theoretically) carriage of goods by sea.

[103] Quatrocchi (2015).

[104] Pacific Telegraph Act of 1860, 18 June.

[105] See *Western Union Telegraph Co. v. Call Publishing Co.* 181 US 92, 98 (1901).

[106] Noam (1994) p. 435, explaining that: 'When historically they [infrastructure services] were provided in the past by private firms, English common law courts often imposed some quasi-public obligations, one of which one was common carriage. It mandated the provision of service of service to willing customers, bringing common carriage close to a service obligation to all once it was offered to some.'

commodities that they will carry. Carriers may refuse to carry dangerous goods, improperly packed goods or those that they are unable to carry (on account of size, legal prohibition or lack of facilities). This definition offers several reasons for refusal of common carriage that can be extended to IAPs – for instance, spam and viruses may be refused. In common law countries such as the UK and US, carriers are liable for damage or loss of the goods that are in their possession as carriers, unless they prove that the damage or loss is attributable to certain excepted causes (e.g. 'Acts of God').[107] That provides several more reasons for loss – one thinks of the loss of undersea cables, or alleged foreign power Denial of Service (DoS) attacks. It might be stretching a definition to suggest that P2P streams can be 'jettisoned' in order to allow other traffic to progress during peak time congestion.

Twenty-first-century IAPs who choose to manage traffic in a discriminatory fashion cannot be considered common carriers. Chapters 1 to 4 deal with the legal implications of abandoning common carriage and the need for net neutrality rules to replace them. If they cease being common carriers, they then open themselves to liability for the 'cargo' they inspect before they agree to carry it on a discriminatory basis. Chapter 5 deals with that issue.

Deep packet inspection and traffic management

In order to manage traffic, new technology allows any of the IAP routers (if so equipped) to look inside an unencrypted data packet to 'see' its content, via DPI and other techniques. Previous routers were not powerful enough to conduct more than a shallow inspection that simply established the header information – the equivalent of the postal address for the packet. An IAP can use DPI to determine whether a data packet values high-speed transport – as a television stream does in requiring a dedicated broadcast channel – and to offer higher speed dedicated capacity to time-dependent content such as HD video or voice calls using VoIP. That could make a good business for IAPs that wish to offer higher capability for 'managed services' via DPI.[108] Not all IAPs will do so, and it is quite possible to manage traffic less obtrusively by using the DiffServ protocol to prioritise traffic streams within the same Internet channel.[109]

[107] In the wonderfully descriptive language of the common law: 'Fault of the shipper as an excepted cause is any negligent act or omission that has caused damage or loss – for example, faulty packing. Inherent vice is some default or defect latent in the thing itself, which, by its development, tends to the injury or destruction of the thing carried. Fraud of the shipper is an untrue statement as to the nature or value of the goods. And jettison in maritime transport is an intentional sacrifice of goods to preserve the safety of the ship and cargo.' See Longley (1967) and references in Noam (1994).

[108] Frieden (2008).

[109] Brown and Marsden (2013a), p. 144.

DPI and other techniques that let IAPs prioritise content also allow them to slow down other content, as well as speed up content for those users who pay (and for emergency communications and other 'good' packets). This potentially threatens competitors using that content: Skype offers VoIP using normal Internet speeds; uTorrent and BBC's iPlayer have offered video using P2P protocols. Encryption is common in these applications and partially successful in overcoming these IAP controls, but even if all users and applications used strong encryption, this would not succeed in overcoming decisions by IAPs simply to route known premium traffic to a 'faster lane', consigning all other traffic to a slower, non-priority lane (a policy explanation simplifying a complex engineering decision). P2P is designed to make the most efficient use of congested networks, and its proponents claim that, with sufficient deployment, P2P could largely overcome congestion problems.

In 2009 congestion on the Internet was said to be caused by P2P file sharing, and consumer advocates feared DPI leading to pervasive Internet monitoring by IAPs and advertisers. Seven years later, how quaint these fears seem. First, P2P is no longer seen as a significant cause of congestion, but merely an artefact of the midband decade in which consumers struggled with 256 Kbps–4 Mbps connections. Today it is UHDTV video streaming and downloading which is the concern. Video streaming arose as a policy concern with the blocking of both Norwegian and UK state broadcaster video streaming in the mid-2000s, which led both nations to a co-regulatory solution with varying levels of success. In the period since 2010, it has been redefined as a commercial concern arising in the US with NetFlix and in varying manifestations in other nations. In addition, audio streaming on the far more limited mobile bandwidth has been a concern with various IAPs providing 'free' (i.e. positively discriminated) offers for music services, such as Spotify. This is a 'legitimate' successor to the P2P service that accompanied much of the early net neutrality controversy surrounding Napster. Net neutrality may therefore be returning to its roots in the 1990s, when Lemley and Lesssig first identified it as a video over Internet issue – turning the Internet into an on-demand cable TV service where Internet traffic becomes a second-class service to the first-class proprietary video (and audio) offer.

As what of pervasive monitoring? In June 2013 National Security Agency contractor Edward Snowden left his Hawaii home for Hong Kong and then Russia. He had given a hard drive of classified documents to documentary film maker Laura Poitras and investigative journalist Glenn Greenwald. These proved beyond all reasonable doubt that major IAPs had collaborated for years with security agencies in the US, UK and many other countries to provide monitoring in real time and via retained historic browsing data of all Internet users in those territories and whose traffic passed through those territories. Moreover, spyware had been used to infect user machines and to trace user behaviour online. The era of Total Information Awareness and surveillance was shown to have started. This made concerns regarding DPI and advertising both proven and trivial – yes, IAPs clearly had the ability to track all users, but more importantly they had been using this capacity for years and it was funded by both advertisers and in advanced projects by the security agencies. Snowden's revelations revealed how acutely the privacy concerns of advocates were shown to have been under- rather than over-blown.[110] Net neutrality, defined as freedom to use the Internet without interference, may be considered a pre-Snowden anachronism given the knowledge he revealed. Bear in mind that what Snowden knew is historical, dating to 2013, and surveillance has advanced considerably in the intervening period.

Back to the future: *plus ça change ...*

Cast your mind back to the start of 2009. It was three months after the global financial crisis had sent the developed world's economies into near meltdown. This in part secured the election of the first black US president, who pledged to rescue the economy through infrastructure spending, restart US relations with the many countries opposed to the invasions of Iraq and Afghanistan, close the Guantánamo Bay torture facility and secure something called 'net neutrality'. It was barely eight years since the 'dot-com' meltdown that bankrupted Internet and telecoms corporations worldwide, ending the great Internet boom of the late 1990s and the largest consumer boom that the US and UK had ever seen. In spring 2009 I wrote a book about how net neutrality could best be secured by using a mix of co-regulation and pragmatism rather than competition law or engineering alone.

Those conclusions stand, though there is now an enormously greater store of empirical evidence to bring to bear and the solutions are clearer and more challenging than it appeared then. This is no victory lap, nor have my conclusions changed. What has changed is that there is now much more legislation and regulation for a socio-legal scholar to write about. There is also a much larger canvas on which to paint a story about net neutrality, not only because we are now acknowledged to live in a post-Foucaultian nightmare of control and surveillance, but also because the Internet is as ubiquitous as its proponents claimed it would become. In 2009 average Internet download speeds in the UK were 4Mbps. P2P file sharing was threatening congestion on slow broadband networks. Piracy threatened the future of the music and movie industries, claimed their intellectual property lawyers. Facebook was overtaking MySpace as the largest social network in the English-speaking world, with one in ten people in the UK having signed up. Google was emerging as a dominant search

[110] Richards (2015).

engine that would be affected by competition law investigations.[111] No-one had heard of Spotify, Twitter, WhatsApp, SnapChat, Instagram or NetFlix. You could not use a 3G network to download apps from the iTunes Store, because the iPhone was only launched in mid-2007 and data acquisition was so slow and expensive that it was assumed downloads all took place on Wifi. The iPad and all other tablet computers did not exist. No one had heard of Julian Assange or Wikileaks, let alone Edward Snowden. The extraordinary information-sharing capacities of the Internet have created huge problems and capabilities for governments and corporations that invest in trying to control information and how citizens use that power. A purity of net neutrality intentions or declarations is hence impossible, whatever advocates may argue. Internet surveillance is pervasive for all but the most advanced users of military surveillance-strength encryption. The 2009 book was written before Wikileaks; this book is written far after Snowden. We need to exhibit realism not naivety in the legal expectations placed on IAPs, to govern what is now known of the extra-legal powers that were exerted to persuade those IAPs to participate in mass surveillance.

The single most interesting aspect of net neutrality is the fierce fight that has been waged for over 15 years to secure the future of the Internet: the international political economy of net neutrality, and the institutional economic aspect. Telecoms companies and their lobbyists first claimed it is of no relevance, then fought fiercely to oppose it. It was meant to be solved in Europe in 2009, when options for regulation were attached to the 'telecoms package'. It was meant to be solved in the US when Obama's first FCC chairman announced that there would be a consultation, then an 'Open Internet Order' in 2009–10. It was meant to be irrelevant to mobile data or developing countries because net neutrality was a luxury problem, not a question of universal access and human rights. But net neutrality is the policy gift that keeps on giving, encompassing all of these areas in ever greater profundity and detail. It will keep academics in articles for decades. We have moved on from the innocence of the 1990s, when it could be declared that:

> The introduction of the Internet was accompanied by evolving procedures and behavioural patterns among its users. A new field of industry self-regulation has emerged in relation to the Internet: 'Netiquette' was the first informal code of conduct ... Codes of practice are needed to regulate issues like respect for privacy, public decency, and protection of minors, accuracy or the application of filtering software.[112]

IAPs never did netiquette for last mile access; they relied on hard regulation and softer rules for interconnection.[113] While there are informal standards – for

[111] Pollock (2010). Pollock's paper had been presented in drafts since 2007.

[112] Kleinsteuber (2004), pp. 61–75.

[113] BEREC BoR (12) 33, EC (2014).

instance BT has been remarkably generous in not cutting off users of 'spammy' competitor IAPs – hard rules and hard cash rules this field. The elements of self- and co-regulation that prevail in technical standard setting and much technical interconnection do so because there is both 'positive sum' economic self-interest in expanding the Internet market for all IAPs,[114] and also the 'zero sum' backstop of a regulator and ultimately courts enforcing the meta-narrative of a rule-based game. I explained in 2008 that this may be an emerging 'middle mile' net neutrality problem, with IAP discrimination against CDNs and other actors, which is discussed in Chapter 3.[115]

The net neutrality problem is complex and far-reaching: European attempts to dismiss it as a problem that can be overcome by local loop (last mile) telecoms competition fail to acknowledge persistent problems with market failure. The physical delivery of Internet to consumers is subject to a wide range of bottlenecks, not simply in the 'last mile' to the end user. There is little 'middle mile' (backhaul) competition in fixed IAP markets, even in Europe where the commitment to regulation for competition remains, as wholesale backhaul is provided by the incumbent privatised national telecoms provider (in the UK, British Telecom). Even if platforms did compete in, for instance, heavily cabled countries, there would remain 'n-sided' market problems in that there is no necessary direct (even non-contractual) relationship between innovative application providers and IAPs, for instance a Korean games developer and a UK IAP.[116] Platforms may set rules to 'tax' data packets that ultimately impoverish the open innovation value chain, so ultimately causing consumer harm. Thus the archetypal garage start-ups such as Facebook (founded in 2003) and YouTube (founded in 2005) would have had less opportunity to spread 'virally' across the Internet, as their services would have been subject to these extra costs.

We need to dig deeper into why IAPs want to infringe on neutrality in the first place, and how policymakers responded short of legislation in the period to September 2013. These are the foci of Chapters 1 to 3.

[114] D'Ignazioa and Giovannetti (2015).

[115] Candeub (2015) references my arguments at note 55.

[116] Economides and Tåg (2007).

1

A brief history of net neutrality law

> There have been suggestions that we don't need legislation because we haven't had it. These are nonsense, because in fact we have had net neutrality in the past – it is only recently that real explicit threats have occurred.
>
> Sir Tim Berners-Lee[1]

We must begin with a quick note on terminology: we are concerned with access to the last mile, not generic services on the Internet. The network access providers are Internet Access Providers (IAPs), not Internet Service Providers (ISPs). Open Internet protection affects IAPs, not generic ISPs (though IAPs offer services too, and are thus both IAPs and ISPs). In this book I will refer specifically to IAPs (and much less often generic ISPs), though when I quote regulators such as CRTC (Canadian Radio-television and Telecommunications Commission), BEREC (Body of European Regulators of Electronic Communications) and Ofcom (Office of Communications Regulation), I will not correct their use of the more generic term 'ISP' even though they are specifically referring to IAPs. So why do they use this term 'ISP'?

The term 'ISP' has different meanings in different contexts, though it is used much more often than are more legally specific terms. In Europe, a provider of Internet access is an Electronic Communications Network Provider (ECNP), whereas a provider of content and services is termed an Information Society Service Provider (ISSP)[2] or an Audiovisual Media Services (AVMS) provider where it is the editorial controller of video.[3] Under European law:

> 'electronic communications service' means a service normally provided for remuneration which consists wholly or mainly in the conveyance of signals on

[1] Berners-Lee (2006).

[2] Directive 2000/31/EC, Art. 2(a), reiterating Art. 1(2) of Directive 98/34/EC, as amended by Art. 1(2)(a) and Annex V of Directive 98/48/EC.

[3] See Directive 2010/13/EU, Art. 1(a)(i) and Art. 2.

> electronic communications networks, including telecommunications services and transmission services in networks used for broadcasting, but exclude services providing, or exercising editorial control over, content transmitted using electronic communications networks and services.[4]

The definition explicitly excludes ISSPs 'which do not consist wholly or mainly in the conveyance of signals on electronic communications networks'. In the US, the access provider is an Internet Access Provider (IAP), and the service provider an Online Service Provider (OSP) under the Digital Millennium Copyright Act 1998 (DMCA),[5] though a further distinction lies between access providers classified under Title I and Title II of the Telecommunications Act 1934.[6] In Section 512 an OSP is defined as 'an entity offering transmission, routing, or providing connections for digital online communications, between or among points specified by a user, of material of the user's choosing, without modification to the content of the material as sent or received' or 'a provider of online services or network access, or the operator of facilities thereof'.[7] This broad definition includes network services companies such as access providers, search engines, bulletin board system operators and even auction websites. It should be noted that most (if not all) access providers are also service providers, and in fact the largest IAPs are also amongst the largest service providers.

Net neutrality was regulated narrowly in the United States, Canada and Europe in 2009 (the latter via a Declaration and amendments to the 2002 Electronic Communications Package). My 2010 book analysed developments to that 2009 settlement, and detailed US regulation by the FCC (though not its eventual but predictable demise in the 2014 District of Columbia Appeals Court decision in *Verizon v. FCC*). I also critically assessed the 2009 European amendments, and gave (accurate) predictions for their failure in practice. I summarise that long development in the first part of this chapter. Development of European legal implementation of the network neutrality principles has been slow.[8] I explain in the second part of this chapter that, at European Member State level, only Netherlands, Finland and Slovenia had passed laws by the end of 2014. I summarise the outcome of 2014/15 legal manoeuvres in both the United States and European Union. Chapter 3 considers Specialised Services in both areas, and then Chapter 4 examines European law in minute detail. Note that as this book is written with Europe as the focus, US regulation is described more briefly.

[4] Directive 2002/21/EC (Framework Directive), Art. 2(c).

[5] Online Copyright Infringement Liability Limitation Act (OCILLA) 1998, which amended the 1976 Copyright Act, passed as a part of the 1998 Digital Millennium Copyright Act (DMCA) and referred to as the 'Safe Harbor' provision because it added Section 512 to Title 17 of the United States Code.

[6] 47 USC §201(a) and (b). See Chapter 5.

[7] DMCA 1998, s.512(k)(1)(A–B).

[8] Cave (2011).

The development of net neutrality regulation

US regulation of network neutrality has a history dating back to 1999, and was introduced via merger conditions placed on major IAPs. The debate began when academics feared that cable TV's closed business model would overtake the Open Internet in 1999.[9] While issues about potential discrimination by IAPs have been current since at least 1999, the term 'network (net) neutrality' was coined in 2003.[10] The pre-history of United States regulation prior to the 2015 Open Internet Order is well-documented,[11] with the 2010 Order[12] being highly controversial in its exclusion of mobile ('wireless'), resulting in several data caps being imposed, notably by AT&T in 2011,[13] zero-rating plans being adopted and the Order itself becoming incapable of effective enforcement following a litigation which ended in 2014.[14] Only lawyers may take joy that the FCC has spent a decade trying to enforce net neutrality since its original regulatory declaration.[15]

Data caps have been controversial throughout the consumer Internet's history, especially in the United States where dial-up Internet was virtually free to the end user (simply the cost of a local telephone call). The FCC Open Internet Advisory Committee in 2013 noted the move towards capping data especially for mobile users and worried 'whether caps or thresholds that are set too low could lead to a world where the average user carefully monitors her bandwidth use' given uncertainty over data caps as a 'transitory or permanent concern', which appears to be the case in developing (and many developed) nations' mobile data access.[16] While data caps apply in many nations and are applied by many IAPs, the user often has little or no idea that they are approaching their monthly limit until informed by the IAP, and such warnings are often inaccurate. It is at best a blunt weapon for handling congestion, though there is little argument that data caps per se do not infringe net neutrality, as long as the cap gradually increases over time. The OECD states: 'zero rating can clearly be pro-competitive ... [and] becomes less of an issue with ... higher or unlimited data allowances. Regulators need to be vigilant.'[17]

Competition in the US is 'inter-modal' between cable and telecoms, not 'intra-modal' between different telecoms companies using the incumbents'

[9] Lemley and Lessig (2000). In Europe, see Marsden (1999) at Section 5.1.

[10] Wu (2003b).

[11] Marsden (2013a).

[12] FCC, Report and Order Preserving the Open Internet, 2010.

[13] See Kang and Tsukayama (2011).

[14] *Verizon v. Federal Communications Commission* 740 F.3d 623 (D.C. Cir. 2014); 11–1355.

[15] FCC, Internet Policy Statement 05–151, 2005; FCC, *Madison River Communications*, LLC, Order, DA 05–543, 2005.

[16] Open Internet Advisory Committee (2013a), p. 13.

[17] OECD (2015).

exchanges to access the 'Last Mile'.[18] Instead of regulated access to telecoms networks, the US has less regulated broadband 'information', not 'telecommunications' services.[19] In 2004 deregulatory FCC Chair Michael Powell declared:

> I challenge the broadband network industry to preserve the following Internet Freedoms: Freedom to Access Content; Freedom to Use Applications; Freedom to Attach Personal Devices; Freedom to Obtain Service Plan Information.[20]

The 'Four Freedoms' were formalised as regulatory policy in the FCC Internet Policy Statement of August 2005.[21]

In *Madison River*,[22] the FCC enforced these policy principles. Madison River is a small consumer IAP and telephone company, not a large national carrier, which was ordered by the FCC to stop blocking rival VoIP services. Madison River's abuse of its access monopoly was incontrovertible; the vertical integration of the IAP with its voice telephone service meant it had obvious incentives to block its competitor VoIP services, and the practice was intended to degrade its customers' Internet access. It was an example of negative network neutrality: customers signed up for broadband service with the IAP, but it chose to degrade that service in the interest of preserving its monopoly in telephone service.

The 2007 merger of AT&T and BellSouth involved both parties assuaging net neutrality concerns by various commitments not to block other companies' applications directed to their users.[23] The FCC then made a major intervention with its 2008 Order against Comcast, a major cable broadband IAP.[24] Comcast's deposition to the FCC stated that it began throttling P2P file-sharing application BitTorrent from May 2005 until 2007, using Sandvine DPI technology. The FCC ruling against Comcast's attempts to stop P2P by sending phantom reset packets to customers reflects another 'easy' case of abuse of customers, like the VoIP blocking in *Madison River* in 2005.[25] Table 2 shows the most important regulatory developments since 2005 in three overlapping phases: 2005–10; 2009–14 and 2014–16.

Net neutrality regulation was supplemented by legislation after the election of Barack Obama as president in 2009. The infrastructure spending stimulus law titled the American Recovery and Reinvestment Act 2009 included a

[18] Communications Act of 1934 as amended by Communications (Deregulatory) Act of 1996, 47 USC.
[19] Candeub and McCartney (2012).
[20] Powell (2004).
[21] FCC, Internet Policy Statement 05–151, 2005.
[22] FCC, *Madison River Communications*, LLC, Order, DA 05–543, 2005.
[23] FCC, *In AT&T Inc and BellSouth Corp*, 2007.
[24] FCC, *In AT&T Inc and BellSouth Corp*, 2007. FCC, *In AT&T Inc and BellSouth Corp*, 2008.
[25] See Karpinski (2009).

Table 2 United States regulation and litigation on the Open Internet 2007–15[a]

Phase of Regulation	Phase 1: Policy 2005–10	Phase 2: Open Internet under Title I, 2010–14	Phase 3: Title II Open Internet, 2014–
Challenge to Existing Regulation	Internet Policy Statement 2005 following 'Four Freedoms' speech 2004	Open Internet Order	Response to *Verizon v. FCC* (2014)
Commission Proceeding	Comcast investigation 2007–08[b]	Preserving the Open Internet (2009)[c]	Protecting and Promoting Open Internet 2014[d]

[a] An excellent narrative account of the policy and judicial history can be found in Feld (2015).

[b] FCC, Formal Complaint of Free Press and Public Knowledge Against Comcast Corporation for Secretly Degrading Peer-to-Peer Applications, 2008.

[c] FCC, Report and Order, In the Matter of Preserving the Open Internet; Broadband Industry Practices, 2009.

[d] FCC, Notice of proposed rulemaking, 2014.

broadband open access stimulus.[26] This extended broadband into under-served areas via federal grants, with open access and net neutrality provisions built into the grants.[27] The FCC then consulted through 2009–10 on its Net Neutrality Order ruling of 23 December 2010,[28] which was challenged before the courts in 2012–14. The FCC refused several times to intervene in interconnection and peering disputes that were claimed by CDNs to unreasonably impair traffic, contrary to the controversial net neutrality rules in that 2012–14 period.[29]

In 2014 the FCC revisited network neutrality in view of the loss of the court case over the 2010 Open Internet Order,[30] resulting in an inevitable further legal challenge in 2016. It appears unlikely that general net neutrality (as opposed to specific merger) conditions can be made that can survive court challenge before the end of the Obama presidency in 2017.

[26] American Recovery and Reinvestment Act 2009.

[27] FCC, Report and Order, In the Matter of Preserving the Open Internet, 2009, especially footnotes 62–63.

[28] FCC, Report and Order Preserving the Open Internet, 2010.

[29] Frieden (2012).

[30] *Verizon v. Federal Communications Commission* 740 F.3d 623 (D.C. Cir. 2014); 11–1355, 14 January.

Reasonable network management and regulatory consultation

The phrase 'reasonable' in connection with IAP traffic management was first included in footnote 15 to the 2005 Internet Policy Statement. It was designed to ensure that an IAP must demonstrate both that its management purpose is reasonable and that it has used a minimally invasive means of so doing, in language borrowed from the US courts' approach to regulation of speech. It was thus a tough two-part test that the IAP 'practice should further a critically important interest and be narrowly tailored to serve that interest'.[31] The FCC expanded on this principle to explain exceptions in 2009. IAPs:

> may employ generally accepted technical measures to provide acceptable service levels to all customers, such as caching and application-neutral bandwidth allocation, as well as measures to address spam, denial of service attacks, illegal content, and other harmful activities.[32]

This means IAPs can deploy to prevent behavioural advertising by third parties, but not to enhance that advertising in a discriminatory fashion themselves, which we examine in Chapter 5 in the European case, and which is the subject of regulatory proceedings to apply s.222 Telecommunications Act 1934 in the United States. Denial of Service (DoS) is a technique for damaging websites via a flood of traffic that causes congestion. The FCC made clear that it is intended to prohibit all non-critical traffic management, though noting that technologies will differ in criticality as between co-axial cable, copper telecoms, fibre broadband and wireless systems: 'We believe that a bright-line rule against discrimination, subject to reasonable network management and enumerated exceptions, may better fit the unique characteristics of the Internet [than a less clear rule].'[33] It was, however, less tightly drawn than the 2005 Order's language of 'narrowly drawn' and 'critically important', which in 2009 it described as 'unnecessarily restrictive'.

The question is what 'harmful activities' involves, in relation to what is not 'generally accepted technical measures'. This must be subject to change over time, such that industry can agree on particular measures that are commonly used. The definition of harmful measures will depend on both the network's robustness and the particular measure, with DoS an obvious example of harmful activity. It will thus be legal for an IAP to intervene to stop a flood of DoS traffic (though such activities will change over time: a million simultaneous requests in 2003 on dial-up would cripple an IAP; in 2013 on broadband that is less likely). FCC Commissioner Copps stated that:

[31] Karpinski (2009), p. 47.

[32] Broadband Initiatives Program (2009).

[33] FCC, Report and Order, In the Matter of Preserving the Open Internet; Broadband Industry Practices, 2009.

> What constitutes reasonable network management in a 768 Kbps world will likely be different from reasonable network management in a 50 or 100 Mbps world.[34]

Implementation of the technical means for measuring reasonable traffic management had been tested in a self-regulatory forum established after the 2010 Order, the Broadband Internet Technical Advisory Group (BITAG). Its specific duties include offering 'safe harbor' opinions on traffic management practices of parties making formal reference for an advisory technical opinion.[35]

The 26 February 2015 Open Internet Order applied from 12 June 2015 and promised to enforce net neutrality.[36] The FCC claimed that the Order offered three 'Bright Line Rules', as already discussed in the Introduction.

Zero rating controversy

In 2016 zero rating was becoming a common practice in the US. It appeared to reflect paid prioritisation (though details are limited by commercial confidentiality), the final FCC 'bright line', and where it also involves degrading non-zero-rated content, also the second line on throttling. T-Mobile offered 33 zero-rated music services in its Music Freedom Plan since 2014,[37] which avoided any negative regulatory scrutiny in part due to the facts: its offer is non-exclusive, relates to music rather than heavily congesting and expensive video, and T-Mobile itself is the smallest of the national mobile IAPs. Goldstein argues:

> Music Freedom plan is inclusive and supports numerous streaming music services, and since T-Mobile does not receive compensation from any company for not counting music streaming traffic against customers' data limits, such a plan is likely going to be fine by the FCC, since it benefits consumers. However, if a zero-rating plan were exclusive to one company that offers a particular type of service, that likely would draw more scrutiny from the FCC.[38]

The decision by T-Mobile to introduce its BingeOn video service in autumn 2015, following the success of its music service, caused huge controversy, in part because BingeOn not only offered a wide range of video services without affecting the user's data cap, but also automatically throttled (i.e. degraded) video service from all providers down to DVD quality (below HD), whether within the free data offer or not. All T-Mobile users with a

[34] Copps, M., quoted in FCC, Report and Order, In the Matter of Preserving the Open Internet; Broadband Industry Practices, 2009 at 94–5.

[35] Broadband Internet Technical Advisory Group (2011).

[36] FCC, Open Internet Order, 2015.

[37] Northrup (2015).

[38] Goldstein (2015).

monthly 3GB or greater data cap were enrolled, and had to specifically delist in order to view HD video from providers such as Amazon or YouTube. Ammori wrote that:

> Degrading video quality this way violates the FCC's no-throttling part of the net neutrality rule, which forbids reducing the quality of an application or an entire class of applications … As a purely legal matter, T-Mobile cannot easily defend its actions by arguing that this discrimination is good for its users. The FCC has already rejected that argument in advance by adopting a 'bright-line' rule for all technical forms of discrimination absent some special technical justification … The FCC made it clear that throttling was 'inherently … unjust and unreasonable', so it 'bann[ed] conduct that … inhibits the delivery of … particular classes of content, applications, or services'. Said another way, and said again by the FCC, 'if a broadband provider degraded the delivery of a particular application … or class of application … it would violate the bright-line no-throttling rule.'[39]

The FCC announced that it would be asking companies to respond by 15 January 2016 to such complaints, including not just those about T-Mobile and AT&T's free mobile data, but others, such as Comcast, who claimed their Stream TV video offer was a Specialised Service, with an expectation that an investigation would be opened in February.[40] Such services are considered in Chapter 3.

The US Federal Appeals Court heard the oral pleadings about the legitimacy of 2015 Open Internet Order on 4 December 2015, and was expected to deliver judgment in spring 2016. Frieden, the foremost academic chronicler and analyst of FCC Open Internet policy, analysed the 2016 litigation possibilities:

> FCC has acted in a manner predicted by Justice Scalia in 2002.[41] The Commission succeed in convincing a majority that it needed to ignore the telecommunications component to support a deregulatory regime. Now the Commission needs to convince an appellate court that the telecommunications component has become so important that it must be pulled from the deregulated safe harbor the FCC previously created. The Commission may not have sufficient persuasive power to finesse a changed regulatory classification based on a collection of conflicting factual and legal rationales.[42]

US regulation of net neutrality remained in a legal no man's land in 2016.

[39] Ammori (2015).

[40] Shepardson (2015).

[41] *National Cable & Telecommunications Association et al. v. Brand X Internet Services et al.* (2005) 545 U.S. at 1013–1014; see also Frieden (2006).

[42] Frieden (2015a, 2015b).

European Union law 2009–12

This section explores European legislative and regulatory responses to net neutrality in more detail. US net neutrality reforms were to be followed slowly in Europe, in reform of its 'Telecoms Package' completed in 2009 and implemented from 2011.[43] This difference led many political scientists to investigate the dynamics of the debate in both the US and Europe.[44] European law upheld transparency on a mandatory basis, and minimum QoS on a voluntary basis, under provisions in the 2009 framework. In its initial 2006 explanation of its reasons to review the 2002 Directives, the Commission noted the US net neutrality debate but did no more than discuss the theoretical problem.[45] Over 2007–08, the volume of regulatory reform proposals in the US, Japan, Canada and Norway had grown, along with consumer outrage at IAP malpractice and misleading advertising, notably over notorious fixed and mobile advertisements which presented theoretical laboratory maximum speeds on a dedicated connection subject to 'reasonable terms of usage' – which meant capacity constraints on a monthly basis, some of these on mobile as low as 100MB monthly download totals.[46]

Net neutrality became a significant issue, together with graduated response to copyright infringement (notoriously giving rise to the HADOPI law in France), in the European Parliament vote to reject the reforms at the First Reading in May 2009, prior to elections in June 2009. Amendments on consumer transparency and network openness were then offered to the European Parliament in the conciliation process. The new rules in two Directives were published in the Official Journal on 18 December and gave Member States 18 months to implement them (by 18 June 2011).[47] The Commission 'Declaration on Net Neutrality' called for reporting by end-2010:

> The Commission attaches high importance to preserving the open and neutral character of the Internet, taking full account of the will of the co-legislators now to enshrine net neutrality as a policy objective and regulatory principle to be promoted by national regulatory authorities(1) alongside the strengthening of related transparency requirements(2) and the creation of safeguard powers for national regulatory authorities to prevent the degradation of services and the hindering or slowing down of traffic over public networks.(3) The Commission will monitor closely the implementation of these provisions in the Member States, introducing a particular focus on how the 'net freedoms' of European citizens are being

[43] Marsden (2012a).

[44] Cooper and Powell (2011).

[45] COM(2006) 334.

[46] Leading to a significant emphasis in EC SEC(2007) 1472, pp. 90–102.

[47] Directive 2009/136/EC (Citizens' Rights Directive) and Directive 2009/140/EC (Better Regulation Directive).

safeguarded in its annual Progress Report to the European Parliament and the Council. In the meantime, the Commission will monitor the impact of market and technological developments on 'net freedoms' reporting to the European Parliament and Council before the end of 2010 on whether additional guidance is required, and will invoke its existing competition law powers to deal with any anti-competitive practices that may emerge.

(1) Article 8(4)(g) Framework Directive.
(2) Articles 20(1)(b) and 21(3)(c) and (d) of the Universal Service Directive.
(3) Article 22(3) of the Universal Service Directive.[48]

What was intended with regard to EU net neutrality was actually a very 'lite' approach, ensuring services are not blocked and/or degraded beyond usefulness:

(34) A competitive market should ensure that end-users enjoy the QoS they require, but in particular cases it may be necessary to ensure that public communications networks attain minimum quality levels so as to prevent degradation of service, the blocking of access and the slowing of traffic over networks.

In order to meet QoS requirements, operators may use procedures to measure and shape traffic on a network link so as to avoid filling the link to capacity or overfilling the link, which would result in network congestion and poor performance.

Those procedures should be subject to scrutiny by NRAs [National Regulatory Authorities] ... in particular by addressing discriminatory behaviour, in order to ensure that they do not restrict competition.

If appropriate, NRAs may also impose minimum QoS requirements on undertakings to ensure that services and applications dependent on the network are delivered at a minimum quality standard, subject to examination by the Commission.

NRAs should be empowered to take action to address degradation of service, including the hindering or slowing down of traffic, to the detriment of consumers.

However, since inconsistent remedies can impair the functioning of the internal market, the Commission should assess any requirements intended to be set by NRAs for possible regulatory intervention across the Community and, if necessary, issue comments or recommendations in order to achieve consistent application.[49]

The net neutrality provisions were transparency in new Article 20, and Article 22 (showing that if NRAs did anything, it should be subject to an effecttive veto by the Commission):

[48] Directive 2009/140/EC (Better Regulation Diective).
[49] Directive 2009/136/EC (Citizens' Rights Directive), Recital 34.

1. Member States shall ensure that NRAs are, after taking account of the views of interested parties, able to require networks and/or services to publish comparable, adequate and up-to-date information for end-users on the quality of their services ... That information shall, on request, be supplied to the NRA in advance of its publication.
2. NRAs may specify, inter alia, the quality of service parameters to be measured and the content, form and manner of the information to be published, including possible quality certification mechanisms, in order to ensure that end-users ... have access to comprehensive, comparable, reliable and user-friendly information.
3. In order to prevent the degradation of service and the hindering or slowing down of traffic over networks, Member States shall ensure that NRAs are able to set minimum quality of service requirements on an undertaking or undertakings providing public communications networks.

NRAs shall provide the Commission ... with a summary of the grounds for action, the envisaged requirements and the proposed course of action. This information shall also be made available to [BEREC]. The Commission may ... make comments or recommendations ... NRAs shall take the utmost account of the Commission's comments or recommendations when deciding on the requirements.

The 15th (and final) Implementation Report on the telecoms single market in May 2010 stated that VoIP blocking is the single biggest issue, and that the Commission will report by end-2010:

> In Italy, consumers are now able to withdraw from their contracts in case of divergence with the declared connection speed. The Slovenian NRA issued a recommendation on the provision of broadband speeds, and the Portuguese NRA published a report on the quality of service for access to internet services, highlighting the upload speeds and network latency as the main differences between fixed and mobile networks. The United Kingdom NRA carried out a broadband speeds survey comparing the service provision of the largest internet service providers. In Hungary, several operators were subject to fines for a failure to provide correct information.[50]

It also suggested that Member States should consult by end-2010 on national plans for net neutrality, with the result that BEREC,[51] Sweden,[52] UK,[53] France[54] and others did so,[55] with Italy following in 2011.[56] Non-members

[50] COM(2010) 253, p. 56.
[51] BEREC, BoR (10) 42.
[52] Swedish Post and Telecom Agency (2009).
[53] Ofcom, Traffic Management and net neutrality, 2010.
[54] ARCEP (2010b).
[55] IP/10/860.
[56] AGCOM (2011).

Canada,[57] Norway[58] and the United States[59] had already done so. The European Commission rather lamely concluded in November, from 318 responses, that everyone agreed the Open Internet was important, and presumably also that the Pope was Catholic.[60]

The Declaration, and the more legally relevant Directive clauses, rely heavily on implementation at national level and proactive monitoring by the Commission itself, together with national courts, and privacy regulators where content discrimination contains traffic management practices which collate personal subscriber data.[61] The Commission promised 'a particular focus on how the "net freedoms" of European citizens are being safeguarded in its annual Progress Report to the European Parliament and the Council'.[62] By the deadline for implementation of 18 June 2011, 20 NRAs had not implemented any part of the 2009 Directives, and the Commission opened infringement proceeedings (this is quite normal in the European Union).[63]

The European Commission consulted on implementation of its 2009 network neutrality laws in September 2010. It referred much of the detailed work to the new BEREC, which has developed an extensive work programme on net neutrality since 2010.[64] BEREC's 2010 response to the EC Consultation[65] concluded that mobile should be subject to the net neutrality provisions, although 'mobile network access may need the ability to limit the overall capacity consumption per user in certain circumstances (more than fixed network access with high bandwidth resources) and as this does not involve selective treatment of content it does not, in principle, raise network neutrality concerns'.[66] BEREC explained some breaches of neutrality: 'blocking of VoIP in mobile networks occurred in Austria, Croatia, Germany, Italy, the Netherlands, Portugal, Romania and Switzerland.'[67] They explain that though mobile will always need greater traffic management than fixed ('traffic management for mobile accesses is more challenging'), symmetrical regulation must be maintained to ensure technological neutrality: 'there are not enough arguments to support

[57] Privacy Commissioner of Canada (2009).

[58] See Norwegian Communications Authority, Net neutrality guidelines, 2013.

[59] FCC, Report and Order, In the Matter of Preserving the Open Internet; Broadband Industry Practices, 2009.

[60] IP/10/1482.

[61] See Directive 95/46/EC, Directive 2002/58/EC (E-Privacy Directive), Directive 2006/24/EC (Data Retention Directive).

[62] Directive 2009/140/EC (Better Regulation Directive).

[63] IP/11/905.

[64] See generally BEREC Net Neutrality Expert Working Group, available at http://berec.europa.eu/eng/about_berec/working_groups/net_neutrality_expert_working_group_/282-net-neutrality-expert-working-group.

[65] BEREC, BoR (10) 42.

[66] *Ibid.*, p. 11.

[67] *Ibid.*, p. 3.

having a different approach on network neutrality in the fixed and mobile networks. And especially future-oriented approach for network neutrality should not include differentiation between different types of the networks.'[68]

In December 2011 BEREC published to member NRAs its detailed guidelines on transparency and QoS,[69] including for instance Network Performance (what IAP monitoring is required for effective detection of discrimination).[70] NRAs have to implement net neutrality in 2014 with such detailed guidance. However, on transparency, 'BEREC states that probably no single method will be sufficient'[71] and points out the limited role of NRAs. Governments' consumer and information commission bodies are also likely to play a key role.

BEREC note that legal provisions in the Directives permit greater 'symmetric' regulation of all operators, not simply dominant actors, but ask for clarification on these measures:

> Access Directive, Art 5(1) now explicitly mentions that NRAs are able to impose obligations on undertakings that control access to end-users to make their services interoperable.[72]

The 2002 framework did not permit formal complaints to be made by content providers regarding treatment by IAPs, which meant no 'formal complaints' could be made to NRAs about content blocking until 2011. BEREC analysed how to define 'reasonable' and concluded it is:

> more reasonable to simply throttle P2P applications in times of congestion to the benefit of, for example, time-sensitive applications ... Those practices would be considered more reasonable than totally blocking special applications because they induce fewer side effects.[73]

Furthermore, the new wider scope for solving interoperability disputes could be used in France, which is explored in the following section.

National law and regulation since 2012

One of the several principles of network neutrality promulgated by both the FCC and European Commission in 2009/10 is that only 'reasonable network management' is permitted, and that the end user be informed of this reasonableness via clear information. Both the FCC in the US and the European Commission have relied on non-binding declarations to make clear their intention to regulate the 'reasonableness' of traffic management practices. Little was

[68] *Ibid.*
[69] BEREC, BoR (11) 53 and BoR (11) 67.
[70] See BEREC, BoR 53 (11), p. 3.
[71] See BEREC, BoR 67 (11), p. 5.
[72] BEREC, BoR (11) 53.
[73] BEREC, BoR (12) 132, p. 56, para. 265.

done to define reasonableness and transparency by the European Commission prior to the implementation deadline. This has led to extensive and prolonged criticism by the European consumers' organisation, and a substantial package of measurement, consumer empowerment and regulation for greater transparency and consumer rights in the proposed reforms (discussed below).

The 28 Member States, the European Economic Area members and the 47 members of the Council of Europe (CoE) must also conform to the human rights law of the European Convention on Human Rights and Fundamental Freedoms 1950 (ECHR).[74] This is supplemented in the European Union by legal instruments on data protection which are implemented using both the decisions of national and European courts,[75] and taking account of the advice of the group of European Union privacy commissioners. In 2011 the European Data Protection Supervisor (EDPS) expressed his concern that traffic management would result in exposure of users' personal data including IP addresses, repeating these concerns in November 2013.[76] The CoE also issues various soft law instruments to guide member states in observance of citizens' rights to privacy and free expression.[77] In 2016, conveniently just after the publication of the European Union Regulation on the Open Internet, the Committee of Ministers of the 47 CoE Member States issued their own Declaration,[78] a soft law instrument which Member States are expected to follow but which has no binding force. It appeared equally as full of exceptions as Regulation 2015/2120, at the time of writing. Telecommunications regulators are aware that net neutrality is a more important issue than they are equipped to explore, as the technologies at stake are technologies of censorship. Private Internet censorship, consistent with Article 10(2) ECHR, may only be acceptable in limited circumstances. Note that the introduction of network neutrality rules into European law was under the rubric of consumer information safeguards and privacy regulation, not competition policy.

Norway

Norway put in place co-regulation for net neutrality in 2009,[79] which had been in negotiation since 2008.[80] The need for net neutrality resulted from an

[74] See Koops and Sluijs (2012); Sluijs (2012).

[75] See Case C-461/10 *Bonnier Audio AB and others v. Perfect Communication Sweden AB*.

[76] EDPS, Opinion on net neutrality, 2011; EDPS, Opinion on the Proposal for a Regulation to achive a Connected Continent, 2013.

[77] See Council of Europe (2010) Declaration of the Committee of Ministers on network neutrality adopted 29/9/2010: 1094th meeting of the Ministers' Deputies.

[78] Council of Europe (2016).

[79] The Norwegian guidelines are available publicly, unchanged since 2009 (Norwegian Communications Authority, Net neutrality guidelines, 2013).

[80] In addition to semi-regular (annual) meetings with the regulator in Oslo, Dublin, Edinburgh, Brussels and Barcelona, I am grateful to representatives of the Norwegian consumer council, Opera software

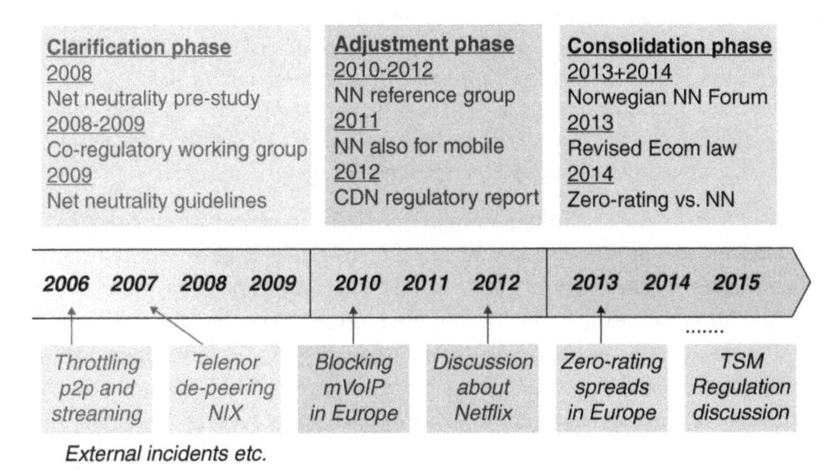

Figure 1 Timeline of net neutrality in Norway

IAP choosing not to carry the video traffic of the state broadcaster, resulting in strong political pressure for neutrality. Sørensen for the regulator explains that: 'In 2009 the Norwegian guidelines for net neutrality were launched and there have since been annual stakeholder meetings to monitor the status of net neutrality in Norway.'[81] A short description of the co-regulatory approach has been published on the regulatory website[82] and it has been extensively described by Sørensen. The Norwegian Electronic Communications Act 2013 did not introduce any (hard) law for net neutrality, but the corresponding Communication (Proposition) to the Norwegian Parliament (Stortinget)[83] confirmed the running soft law approach of Nkom (NPT at the time):

> in the proposals for amendments to the Electronic Communications Act the Norwegian Post and Telecommunications Authority and the ministry are following developments and will if necessary consider regulatory measures if it proves that a voluntary arrangement is insufficient to ensure good development in line with the overall goals.[84]

and Telenor for their comments. In Oslo, I thank in particular Professor Lee Bygrave of the University of Oslo for hosting the various meetings of the iGov and iGov2 projects, to which he invited me to meet regulators and ministry officials.

[81] Sørensen (2013).

[82] *Ibid.*

[83] Email to author from Frode Sørensen, 23 April 2015, on file with the author.

[84] See Sørensen (2014a).

Sørensen states that 'market players that have not formally endorsed the guidelines follow the guidelines in practice'. He explains that:

> CDN servers that are connected to dedicated transmission lines or that use a higher priority level than 'best effort' will not be considered net neutral. IPTV provided on a closed network (i.e. not over the Internet) can, in principle, be considered a modern form of cable TV. These types of services are often referred to as 'specialised services' and as long as these are not provided at the expense of the Internet service, net neutrality will not apply for them.'[85]

Neutrality was defined as excluding zero rating in 2014, in order to ensure that IAPs did not attempt to introduce such a practice: 'zero-rating lead to selected traffic from the Internet service provider itself or affiliated providers being favoured above other traffic. And this is exactly the kind of situation net neutrality aims to avoid.'[86] Norway is unique in that its co-regulatory net neutrality approach was agreed prior to other European nations, yet remains in place unchallenged by affected companies (although Telenor actively zero rates in Asian nations such as Myanmar and Bangladesh). It may therefore prove an exception to the general rule of litigious companies and captured regulators. Note that Norway practises an advanced form of Scandinavian social democracy, supported by strong and independent bureaucracy and government, a social compact between companies and society, and economic growth fuelled by North Sea oil wealth. It is an atypical example.

Netherlands and Slovenia 2012

In mid-June 2011 the Netherlands moved to implement the powers to require Quality of Service guarantees without discrimination.[87] Netherlands network neutrality regulation was voted on by its Senate on 6 March 2012,[88] which made it the first European nation to formally introduce mandated network neutrality. Implementation of the law was delayed until spring 2013 by the need for secondary legislation from the ministry mandating the regulator to implement the law, and the regulator was merged into the competition authority in April 2013, delaying implementation by over two years.[89] By late 2014 it was issuing regulatory decisions to enforce net neutrality and prevent discrimination.

[85] Sørensen (2013).

[86] Sørensen (2014b).

[87] Bits of Freedom (2015).

[88] Article 7.4a (3) of the Netherlands Telecommunications Act 2012.

[89] From 1 April 2013, OPTA (Onafhankelijke Post en Telecommunicatie Autoriteit/Independent Post and Telecommunications Authority) merged with the Competition and Consumer Authorities into the ACM (Autoriteit Consument en Markten/Authority for Consumers and Markets). In 2012–14 I conducted interviews in the Netherlands with Robert Stil and Mark de Hek of ACM, Professor Nico van Eijk and Mariejte Schaake MEP, and had many conversations with researchers at IVIR,

Slovenia also passed a law mandating net neutrality, on 28 December 2012, which is on its face more restrictive than the Netherlands law.[90] This was also implemented in 2013. Field research reveals the effectiveness of such laws and their operator and consumer effects.[91] The new 2012 Netherlands and Slovenian laws prohibit traffic management that discriminates, with few exceptions. In the Netherlands, these exceptions are:

> [a] to minimize the effects of congestion, whereby equal types of traffic should be treated equally;
> [b] to preserve the integrity and security of the network and service of the provider in question or the terminal of the end-user
> plus to stop spam and enforce legal requirements.[92]

In Slovenia they are:

1. applying necessary technical measures in order to ensure a smooth use of the Internet network (e.g. to avoid traffic congestion),
2. applying necessary precautions to preserve the integrity and security of networks
3. plus spam/legal requirements.[93]

Slovenia's law makes clear that these must be temporary fixes: 'proportionate, non-discriminatory and time limited and applied only to the extent necessary'. They both prohibit 'limited Internet offers' which block certain traffic, for instance zero rating by mobile providers, as the Netherlands law commands IAPs: 'do not make the price of the rates for Internet access services dependent on the services and applications which are offered or used via these services.' The 2015 European Regulation adopts less clear language on what is 'reasonable', as we will see in Chapter 4.

Amsterdam, Bits of Freedom and Oxford researcher Ben Zevenbergen. My former co-blogger Dr Jasper Sluijs was also a source of informed comment.

[90] Slovenia, Law on Electronic Communications, No. 003-02-10/2012-32, 20 December 2012, Article 203(4). A helpful translation of key aspects is available at https://wlan-si.net/en/blog/2013/06/16/net-neutrality-in-slovenia/ (Accessed 24 September 2016).

[91] The author has conducted personal interviews with the relevant national experts in April 2013 (Netherlands) and June 2013 (Slovenia), as well as the Minister responsible in Slovenia (August 2013) and consumer representatives (June 2013). More such research with operators and consumer groups is needed.

[92] Netherlands Telecommunications Act 2012, official translation by the Dutch government. Netherlands regulators were not required to implement net neutrality until summer 2013, a deadline delayed by the need for the Ministry to issue secondary legislation and guidance to the regulator on the form that such implementation should take. It is therefore too soon to draw firm conclusions about the efficacy of the Netherlands law.

[93] Slovenian Law on Electronic Communications, No. 003-02-10/2012-32, 20 December 2012, Article 203(4).

The practice of zero rating was outlawed by the Netherlands in its 2015 Guidelines clarifying application of its 2012 net neutrality law.[94] The four issues dealt with by the Netherlands regulator once its net neutrality law came into effect in 2013 have caused van Eijk to caution that 'hard cases make bad laws', including for zero rating: 'the new net neutrality rules ... led to a new subscription structure, with a substantially increased emphasis on data traffic. Data bundles are priced more specifically, and existing packages with unlimited data access have been replaced by packages with a specific size (data caps) and specific speeds.'[95] He cautions that 'it is too early to tell whether net neutrality has had an effect on the overall costs for mobile broadband'. The new Netherlands rules in practice only affect mobile IAPs: 'The new neutrality rules had no effect on the fixed market.' He explains: 'In two cases, the Authority investigated the bundling of data packages with free services (i.e. a mobile subscription with 'free' access to Spotify). To deal with these cases, a new guideline has been drafted by the ministry involved.'[96] This clarifies that zero rating is illegal in the Netherlands, though it may not be a ruling that is compatible with the new European law.

Slovenian regulation 2015

Due to language barriers, limited regulator size and the rather obscure position of Slovenia as a small Member State of the EU (population 2 million), Slovenia's very strict net neutrality law has been analysed very little by non-Slovenes. The net neutrality law is Article 203 of the wider Electronic Communications Law 2012 (ZEKOM), drafted as an innovation measure in response to hostility from the dominant IAP and trades unions towards competition in Internet supply. The regulator is the Communications Networks and Services Agency of the Republic of Slovenia (AKOS). The law's author, Professor Ziga Turk, when Minister for Communications, examined its genesis and implementation in a publication for the European Commission.[97] His main conclusion was that implementing net neutrality in a nation with such a weak regulator would prove very difficult. Caf agreed with this analysis, arguing that AKOS 'led by a

[94] Netherlands Department of Economic Affairs, Net Neutrality Guidelines, 2015 (official translation). In summary: 'Pursuant to the Act, providers of Internet access services may not block or obstruct services and applications on the Internet (with limited strict exceptions). Furthermore, providers may not differentiate between tariffs for Internet access services, and services and applications provided or used through these services.'

[95] van Eijk (2014).

[96] *Ibid.* The other two cases in 2013/14 concerned public Wifi and mobile ISP throttling. See van Eijk (2014): 'The regulator in charge – the Authority for Consumers and Markets – took a first decision on applying the new rules in a case where Internet access in trains was blocked for congestion reasons. In another case, a service similar to WhatsApp was inaccessible via wireless networks.'

[97] Turk (2015). I declare an interest as co-author.

former industry executive, has not been an advocate of net neutrality. Instead, it has taken a pro-industry stance on net neutrality and has not opposed attempts to weaken or even remove net neutrality provisions from the law.'[98]

While the ZEKOM law dates to the start of 2013, its regulation by AKOS was slow to arrive, with the main four rulings – those of 24 January and 20 February 2015 – against zero rating. AKOS confounded its critics with a strong zero-rating decision when forced to investigate by the Electronic Communications Council (SEK), which filed a complaint in July 2014 alleging Telekom Slovenije violated net neutrality with zero-rated products. From 2013 Telekom Slovenije provided free data for video channel HBO and UEFA Champions League football, then later the music streaming service Deezer. AKOS also found against Si.mobil (the largest mobile IAP) for zero rating cloud storage service Hanger Mapa. Telekom Slovenije and Si.mobil were instructed to stop zero rating. In the second pair of cases, bans were imposed against a zero-rated mobile TV service and web portal provided by AMIS (Mobia TV) and Tušmobil (Tuškamra), respectively. These were the only rulings against all major IAPs in Slovenia, all of whom had zero-rated affiliated content, and were given 60 days to comply. The issue was fought for by AKOS against substantial industry lobbying and the huge asymmetry in personnel between the IAPs and the very small regulator.

A remaining issue is that football and cloud storage on Telecom Slovenije remains zero rated, though this practice was stopped with HBO, whereas AMIS and Si.mobil were banned from video and cloud zero rating. The importance of Champions League football to many users means it may be politically impossible to deprive viewers of that stream by capping downloads in Slovenia. As a result of these bans, 'Telekom Slovenije and Si.mobile have both come up with special offers and packages with larger data caps or inexpensive data cap options'[99] to expand the cap, presumably to try to include their formerly zero-rated services.

Just as in the US, Slovenian operators and the regulator are highly litigious and all cases were on appeal at the time of writing.[100]

French principles 2010/14

In 2010 French regulator ARCEP released a '10 point' set of principles for net neutrality,[101] having consulted extensively over an entire year on how to implement the 2009 framework on net neutrality.[102] ARCEP updated their '10

[98] Caf (2014).
[99] Caf (2015).
[100] *Ibid.*
[101] ARCEP (2010a).
[102] See further Curien and Maxwell (2010) and Sieradzki and Maxwell (2008).

points' in a report to the French parliament in September 2012, which concluded that competition and transparency was insufficient to deal with potential long-term detriments to consumers from anti-neutrality behaviours.[103] It concluded that further legislation of the type passed in the Netherlands and Slovenia would be required in order to stop blocking and throttling, especially of VoIP over mobile networks, but that this was of course within parliament's competence.

ARCEP decided that Specialised Services would be permitted to be offered alongside Open Internet access, 'provided that the managed service does not degrade the quality of Internet access below a certain satisfactory level, and that vendors act in accordance with existing competition laws and sector-specific regulation' (Principle 4 of 2010). It confirmed this stance in permitting an agreement for preferential access to France Telecom/Orange and Free's services by Google's YouTube CDN in early 2013.[104] It is important to note that this is a non-neutral provision for a higher speed 'managed service' (Specialised Services), to which we return in Chapter 3. Furthermore, in September 2012 the competition authority demanded that France Telecom clarify the relationship between its wholesale and retail operations in order to ensure it did not cross-subsidise and margin-squeeze competitors, notably Cogent Communications.[105] This has been noted with approval by expert telecoms analysts, with Robinson stating 'ARCEP is therefore calling for the elimination of the blocking of VoIP and P2P traffic. The regulator concludes that QoS is a crucial long-term issue that must be monitored in order to "strengthen competitive emulation".'[106]

Through its decision dated 10 July 2013,[107] the Conseil d'Etat denied the appeal of US IAPs Verizon and AT&T and their French subsidiaries, thus confirming ARCEP's decision of 29 March 2012 on gathering information on the technical and pricing conditions governing interconnection and data routing.[108] This decision was itself challenged by US operators active in the French market, who did not wish to reveal their traffic data. ARCEP explains: 'The information gathering system that ARCEP introduced concerns the interconnection and data routing markets. These markets are home to complex and potentially strained relationships between internet service providers (ISP), providers of public online communication services (PPOCS) and technical intermediaries such as transit operators and content delivery networks (CDN).' ARCEP considered that 'regular, twice-yearly information gathering campaigns were vital

[103] ARCEP (2012a).
[104] DSL Prime (2012).
[105] Autorité de la concurrence (2012).
[106] Robinson (2012).
[107] France, Conseil d'Etat, Decision No. 360397/360398 of 10 July 2013.
[108] ARCEP (2012b).

to the regulator's ability to ensure that these markets run smoothly over time from a technical and economic perspective, particularly in relation to ARCEP's ability to settle any possible disputes that might arise between ISPs and providers of public online communication services'.[109]

The decision to uphold the information-gathering demands of ARCEP means that the French regulator was able to gather more information on the traffic management practices ofTier 1 IAPs and CDNs such as Google than any other national regulator, including those outside the European Union.[110] Arguably, it also means that ARCEP will be placed in the best position in Europe to assess the state of competition in the backbone IP interconnect market.

Conclusion

While both European and United States legislation and regulation affecting network neutrality were in peril of failing in 2014, the adoption of regulation and legislation in both regions in 2015 suggested that net neutrality was a permanent feature of the Internet law landscape. The European law proposed in 2013 was being rapidly overtaken by events in the Netherlands, France, Slovenia and Finland (see Introduction).

Net neutrality discourse has seen a sea-change in terminology since 2010, with governments keen to term the debate as being about 'the Open Internet', while telecoms companies and the European Commissioner,[111] who were firmly opposed to net neutrality in 2010, claimed in a Janus-faced manner to favour that term, as redefined and watered down to allow Specialised Service loopholes. These will be explored in Chapter 3, but first we turn to the reasons why competition is not the solution to net neutrality in Chapter 2.

[109] ARCEP (2012b).

[110] See ARCEP (2013). The Conseil d'Etat backs up ARCEP's powers in interconnection and data routing markets, and confirms its ability to query all of the players in theses markets, including those located outside the European Union, stating 'The Conseil d'Etat thereby also upheld ARCEP's power to query all market undertakings, including those located outside the European Union whose business and/or activity could have a significant impact on internet users in France ... ARCEP's information-gathering campaigns were necessary and proportionate.'

[111] See my comments at 31:00 on Commissioner Kroes in Marsden, Chris (2013b).

2

The limits of competition law and communications regulation

> Essential layers of this new infrastructure are either still under bottleneck control, e.g. local telecommunications access, or threaten to fall under such control, e.g. access to top-level Internet connectivity.
>
> Herbert Ungerer[1]

Net neutrality is not simply a competition problem, and viewing it through that lens leads one to the quite erroneous conclusion that no problem is proven to exist: that 'net neutrality is a solution in search of a problem'.[2] But competition policy is useful in helping us to understand both the limits of net neutrality as a problem, the limits of competition law's ability to explore the problem and deliver behavioural or structural remedies, and the limits of competition policy itself in the Digital Age. As the net neutrality principle refers to all access providers, large or small, it is a consumer protection measure that goes far beyond standard competition law. As network neutrality extends to all consumer IAPs symmetrically, it is not subject to competition law assessments of dominance, as abuse of dominance is not necessarily an accurate analysis of the network neutrality problem, at least in Europe.[3] Dominance is neither a necessary nor a sufficient condition for abuse of the termination monopoly to take place, especially under conditions of misleading advertising and inevitable consumer ignorance of potential abuses perpetrated by their IAP.[4]

The main justifications for the net neutrality principle are consumer protection, free speech and innovation by those creating Internet content,

[1] Ungerer (2000).

[2] Language used repeatedly by telecoms companies opposed to net neutrality from 2005 onwards, as documented in Hart (2011).

[3] See Marsden (2010), p 1.

[4] Some authors questioned the the need to distinguish between degrading and prioritising traffic, as they found that the latter naturally presupposes the former. See e.g. Chirico *et al.* (2007).

applications and services: the clichéd garage-based start-ups, which include those behemoth content providers Google, Facebook, Microsoft-owned Skype and Facebook-owned WhatsApp. Scholars and policymakers should begin by reading the many analyses by Ungerer, the architect of European communications policy, cited above, to understand how pervasive and enduring bottlenecks in the emerging Internet access value chain have been.[5] He explained in 2000:

> Many segments of the new Internet- based economy could develop, driven by the requirement [of] world- wide presence to reach scale economies, towards structures controlled by highly dominant enterprises, quite contrary to beliefs of internet libertarians. Potential anti-competitive behaviour becomes more difficult to regulate & check in the different geographical markets and jurisdictions, the behaviour itself can only be judged on a global level.[6]

It is worth noting that in late 2014 the EC closed a long competition investigation into Internet connectivity, yet 'found no evidence of behaviour aimed at foreclosing transit services from the market or at providing an unfair advantage to the telecoms operators' own proprietary content services'.[7]

In this chapter, I relatively briefly outline competition policy's purpose, referring to the exceptionally rigorous recent analysis of competition law suitability to regulate net neutrality by Maniadaki. Having analysed regulatory tools with little chance of success, I then examine what communications regulators actually do: regulating telecoms access based on the UK case study. This provides insights into how difficult net neutrality regulation will prove in practice, a subject to which we return in Chapter 6 and the concluding Chapter 8. I then consider the possibility of platform neutrality or some other form of platform regulation. I consider both competition law as a net neutrality tool and platform regulation only briefly, as my previous books[8] considered these issues in depth. I assess the possibilities of behavioural regulation to overcome some of the consumer detriments identified in nascent net neutrality regulation, and the wider use of behavioural 'nudge regulation' in Internet policy. First, I examine competition policy.

[5] Ungerer (2013). Ungerer wrote the 1987 Telecoms Green Paper, rising over a four-decade Commission career to become Deputy Director General of DG Competition before retiring in 2011. He is a physicist and economist: 1982 Doctorat in Economics, École des hautes etudes en sciences sociales, Paris; 1981: MBA, INSEAD; 1973 PhD, Theoretical Physics, Tübingen; 1969 MSc Physics, Technical University Munich. See www.acer.europa.eu/en/The_agency/Organisation/Board_of_Appeal/CVs/Herbert%20Ungerer%20CV%202016.pdf (Accessed 16 September 2016).

[6] Ungerer (2000).

[7] IP/14/1089.

[8] Marsden (2010), pp. 6–10, 42, 131, 153–164 and references therein; Brown and Marsden (2013a) pp. 17–32, 123–140 and references therein.

Competition policy: origins and purposes

It is worth reminding ourselves why competition policy exists: it was created to tackle the political and economic problems caused by monopolists abusing their dominance of certain trades, largely caused by the award of patronage by their political allies who were in turn funded by bribes or other 'political donations'.[9] This applied from the Statute of Monopolies 1623,[10] and more recently and particularly inspired the Sherman Act 1890 and Clayton Act 1906 in response to the abuses of the 'Gilded Age' of industrial capitalism in the United States in the late nineteenth century.

In Europe, the legal response to the terrible crimes committed by the cartels and monopolies of corporatist Germany and her allies in the Second World War was the so-called ordoliberalism of post-war competition policy, in particular as constituted in Articles 101–106 of the 1957 Treaty establishing the European Economic Community (Treaty of Rome) (now the European Union or EU). Herrera Anchustegui explains that: 'Ordoliberalism … proposes an alternative method to pure laissez-faire and state planned economy for the better regulation of the market economy by having as goals the protection of the competitive process and individual freedom.'[11] Vatiero reminds us that ordoliberalism is a political aim to be achieved by the economic means of preventing dominant companies capturing the economic and political process:

> The ordoliberal distinction between performance competition and impediment competition may improve the understanding of the European distinction between a 'dominant position' and an 'abuse' of that position … such an ordoliberal standard leads to a wider concept of dominance that not only includes the economic domain but also considers the impact of private economic power on the political sphere.[12]

Pasquale explains that communications policy is about cultural and political impact as much as economic impact, and that it is essential due to a special exceptional position within political and cultural debate: 'It is now time for scholars and activists to move beyond the crabbed vocabulary of competition law to develop a richer normative critique.' He sums up the issue: 'As robust American competition law fades into a secluded corner of legal history, essential facilities doctrine still remains, for some scholars, a ray of hope for intermediary responsibility.'[13]

[9] See Whish and Bailey (2015), p. 513.

[10] Statute of Monopolies 1623 c. 3, placing on a stautory basis the common law Case of Monopolies decision: *Edward Darcy Esquire v. Thomas Allin of London Haberdasher* (1599).

[11] Herrera Anchustegui (2015).

[12] Vatiero (2015).

[13] Pasquale (2010), p. 401, citing Lynn (2010) and Frischmann and Weber Waller (2008).

The brilliant minds who married economic analysis to law and were to dominate antitrust analysis from the 1970s to date (with a brief interregnum under the Clinton Presidency in the mid-1990s) included academics who became practitioners and leading judges. These economists and lawyers pursued the Chicago School approach based on neoclassical economic modelling, which has proved astonishingly adept at demonstrating no abuses exist in markets where the anecdotal evidence appears overwhelming.[14] Lessig's 'New Chicago School' paper, incorporating behavioural analysis, was presented in Chicago to the leading judge and scholar of that School, Richard Posner.[15] The US move towards a post-Chicago School from the mid-1990s reflects both a wider distrust of the apparent corporate takeover of antitrust, not least in the wave of predatory mergers from the 1980s onwards, as well as the examples discussed in the introductory chapter that demonstrate that *homo economicus* is a poorly motivated actor who often makes decisions that are not only inimical to the wider public interest but even to his own interest.

Particularly powerful models of inferior decision making have been developed in the economics of privacy, where the price of 'free', in particular when divulging personal information to marketers and/or government via the Internet, has been exposed. Economic actors are very imperfectly rational.[16] Data protection throws up substantial issues for competition analysis,[17] and European proponents of the disciplines imposed by economics on competition law now agree with Lamadrid that:

> over the years we have also come to realize that competition law is not the answer to every problem ... This is a message in which I have insisted repeatedly in the context of the debate on the role of competition policy in addressing data protection/privacy concerns.[18]

Abusive discrimination in access to networks is still characterised in telecoms as a monopoly problem, manifested where one or two IAPs have dominance, typically in the last mile of access for end users.[19] Even if platforms did compete in, for instance, heavily cabled countries, there would remain 'n-sided' market problems in that there is no necessary direct (even a non-contractual) relationship between innovative application providers and IAPs, so that platforms may set rules to 'tax' data packets that ultimately impoverish the open innovation value chain, and so ultimately cause consumer harm.[20] If you are a Korean

[14] The leading literature includes: Posner (1974, 1979); Areeda and Turner (1975); Landes and Posner (1981); Easterbrook (1984).

[15] Lessig (1998). For comment, see Tushnet (1998).

[16] See Fatas *et al.* (2013).

[17] Costa-Cabral and Lynskey (2015).

[18] Lamadrid (2015).

[19] I argued that the real problem lies in the 'middle mile' of interconnection, in Marsden (2010).

[20] Economides and Tåg (2007). Economides (2015).

game developer, you do not know BT's traffic management policies. The archetypal garage start-ups such as Facebook (founded 2003) and YouTube (founded 2005) would have had less opportunity to spread 'virally' across the Internet, as their services would be subject to these extra costs. Many commercial content providers, such as Google, use CDNs and other caching mechanisms to accelerate the speed of delivery to users, in essence reducing the number of those 'hops' (see Chapter 3). Content is therefore already delivered at different speeds depending on the paid priority the content provider assigns to it, but not on the IAPs' policies.

Maniadaki on competition and net neutrality

Maniadaki provides a magisterial examination of discrimination by termination monopolists in Internet access, explored from a competition law perspective.[21] She does not consider illegal content filtering (for copyright violation, sundry criminal law matters, national security),[22] nor does she enter into broad debate as to whether *ex ante* or *ex post* intervention is more effective. She focuses on single-company dominance (Treaty of Rome Article 102 rather than Article 101 cartel) violations, given that most of the potential abuses highlighted are carried out by a single access provider, whether mobile or fixed (telecoms/cable providers). Her chapter 3 performs a far more complete dismantling of competition law's claims to address the main net neutrality claims. Chapter 4 goes to the heart of the tension between an *ex ante* principle founded on innovation and consumer protection and her preferred competition law lens for inquiry: the 'Inherent Potential and Limitations of Competition Law in Protecting Net Neutrality'. She discusses at length the issue of what she terms pluralism and diversity,[23] European Commission new media policy 'weasel words' for the enormous flourishing of content made possible by the Internet, which we should note is the most powerful interactive communications platform in history. It is clear that pluralism, a diversity of views, is a valid reason to impose conditions on, for instance, mergers under EU law, an inheritance from newspapers and television stations. It is rather inadequate, as she acknowledges, for the Internet's variety of content, applications and services, in particular as privacy is as important to consumers as is free expression in the regulation of net neutrality, a subject of increasing interest to regulators on both sides of the Atlantic.

[21] Maniadaki (2015).

[22] Maniadaki does, however, refer to those vital standards documents produced by the Internet Engineering Task Force, the Request For Comments (IETF-RFC), which are rather closer to setting out physical laws than anything lawyers encounter in court.

[23] Maniadaki (2015), pp. 137–158.

Her chapter 5 then returns to more familiar territory for competition lawyers: definition of relevant market and market power assessment. Here again the tension between a legal mechanism designed to protect all consumers interacting with competition law, which only regulates dominant actors, comes to the fore, particularly given the notorious difficulty in proving collusion between entities in markets such as mobile telephony. Maniadaki makes a strong plea for better use of behavioural analysis in this case and more broadly, based on the Microsoft cases and the merger cases involving MCI with successively British Telecom, WorldCom, Sprint, Verizon and succeeding interconnection cases.[24] It is particularly noteworthy in a meticulously researched work that she identifies such a body of economic and technological literature which is less well known to the competition law academy. Behavioural analysis is a subject to which she returns in her final chapter, and which I consider in the final part of this chapter, dealing with platform regulation.

Maniadaki considers net neutrality violations as refusal to deal, as for instance in infamous cases where the access providers block access to VoIP or IM applications such as Skype or WhatsApp. Analysis here takes on the argument between David Evans and Nicholas Economides, economics professors with very different views of the imperfections of Internet access markets and the manner in which net neutrality leads to innovation.[25] In her chapter 7 she explores discrimination and unfair pricing as aspects of the problem, though concludes that, however open-ended the case law may be on the potential for abuses to include, for instance, over-pricing for so-called 'fast lanes' on the Internet, 'a large proportion of practices … would generally fall outside the scope of competition law'.[26] Analysts who find that their preferred legal approach is largely useless should not, however, despair. Maniadaki skilfully establishes the limits of competition law to date, as well as offering the possibilities to extend that corpus of legal application to make it more relevant for dynamic digital services. It will not be easy for competition lawyers to move from widgets to bytes, but without such a transformation of their skillset, they will be excluded from Internet law debates such as that involving net neutrality.

Maniadaki has performed an extraordinary service, which all competition lawyers should study. She has shown how much competition law can address the critical issue of net neutrality (not much) and how much it needs to transform itself to have relevance for digital services markets (a great deal). She also shows that the tools of competition law are used a great deal in informing

[24] *Ibid.*, pp. 179–188. The academic literature in that field is particularly sparse, although she located an obscure 1997 paper by Kenn Cukier, and later more mainstream work by Giovanetti (2015), David (2001), and Clark and Claffy (2015).

[25] Evans is in favour of self-regulation by markets; Economides (2015) thinks that regulation is needed due to players' gaming of n-sided markets. See Maniadaki (2015), Chapter 6.

[26] Maniadaki (2015), p. 297.

both sector-specific communications law, as well as the specific regulation of net neutrality. Maniadaki herself is an official in UK communications regulator Ofcom, and it is to Ofcom's regular task of regulating communications networks that we now turn, to see how economic regulation works in practice.

Telecom regulation: the UK case study

To understand UK communications regulation, we need to examine the market structures and legal challenges to regulation of the infrastructure. I examine first fixed then mobile regulation, including fixed-to-mobile termination, which we will see is vital to the communications ecology and explains many of the attempts to breach net neutrality by blocking rivals to the incumbents.

Fixed network regulation

I now briefly explain the fixed network market of UK communications. It is characterised by wholesale duopoly between the Virgin cable network (owned by US investor John Malone via Liberty Global) and British Telecom (BT), the former domestic monopoly wholesale network (telco) in the entire UK except the city of Kingston-upon-Hull, in fixed local ('last mile') access to telecoms and thus Internet communication.

The UK was a notable early example in the wave of neoliberal privatisations of communications. BT was privatised in 1984, and is the successor to Post Office Telecommunications, itself the successor to the nationalised telegraph companies, to which BT traces its origins in 1846.[27] The nationalisation of communications from the origins of telephony in the 1870s to the 1980s is a feature shared with almost every country in the world outside the United States[28] (UK colonial monopoly Eastern Telegraph Company, renamed Cable & Wireless in 1932, nationalised in 1947, had about 50 national monopolies itself in former colonies and protectorates including the West Indies, India, Panama, Bahrain, Hong Kong, Macau[29]). The UK government sold all of its interest in BT in three stages over the period 1984–93, whereas most major European former monopoly telcos retain the government as the largest shareholder: for instance, in 2016 22 per cent in France Telecom (FT: since 2006 branded as Orange) and 16 per cent in Deutsche Telekom. An unusual historic exception is Telefónica de España, formed by US multinational IT&T in 1924, which came under partial state ownership in the fascist period from 1945,[30] but was wholly

[27] See www.btplc.com/Thegroup/BTsHistory/.
[28] Waverman and Trillas (2002).
[29] Hills (2002).
[30] Sampson (1973). See also Garcia-Algarra (2010).

privatised in the period 1987–97. In 2014 it briefly controlled Telecom Italia, amongst its many foreign subsidiary interests.[31] It is noteworthy that on completion of the £12.5 billion BT purchase of the largest UK mobile network EE, Deutsche Telekom (largest shareholder the German government) became BT's largest shareholder with about 12 per cent of its shares.[32]

A radical UK departure from standard market structure is that there were five mobile network operators (MNOs) until 2011, but none of them was the former telecoms monopolist. BT halved its net debt from £27.9 billion in 2001/02.[33] The incumbent had become severely financially compromised due to its international expansion, notably its Concert joint venture with AT&T, loss-making European subsidiaries and its BT Openworld content platform (it also had to divest its yell.com directory business to venture capitalists for £2 billion). This included the demerger of BT's wholly owned mobile subsidiary O2 in 2002 (bought by Telefónica for £18 billion in 2005), accompanied by a hugely discounted rights issue. This fragmentation of mobile and fixed operators, and minority position for the cable operator, means that there is no integrated large investor in telecoms (unlike with Deutsche Telekom in Germany, Verizon and AT&T in the US, FT/Orange in France and so on).

It should be noted that BT's pensioners comprise over 300,000 individuals, the largest private pension in the UK – the second largest is for universities with 200,000 individuals. Telco pensioners and the legacy funding of their pensions are a vital element in regulation, together with maintenance of the local access lines – the price of broadband is relatively trivial by comparison. In 2014 the BT Pension Scheme deficit was estimated at £7 billion on assets of over £40 billion, with a 16-year plan in place to reduce that deficit, including £1.5 billion in 2015.[34] (By contrast, in 2014 the University Superannuation Scheme deficit stood at £5.3 billion on assets of £41.2 billion.)[35] To give a sense of scale, BT's revenues in 2015 were approximately £17 billion.[36] BT's pension obligations on privatization under Telecommunications Act 1984, s.68(2) included a Crown Guarantee to back the pension fund should BT become insolvent.[37] This was alleged by an anonymous complainant in 2006 to be illegal state aid breaching Article 87 of the Treaty of Rome (now Article 107), confirmed in

31 Bela and Trillas (2005).

32 Hall (2016). See also www.productsandservices.bt.com/mobile/phones/apple/?s_intcid=con_int-ban_dmedia_content_t017_iPhone7_preorder_promo616_org (Accessed 16 September 2016).

33 See www.btplc.com/report/report03/Businessreview/Restructuring.htm (Accessed 16 September 2016).

34 BT Press Releases (2015).

35 Universities Superannuation Scheme (2014) Actuarial Valuation March 2014, available at www.uss.co.uk/Actuarial%20Valuation/ActuarialValuationMarch2014.pdf. University professors aged 45 in 2014 are expected to live to 91.2 years (male) and 93.6 years (female).

36 Jackson (2015).

37 Summarised in C(2007) 5617 State aid C 55/2007 (ex NN 63/2007 (ex CP 106/2006)).

2009 by the European Commission and then by the European Court of Justice after litigation that was not concluded until 2014.[38]

Why do pensioners matter to net neutrality? If you intend to regulate access providers as regulated actors, you had better understand what their motivations are. Such cash deficit payments by BT affect the wholesale price of broadband to its competitors, with an Ofcom inquiry in 2009–10 explicitly focused on that impact.[39] Broadband access involves the historic legacy of pensions, of utility and common carriage regulation, and of the pensioners, widows and families from the nationalised monopoly. Regulators may be accused of considering net neutrality a Friday afternoon job (five minutes before going home), and policymakers should realise how it is linked to what happens from Monday to that point on Friday.

UK broadband penetration was under 5 per cent of all households in 2003, when it was already ubiquitous in South Korea, and well over 10 per cent in the Netherlands and the United States. It is now equivalent in penetration, but not speed, with more advanced economies. The average download speed is over 20Mbps, though upload speeds were under 2Mbps in 2013 for at least 70 per cent of the population.[40] The download speed is about average for the advanced economies in Western Europe.[41] DOCSIS3.0 cable customers averaged over 30Mbps (34.9Mbps), and VDSL (very high bit rate digital subscriber line) customers 43Mbps. There is very little fibre to the premises. BT trialled vectoring from late 2013. Maximum VDSL speed is therefore in the 40 to 50Mbps range, with cable in the 30 to 60Mbps range. BT set a completion date of spring 2014 for the Openreach commercial roll-out of FTTC/FTTP (fibre to the cabinet/fibre to the home) products, with 24 million premises within its fibre (FTTC/P) footprint from 1,725 exchanges. The UK has 26.4 million households and around 4.8 million businesses (many micro-businesses are home based), which is circa 30 million premises. Openreach fibre in 2013 passed 63 per cent of UK premises, 30.8 per cent of all UK exchanges, after five years of 300,000 premises passed per month, at £131 per premises passed (£2.5 billion/£19 million).[42]

The remaining households depended on the government subsidy of BT's extended fibre roll-out and other means – BT is the only fixed broadband supplier

[38] Decision 2009/703/EC concerning the State aid C 55/07 (ex NN 63/07, CP 106/6) UK Crown guarantee to BT Pension Scheme; Case C-620/13 P – *British Telecommunications v Commission*, appealing Cases T-226/09 *British Telecommunications v. Commission* and T-230/09 *BT Pension Scheme Trustees v. Commission*. For analysis see Lokhandwala (2014). See BT Pensions (2014) Litigation continued on the issue of non-UK dividends: see *Trustees of the BT Pension Scheme v. HMRC* FTC/91 & 92/2011 [2013]; *Trustees of the BT Pension Scheme v. HMRC* [2015] EWCA Civ 713.

[39] Ofcom, BT Pensions Statement, 2010.

[40] Ofcom, Average UK broadband speed continues to rise, 2013.

[41] Source: OECD Broadband Portal: www.oecd.org/sti/broadband/oecdbroadbandportal.htm.

[42] Ferguson (2013). BT claims 1.7 million customers for FTTC/FTP, see BT (2013) First Quarter results.

to most of these households. State-subsidised non-commercial roll-out is subject to government subsidy and only one supplier is qualified to provide roll-out – the former monopolist BT. The first tranche of UK government state aid to BT was delayed, only being approved by the European Commission with amendments to allow for greater competition where feasible, resulting in contracts signed in 2013.[43] Significant concerns have been expressed, including by both parliamentary audit bodies, that BT was reimposing a rural/semirural monopoly with the government subsidy of £1.2 billion.[44] The government's Broadband Delivery UK (BDUK) office's initial aim was to reach 90% of the country by the end of 2015 (with the last 10% having a download speed of at least 2Mbps). The word 'nearly' was added in front of the 90% target (i.e. to allow it to slip into 2016) and an extra £250 million was set aside to push the end goal out to 95% by 2017.[45] The government also confirmed that 'superfast' (VDSL) broadband would cover 88% by end-2015. BDUK put out a new tender in September 2013, seeking 'as many suppliers as possible' to help it spend £250 million to extend fixed line broadband coverage to 95% by 2017 (99% by 2018 including wireless where other companies are active in a subsidised roll-out).[46]

There is some competition in the retail and resale DSL (Digital Subscriber Line) market, with two main mass consumer resellers of BT's services in whole or in part: TalkTalk (a cut-price reseller) and Sky (the dominant pay-TV operator via satellite). These compete with BT Retail. Fifty per cent of the UK is cabled by the monopoly cable provider Virgin (US-owned), though only about 19% of the market subscribes to cable broadband. In September 2015 the market for broadband lines was:

- BT Retail: 7.88 million retail broadband subscribers (33% of the market)
- Virgin (Liberty Global): 4.63 million (19%)
- resellers of unbundled BT lines (mainly TalkTalk and Sky): 8.8 million fully unbundled, 1.1 million shared unbundled and 1.85 million BT Wholesale operated lines
- total of 11.75 million BT lines branded under other resellers (48% of the market).

[43] IP/12/1244 and C(2012) 8223 final, State aid SA.33671 (2012/N). In particular, the EC drew attention to the need to ensure regulated wholesale pricing, supervised by Ofcom. See Broadband Delivery UK (2012). Paragraph 52 is the most relevant and states (in part) that subsidised schemes have an 'obligation to allow effective wholesale access to all parts of the subsidised broadband infrastructure ... Where the supplier operates in the downstream markets it is also required to supply the upstream wholesale inputs on an equality of access basis to its own downstream retail divisions and to competing communications providers. This requirement therefore ensures that there is no discrimination in the supply of key wholesale access.'

[44] National Audit Office (2013). See also Public Accounts Committee (2013).

[45] Jackson (2013a).

[46] Jackson (2013b).

Disagreements over BT's wholesale pricing from 2013 resulted in 2016 in a full reference to the Competition Appeal Tribunal (CAT) on 'virtual unbundled local access' (VULA) lines.[47] The CAT will report in autumn 2016.

The two fixed wholesale urban operators –Virgin Media, the cable TV operator, and BT – become one in the more remote parts of the UK that are not cabled, but which include some urban areas such as Fulham, only four miles from Aldwych, the centre of London. This is in large part because of the weakness of cable TV given the monopoly over live English Premier League (EPL) TV football broadcast rights held by Sky (controlled by Murdoch) in 1992–2013. BT bought some of those rights (2013–16 and 2016–19 contracts) to offer a bundled broadband–IPTV sports package to compete with Sky. Virgin resells both companies' football rights. As it matures, the DSL market is characterised by increasing oligopoly: bundled service offers with little real consumer choice except in the further bundle of premium TV – notably exclusive football – channels. Consumers can now buy bundled TV/broadband offers from all four main operators. The IPTV offer is zero rated and is not included in the prevalent monthly data caps.

The perceived erosion of revenues from fixed line telephone calls, lost to mobile and to VoIP offers (especially for the previously vastly profitable international call market), has led fixed IAPs to charge more for line rental once that was liberated from regulatory control in the 2000s, with Ofcom stating that 'the basic fixed line rental fee has risen by an average of over 25% in real terms since 2010'.[48] While wholesale line rental has fallen over the period, retail line rental has increased, subsidising broadband offers. This disadvantages the (mainly) older landline-only customers, of whom 69 per cent are over 65 and who make up about 10 per cent of BT OpenReach's user base of fixed line customers,[49] who are also poorer, as Ofcom explains: '[Poorer] households (over half of landline-only homes fell into this category)'.[50] The cost of broadband service (including line rental) from the major four IAPs varied in June 2015 from £21.49 to £34.49 per month, depending on IAP, line speed and monthly data cap.[51]

Average usage was accelerating with consumption of real-time video, up from 17GB/month in 2011 to 82GB/month, a 41 per cent increase in the year to 2015. This usage is split between relatively low-volume downloaders with

[47] Competition and Markets Authority, CAT refers superfast broadband price control appeals to CMA, 2016.

[48] Ofcom, 12th Annual Communications Market Report, 2015, p. 273.

[49] Ferguson (2015).

[50] Ofcom, 12th Annual Communications Market Report, p. 277.

[51] *Ibid.*, p. 315.

speeds below 40Mbps (typically ADSL – Asymmetric Digital Subscriber Line) at 75–100GB/month and much higher volume downloads by higher speed users (140–160GB/month).[52]

Mobile network regulation

CommissionerVestager summarised the state of EU telecoms competition in 2015:

> In contrast to the US, a pan-European telecoms market does not yet exist – even though we have far fewer network operators in the EU than in the US. In Europe, we have around 35 mobile network operators at group or company level. The two biggest players are present in eleven and twelve EU countries respectively. The four biggest operators serve around 60% of EU subscribers ... while the biggest companies in the EU are present in multiple territories, consumers in each territory are captive in national markets. They cannot access the same offerings as their neighbours across national borders.[53]

The UK mobile market was traditionally vied for by duopolists Vodafone and O2, but in 2011 two smaller later entrant GSM (Global System for Mobile Communication) operators, owned by France Telecom (branded Orange after the successful UK-founded Hutchison- owned brand of the early 1990s) and Deutsche Telekom, merged to form the ludicrously named Everything Everywhere (EE), which network shares its network with 3G entrant Three. In 2016 Three was trying to merge with O2, reducing the number of players from five in 2011 to three, though with opposition from the new pro-consumer Chief Executive of Ofcom[54] and a Phase 2 investigation by the newly vigorous DG Competition Commissioner.[55] There are three wholesale mobile networks, which in 2013 introduced metropolitan LTE (Long Term Evolution), several years after advanced markets such as Sweden and Germany.

Note that mobile shares post-mergers once BT acquired EE are BT 29 per cent and Vodafone 23 per cent.[56] BT's fixed share rose to 36 per cent.[57] The Three mobile market share would have risen to 45 per cent had it been permitted to merge with O2 (though this was not approved by DG

[52] Ofcom, ConnectedNations, 2015, p. 23, Figure 13.
[53] Speech on 2 October 2015 Competition in telecom markets.
[54] Rushton (2016).
[55] Speech by Margrethe Vestager: Competition in telecom markets.
[56] Thomas (2015).
[57] The merger of BT and EE was referred to a Phase 2 investigation by the CMA on 9 June 2015, and will report by 18 January 2016. See Competition and Markets Authority, BT Group/EE merger inquiry, 2015.

Competition, whose Phase 2 investigation into Three/O2 concluded in May 2016).[58]

Network interconnection

Just as Internet content must 'pay' to transit or directly interconnect (unless its network is so large it can peer) on the access network, so too must telephony. The exception is obviously telephone calls between two subscribers to the same mobile network within the same country, described as on-net calling. It is expected that there are both efficiencies in routing over the same network and incentives to offer lower prices for competitive networks, to encourage users to concentrate their 'friends and family' on the same network. For many mobile monthly subscribers, it is effectively free to call on-net. Calls between networks are described as 'off-net', and here pricing can be set by the network or regulated if pricing is too high. In fixed network calling, the former incumbent was obliged to offer the same interconnection prices to all its competitors, which became described as the most exotic-sounding process in telecoms: RIO. This actually means the Reference Interconnection Offer, and is only exotic in accounting terms. This is not particularly economically controversial, but litigation over two decades by the mobiles has delayed setting mobile off-net prices at the same level as fixed calls. The RIO level is more important for mobile because in Europe and most other regions, the system of charging is Calling Party Pays (CPP). This means that when a user makes a call, it is her network that pays the network that transits (if any) and the receiving party network. This makes the receiving network a monopoly with little incentive to lower costs except for reciprocal value. While that might put cost pressures on calls between mobile networks of near-equal size, it certainly does not for small market-entrant mobile companies. It also does nothing for fixed networks who have to offer RIO regulated rates, which in the late 1990s were 30 times cheaper than the wholesale rates the mobiles offered the fixed operators (and each other). Over the period since 1995, when mobile penetration first exceeded 10% of the UK adult population, this distorted the market significantly.[59] The greater part of this mobile growth took place in the period 1998–2000 with penetration increasing from 25% to 73% in two years (Internet use grew from 14% to 27% in that time).[60]

Markets for both mobile and fixed telephony are a relatively stable split, which some have described as oligopoly. This obscures the fact that the mobile

[58] COMP/M.7612 *Hutchison 3G UK/Telefonica UK.*

[59] Oftel, Mobile Numbering Consultation, 1996.

[60] Oftel, Consumers' use of mobile telephony, 2002.

Table 3 Market shares in fixed and mobile access in the UK

Fixed 2014	**Mobile**
BT 32%	EE 29%
Sky 22%	O2 29%
Virgin 20%	Vodafone 23%
TalkTalk 14%	Three 12%
Others 12% (EE 4%)	Others 7%

Source: Ofcom (2015a), Figure 4.44.

companies in each sector compete for consumers, but receive income from their higher termination rates charged to fixed operators. This is strongly challenged by the fixed operators, especially the regulated former monopolist BT. It must be borne in mind that the UK is a unique market in Europe and globally, as the incumbent telco does not own a mobile operator.[61] In return, mobile networks claim that BT overcharges them for the fixed lines they need to transport their mobile traffic – bear in mind that a mobile network is actually a fixed network with 8,000 (or so) mobile towers for local connectivity attached to that network.

Mobile termination regulatory battles

The Open Internet element of Regulation 2015/2120 will be examined in Chapter 4 and forms the central element in this book. Note the context: it will operate alongside the mobile roaming element, politically a much greater prize for the European institutions that will reduce international roaming charges within the European Digital Single Market by 2017. International roaming charges had already been significantly reduced in the period after the previous Regulations in 2007[62] and 2012.[63] Vodafone had appealed the 2007 Regulation to the General Court of the EU.[64] As Recital 22 of the 2012 Regulation pithily put it, 'retail and wholesale roaming prices are still much higher than domestic prices and continue to cluster at or close to the limits set by Regulation (EC) No 717/2007, with only limited competition below those limits' – in other words, mobile companies were still screwing their customers on international roaming.

[61] An excellent comprehensive analysis of fixed and mobile competition is provided in Competition and Markets Authority, BT–EE Provisional findings report, 2015.

[62] Regulation (EC) 717/2007, p. 32.

[63] Regulation (EU) 531/2012.

[64] Case C-58/08 *Vodafone and Others v. Secretary of State for Business, Enterprise and Regulatory Reform.*

International roaming is, however, an 'affluenza' type of regulatory reform for the cross-border traveller (and European parliamentarian), compared to the much larger cost to consumers of national mobile roaming. In 2015 Ofcom announced that national regulated prices would be reduced below 0.05p by 2017, a 98 per cent decline since 1995 when price regulation was first introduced.[65] This enormous loss of per-minute revenue is compensated for by the ubiquity of mobile phones in 2016 and the substitution of mobile for fixed calls – but also by attempts to cling to text and international call revenues by some MNOs trying to breach net neutrality and block rival services such as Skype and WhatsApp, as we have seen. Around 10 minutes of calls are made from mobiles for every 4 minutes of calls made to them, showing in part the market-distorting effect of their high CPP prices (and the convenience of calling from mobiles compared to landlines).[66] Mobile text and voice revenues per UK user fell by 60 per cent in the period 2008–13 due to this regulatory regime.[67] The next threat to mobile revenues is the elimination of texting revenues, as WhatsApp and other IM clients are used ubiquitously, which is where mobile termination becomes directly a net neutrality issue. Why? Mobiles blocked WhatsApp, Skype and others throughout this 'rate war' with their regulators.

The MNOs had maintained the fiction that on-net prices could be almost free but off-net prices 30 times higher than for fixed networks – an obvious nonsense. This is shown by the US experience, where almost uniquely the Receiving Party Pays (RPP), meaning that the mobile companies have every incentive to maintain prices at the level of fixed operators, which they do. The extreme version of this is shown at the Canada border, where you might have paid a fraction of a cent per minute on the US side under RPP, but 20 cents or more on the Canadian side under CPP. Such huge differentials have led economists such as Dewenter and Kruse to argue:

> implementing RPP instead of CPP would not reduce penetration rates, irrespective of whether a country's penetration process has just begun or has nearly reached saturation levels. Moreover, we expect that adopting RPP instead of CPP seems to be a possible way to reduce the market power of terminating networks as well as of mobile termination rates.[68]

However, few nations have done so and the possibility within the EU seems vanishingly small, especially as regulation has driven prices towards rough parity between mobile and fixed.

[65] Ofcom, Mobile call termination market review 2015–18: draft decision, 2015.

[66] *Ibid.*, Annex 3, p. 42, Figure A5.3. High-value subscription callers proportionally make more calls than receive them; pre-pay low-value callers receive more calls than they make.

[67] *Ibid.*, Annex 3, p. 46, Figure A5.5.

[68] Dewenter and Kruse (2011).

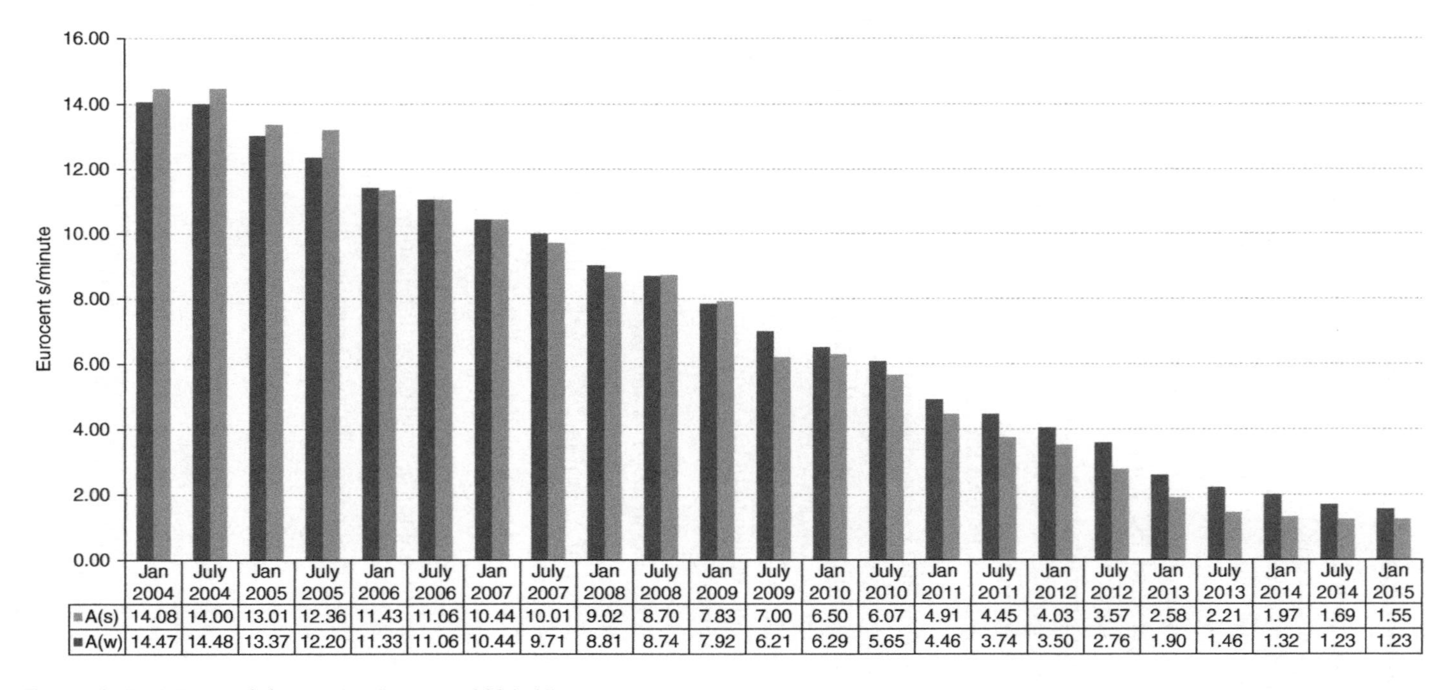

	Jan 2004	July 2004	Jan 2005	July 2005	Jan 2006	July 2006	Jan 2007	July 2007	Jan 2008	July 2008	Jan 2009	July 2009
A(s)	14.08	14.00	13.01	12.36	11.43	11.06	10.44	10.01	9.02	8.70	7.83	7.00
A(w)	14.47	14.48	13.37	12.20	11.33	11.06	10.44	9.71	8.81	8.74	7.92	6.21

	Jan 2010	July 2010	Jan 2011	July 2011	Jan 2012	July 2012	Jan 2013	July 2013	Jan 2014	July 2014	Jan 2015
A(s)	6.50	6.07	4.91	4.45	4.03	3.57	2.58	2.21	1.97	1.69	1.55
A(w)	6.29	5.65	4.46	3.74	3.50	2.76	1.90	1.46	1.32	1.23	1.23

Figure 2 BEREC mobile termination rates 2004–15

Regulating the mobiles: delay, degrade, confuse

Ofcom is obliged to follow the relevant EC Recommendation on market definition,[69] to provide consultation and gather evidence according to set processes, perform market tests in accordance with recognised economic modelling[70] and to notify the EC of draft decisions. A new Recommendation on Market Definition was issued in October 2014,[71] replacing the previous version of 2007, itself superseding the original 2003 Recommendation. In addition to any potential EC hurdles, it also has to offer its regulatees the opportunity to challenge decisions before domestic and EC courts. This can become extremely protracted (and tedious) with the fixed-to-mobile termination saga taking decades in its various episodes. I examine first UK appeals against regulatory decisions, then those at European level. Do not be surprised if net neutrality regulation follows the same agonised litigation path between now and 2025.

UK regulator Ofcom had been tied up in litigation by MNOs almost continuously throughout the twenty-first century, with 'delay, degrade, confuse' the results of the continual consultation, litigation and settlement. Note that the three major mobile companies (in the UK since 1992 there have been four mobile companies with the exception of the late 2000s when there were five) employ the best competition lawyers in London to fight Ofcom and the other regulators, with one former regulator expressing the view that Ofcom 'has to fight the entire London competition Bar' (specialist barristers).

A new epic commenced in 2005 as the previous version ended: Ofcom consulted, a regulation was set in March 2007, it was referred to the CAT in 2007, then to the Court of Appeal in 2009, with a decision in 2010. Meanwhile, the next three-year period began, with a very similar outcome of a visit eventually to the Court of Appeal in 2013.[72] The EC issued a 2009 Recommendation,[73] which the second process followed. All these delays were obviously beneficial to those who gained from imposing higher or lower charges, which made this a multi-million-pound game of litigation. Also note that this meant that – with the exception of 2004 – Ofcom's legal department and its predecessor spent the entire period 2000–13 fighting the mobiles to reduce termination charges.

[69] Recommendation of 17 December 2007.

[70] The most recent is Recommendation C(2013) 5761.

[71] Recommendation 2014/710/EU.

[72] *Everything Everywhere Limited v. Competition Commission, Office of Communications, Hutchison 3G (UK) Limited, British Telecommunications plc* [2013] EWCA Civ 154 of 6 March.

[73] Recommendation C(2009) 3359.

Table 4 Mobile termination sagas in the 2000s

2000–03 (Oftel)	2005–10 (Ofcom)	2009–13 (Ofcom)
Review of the mobile sector began in September 2000, culminating in Oftel regulation statement of the mobile sector 26 September 2001.[a]	Review of the mobile sector began in 2005 ending in 27 March 2007. Mobile call termination regulation was set as all five national MNOs possessed SMP in calls to their own customers.	From May 2009, Ofcom conducted a market review with three consultations: statement published in March 2011.
7 January 2002 Oftel asked Competition Commission to investigate, indicated 12 December 2001 after mobiles reject control.[b]	May 2007 BT and H3G appealed Ofcom's 2007 MCT Statement to the CAT. BT's appeal upheld.	March 2011: Three of four mobiles and BT appealed to CAT. Decision issued 12 March 2012.
Competition Commission inquiry 2002 reports 18 February 2003.[c]	2 April 2009 CAT direction to Ofcom to revise charge control as determined by Competition Commission. Ofcom published revised SMP conditions, took effect 3 April 2009.	Tribunal referred to Competition Commission; Commission agreed with Vodafone and EE not Ofcom.
7 March 2003 appeal by all four major mobiles to High Court, denied in June 2003.[d]	CAT's judgment appealed by T-Mobile, Vodafone, Orange and O2 to Court of Appeal. Upheld: CAT did not have power to direct Ofcom to reset the charges for years 2007–09.	CAT's final judgment appealed by Everything Everywhere, to Court of Appeal. Decision in favour of Competition Commission and Ofcom on 6 March 2013.

[a] Oftel, Statement on mobile termniation rates, 2001.

[b] Oftel, Press Statement Ref 02-01, 2002.

[c] Competition Commission, Report on the charges made by mobile operators for terminating calls, 2003.

[d] *R (T-Mobile, Vodafone, Orange) v. Competition Commission* [2003] EWHC 1566 (Admin) [2003] EuLR 769.

A 2013 Colt Technology Services case before the Competition Appeal Tribunal[74] shed more light on when and how decisions of Ofcom can be appealed.[75] Colt challenged Ofcom's decision to implement a requirement on BT to offer 'active' wholesale services, but not passive infrastructure access, in the Business Connectivity Market Review (BCMR) Statement.[76] Ofcom had reviewed the Wholesale Local Access market in 2010 and imposed limited passive infrastructure access remedies requiring BT to provide residential access to its ducts and poles, but not to its dark fibre. Ofcom imposed less rigorous requirements on business in BCMR, concluding passive infrastructure access remedies could lead to a duplication of investment, discourage future investment, encourage inefficient market entry and disrupt BT's recovery of common costs by encouraging arbitrage ('cherry picking' by competitors such as Colt). The CAT cited observations from two Court of Appeal decisions which ended the mobile termination saga: the applicant (Colt) must show that the decision of Ofcom itself is wrong;[77] if Ofcom addressed the right question by reference to relevant material, any value judgement on its part must carry great weight.[78] CAT noted that Ofcom had conducted a thorough market review process, consulting with all stakeholders, BEREC and the EC, and publishing a number of consultation documents over the period 2011–13, as well as holding meetings. Ofcom's standing as an expert regulator meant that its value judgement in assessing what would be the appropriate remedies must carry great weight.[79]

If a telco cannot easily overturn Ofcom, although it can delay it, what chance might it have in arguing its case in Brussels? The EU Telecoms Framework Directive Article 7 process allows the European Commission within a month of notification to veto inappropriate market definitions or findings of significant market power (SMP). If the Commission judges that an SMP definition is incorrect, it can issue a 'serious doubts' letter, which cause the NRA to rethink and withdraw their decision even before an ultimate veto. Minor issues

[74] Now governed by the Competition Appeal Tribunal Rules 2015 (SI 2015/1648). Cases commenced before 1 October 2015 were governed by Competition Appeal Tribunal Rules 2003 (SI 2003/1372); Competition Appeal Tribunal (Amendment and Communications Act Appeals) Rules 2004 (SI 2004/2068).

[75] *Colt v. Office of Communications* [2013] CAT 29.

[76] Colt made four complaints that Ofcom made the wrong determination in: viewing passive and active remedies as necessarily alternatives, rather than complementary, and rejecting passive infrastructure access remedies as a result; rejecting passive remedies on the basis of a supposed lack of demand for them; not proceeding from the starting point that it should regulate as far 'upstream' as possible; and believing that active remedies were likely to promote innovation and competition at least as effectively as passive remedies. Colt alleged that with respect to the first and third grounds Ofcom had, in effect, taken a prejudicial approach against passive infrastructure access, and not therefore given the issue proper consideration.

[77] *Everything Everywhere Limited v. Ofcom (Mobile Call Termination)* [2013] EWCA Civ 154 at [22].

[78] *Telefónica O2 UK Limited v. Ofcom* [2012] EWCA Civ 1002 at [67].

[79] Conradi and Holley (2014).

are commented upon by the Commission, providing policy guidance on how the Commission intended to handle complex issues in future. Reading these comments became a form of 'Kremlinology' or 'Colasantology',[80] whereby the Commission's future intentions could be gauged. The 2009 EU Directives modified Article 7 to allow the Commission to oversee the remedies imposed, via Phase 2 letters regarding its reasons that the draft decision including remedies would create a barrier to the internal market, or 'serious doubts' as to its compatibility with EU law. BEREC, established in 2009 though not fully operational until 2011,[81] then investigates and must respond within 6 weeks. If it agrees with the Commission, the NRA has two options: to modify or withdraw the measure to take the Commission's comments into account, or to keep its draft decision. 'Taking utmost account of the views' of BEREC, the Commission can then either lift its reservations (which has never happened to date), or much more likely issue a recommendation requiring the NRA to amend or withdraw the draft remedy.

Note that this process applied formally from 2011, but informally only from 2012 once BEREC and NRA processes were put in place. The 28 NRAs, large ones such as Ofcom or small ones such as Cyprus, have to be ready both to respond individually and to coordinate with the other 27 NRAs should an EC letter be issued requiring BEREC response. Allen indicates unsurprisingly:

> Insiders have told us that this is placing a significant burden on the NRAs (which provide the manpower to BEREC); given the restricted duration, the process requires NRA experts to make decisions (about whether they agree with the draft BEREC approach) **within days**, even if the relevant expert in that NRA is on leave or fully busy in their own country. Many NRAs do not have sufficient resources to contribute fully.[82]

In addition to this strain is the problem of BEREC disagreeing with the EC, such as in a case related to mobile termination in France in June 2012, the first real test case for the procedure. This calls for specialists to interpret these decisions in order to work out the relevance for other cases, which has severe resource implications for smaller NRAs, if less so for BEREC and the EC.

If telecoms regulation is bureaucratic and convoluted, how much faster is regulation of Internet platforms which are not IAPs? Not much, if the Microsoft and Google antitrust sagas are a guide. I consider these in the next section of this competition/regulation analysis.

[80] Named after Cold War Kremlin observations for the former, and the thinking of Fabio Colasanti, Director General of DG INFSO 2005–09, for the latter. I am grateful for the latter description to Alexandrina Dospinescu (née Hirtan), when Vice President at the National Regulatory Authority for Communications (ANRC) of Romania in 2006.

[81] Regulation (EC) 1211/2009.

[82] Allen (2012) (original emphasis).

Platform regulation: US cuckoos in the European nest?

While Internet companies were considered until recently the 'goose that lays the golden egg', European policymakers more recently consider them the US cuckoo in the European telecoms nest. This has to be considered in light of the broader platform regulation debate in Europe, which is taking place in 2017. The enormous importance of the Google case as well as various other issues in the 'digital platform economy' affecting such dominant actors as Amazon, Apple and Facebook, are dominating the competition policy debate on both sides of the Atlantic and will grow and grow as litigation follows.[83]

Microsoft was forced to offer consumers greater choice of multimedia players and web browsers in the outcome of its long-standing competition law case brought by the European Union. As a result, consumers have chosen a variety of multimedia players and web browsers, including the far more privacy-invasive offer by Google, which is itself now considered as great an antitrust target as Microsoft. It is also notable that the limits of such nudging are clear: Microsoft's dominance of personal computer (PC) operating systems is almost as great in 2015 as it was in 1995 when the European case commenced, though dominance in mobile operating systems now belongs to Google. Nudging was so familiar to Internet regulatory scholars in the late 1990s that it came to be termed the 'new Chicago School' by Lessig,[84] recognising imperfect information, bounded rationality and thus less than optimum user responses to competition remedies, driven by insights from the Internet's architecture and Microsoft's dominance of computer platform architecture. Thus recent 'nudge' concerns by regulatory scholars and competition lawyers echo 1990s concerns by Internet regulation specialists. It is a mark of Internet regulation's specialisation in Europe, and mainstream regulation and competition law's failure to fully absorb the insights of that scholarship, that in 2016 the debate surrounding nudges and privacy affecting competition outcomes has yet to reinvent the 1990s wheel of nudge limitations.[85] By not learning their Internet regulatory history, competition and regulation are obliged to repeat the lessons of the 1990s Microsoft case.

Maniadaki's work offers not only conclusions, but also 'Extensions', in particular the issue of so-called 'platform neutrality', the telecoms companies' revenge on Facebook and Google in which they argue that network neutrality should be accompanied by similar neutrality for the software platforms which run on the networks – and which are the largest source of consumer traffic for those congested networks. The growth of video, home-workers, industrial

[83] Brown and Marsden (2013b).
[84] Lessig (1998).
[85] See Baldwin (2015) and citations therein.

applications known in Europe as Industry 4.0, and CDNs with interconnection disputes, raises an ongoing issue. As platforms for these various new services interact with monopoly access providers, should platform neutrality also be regulated, and if so on what terms and when? The idea that the bottleneck for physical access to infrastructure should be extended to the 'hyper giants' of Google and even Facebook is considered by Maniadaki in some depth.

Nudge examples abound in net neutrality and software platforms. The architecture of the Internet, and the manner in which companies can influence that to mould user behaviour, was well understood by the founders of Internet regulation.[86] Government regulatory response to companies' abuse of user behaviour itself drives the subsequent choices made by both companies and users. Nudges are thus applied by companies against users, as well as by governments against companies. In advanced cases, it is governments inducing companies to offer consumers a means of nudging towards different platform outcomes.

A further instructive anti-nudge example is the use of cookies in web browsers. The European law on privacy (Directive 95/46/EC on data protection) was designed to offer users an 'opt out' of receiving cookies when browsing websites. In fact, however, the lack of enforcement by various EU Member States and the dubious legality of the US negotiated 'safe harbor' for data processing companies such as Microsoft, Yahoo and Netscape, meant that users' choice was severely constrained by the architecture of the WWW and its prolific use of cookies. Not accepting cookies meant not accessing all commercial and many other forms of web page, with ubiquitous and default cookie downloading a feature (or bug) without which web surfing is all but impossible. European law has been circumvented by both a persistent architecture of control by WWW software developers driven by the commercial advertising imperative of web design, and by the signal lack of enforcement of cookie laws by European regulators. Nudging US-led commercial operators in the direction of cookie choice failed miserably. The continued two-decade pretence that it did not is a salutary lesson in how to learn from past regulatory failures, which will have critical long-term consequences for the architecture of the Internet and of the now massive companies that dominate electronic commerce and advertising.

The strategies of Google, Amazon, Facebook, Apple (the so-called 'GAAF' dominant online retailers often described as the Big Four, or 'Frightful 5' if one includes Microsoft,[87] though noting most of Apple's revenue stems from hardware sales) and others are built on privacy-invasive architectures which are correctly contemptuous of lack of enforcement of European law. In the face of this commercial success where platform markets tip to a single dominant firm,

[86] Brown and Marsden (2013a), pp. 21–27 offers analysis with key references therein.

[87] Manjoo (2016).

it is unsurprising that IAPs are adopting increasingly privacy-invasive behaviour to try to control their users' web habits and therefore commercial revenues from those habits.

Users have adapted to behavioural architectures on the Internet and rejected certain constraints placed by companies and even regulators. In particular, dominant gateway companies have tried to shape and react to the effects of user behaviour throughout Internet history. On occasion, users have outflanked Internet companies, as for instance in the ease with which users decided to step outside walled gardens in the mid-1990s to adopt their own choice of navigation around the Internet. In rejecting the 'AOL model' of walled garden, users in fact adopted an approach that was to become dominated by Google in both search and associated services such as news, maps and shopping. In 2016 the Anglophone country dispute is no longer between IAPs such as AOL (Vodafone through its Live and 360 services, Telefónica through Endemol, and British Telecom through Yahoo! and Openworld tried similar vertical integration) and open search using Navigator and Explorer browsers, but between Google-led open search despite its vertical integration, and the Facebook-dominated closed environment.

Nudge solutions have been derisory in coping with GAAF dominance, as reflected in the reaction to Commissioner Almunia's attempt to settle the Google antitrust case. Nudging has not succeeded in preventing the growth of Amazon's Kindle proprietary grab for books, Apple's iTunes multimedia store (though it is constrained by strong vertical lock-in for Apple hardware), Google's control of the advertising industry, eBay's auction control or Facebook's growing social media monopoly. All of these platforms were subjected to competition analysis through either merger or single company dominance cases (known in Europe as Article 102 cases after the relevant legal provision in the Treaty of Rome as amended). All therefore follow Microsoft down the tried and trusted route subject to traditional competition law analysis and all have succeeded in emerging intact. Privacy invasions have led to greater regulatory control specific to that consumer harm, but the combination of privacy and competition concerns has not been achieved. Both US and European regulators are now examining the case for such analysis, though the separation of consumer harm and competition divisions of regulators is a long-standing practice. Any change is likely to be very long drawn out and ineffectual in grasping effective solutions for Internet markets at this stage in their development.

That returns us to the pressing issue of access provider abuse of control of the end user. If competition authorities and their tools of analysis were left to explore any potential problem and solution, we may be left with an assault that leaves us where Windows is ensconced after two decades: with the WWW totally controlled by a few companies each dominant in their own sector, and users only free to choose where they are super-users sophisticated and motivated enough to explore in the 'dark web' or encrypted parts of the Internet. In the same way, the number of users of the open source free Mozilla Firefox browser, alongside Google proprietary browser Chrome, is sufficient to make the Microsoft control of the browser market somewhat limited compared to 2005, but that is at the price of allowing Google to strengthen its dominance of search. Nudging Microsoft resulted in Google raising its market power given the dynamics of the Internet environment.

Firefox developed rapidly with over 40 versions released, because it finances its developers by agreeing to dominant search provider Google reinforcing its monopoly by becoming the default search engine for Firefox. Google maintained that deal from 2004 to 2014 and paid well over $1,000,000,000 for the privilege.[88] Firefox launched in November 2004 and built a 25 per cent market share in the US by 2010, but this declined to 11 per cent in early 2015.[89] This is in part a reaction to the growth of Chrome, launched in 2008, which became the largest PC browser globally (though still trailing in the US), overtaking Microsoft in 2012/13.[90] In December 2014 Microsoft-supported Yahoo became the default Firefox search engine – a search facility provided to Yahoo by Microsoft, a return to the former monopolist as Chrome became more powerful. In the US, the market shares of Microsoft, Google and Firefox were respectively 52%, 29%, and 11% in 2015. Default search had been split (52% to 41%) until 2015, but with Firefox reverting to Microsoft-powered search, that returns to 64% versus 29%. Note that many Microsoft users actually changed their default browser from its Bing or Yahoo search to Google, and in October 2015 Yahoo switched back to Google search in the United States.

Behavioural regulation: nudging net neutrality

Net neutrality also casts a light on the central issues that are emerging in regulation in a generic sense. Behavioural or 'nudge' regulation has become the flavour of the decade since Thaler and Sunstein's eponymous monograph.[91] The use of behavioural psychology insights to observe changes in regulated outcomes from the 'bounded rational' choices of consumers has been commonplace in Internet regulation since 1998, the same year Sunstein's first article on the subject appeared.[92] The Internet is a network and a real-time laboratory

[88] Murphy (2011) reporting that deal as representing 84 per cent of Mozilla's revenues in 2010.

[89] Keizer (2015).

[90] These statistics are for the United States, and are inherently unstable as the 'market' is for free browsers, and thus they do not measure purchases.

[91] Thaler and Sunstein (2008). See also Sunstein (2011), p. 1349; Johnson and Goldstein (2012), p. 417; Thaler (2015).

[92] Jolls, Sunstein and Thaler (1998). 'Nudging' analysis in neuro-science dates back to at least 1974: see Tversky and Kahneman (1974).

for the distribution and manipulation of information, which is why it is unsurprising that the adaption of that information to affect user behaviour has been a commonplace online throughout the history of the Internet. The competition and regulatory aspect of attempts to direct user and market behaviour are therefore another key theoretical perspective uncovered through examples in this book.

Though clearly platform regulation is not a network neutrality issue – there is no physical access network that contracts with the end user – there are behavioural insights borrowed from the net neutrality debate that can be applied to those dominant firms, notably the extreme level of consumer (and much legal) ignorance of the manner in which these companies trade personal data and network for advertising dollars.[93] UK competition authority chair Lord Currie explains:

> At the heart of behavioural economics is the insight that ordinary consumers do not behave as the so-called perfectly rational consumer of neoclassical economics ... Thus, for example:
>
> - we have limited ability to process and compute information. Faced with complexity we often focus on just a subset of the product's characteristics and so make bad decisions
> - we are very poor at relative probability assessment, not surprisingly since a lot of us don't understand percentages, and we tend to over-estimate the likelihood of small probability events
> - our decisions are often not neutral with respect to how choices are framed: thus we will be unduly influenced in our choice of sofa by notices that say it has been discounted from £1,000 to £500 compared with if it had simply been priced at £500 at the outset
> - we are time inconsistent and exhibit hyperbolic discounting, a lack of self-control and over-confidence. So we will definitely give up smoking and drinking but tomorrow, we won't go overdrawn, and we will go to the gym regularly
> - we care more about losses than gains and so can become inert. For example, fear of making a bad decision by switching outweighs the fact that we might well gain by switching, or makes us over-cautious in our choice of mobile package
> - we care about more than just profit maximisation, and value fairness as well.[94]

Though it would be as inappropriate for Maniadaki to conclude that case, as it was for the retired Competition Commissioner Almunia who failed

[93] This has led to a recent flurry of consultation by the UK competition regulator. See Competition and Markets Authority, Online reviews and endorsements, 2015; Competition and Markets Authority, Commercial use of consumer data, 2015.

[94] Currie (2015).

spectacularly to remedy Google's dominance, she offers interesting thoughts to start the debate, on non-price competition, information failures and behavioural biases, network effects, economies of scale and two-sided markets, innovation and burden of proof.

Ed Richards of Ofcom in 2010 considered 'nudging' and transparency in net neutrality:

> Whether and to what extent the emerging discipline of behavioural economics provides a basis for solid regulatory practice is not yet clear, but it is obviously an area that we should explore if we are serious about ensuring that consumer choice remains central to the way we address these issues ... even if consumers have access to transparent information, they need to understand how traffic management practices will affect their day-to-day experience of a service and be able to assess which product best meets their needs. This may require substantial effort and time, particularly if the information provided about traffic management practices is fairly technical.[95]

He also admits the limits of a transparency approach in consumers' lack of rational self-interest:

> The behavioural economics literature highlights further reasons why consumers may find it difficult to use information to compare products effectively. In particular, studies have shown that consumers can find it difficult to take into account fully different aspects of products when making a decision. For example, in an experiment on purchases made on eBay, researchers found that participants tended to ignore shipping costs, even when they were clearly displayed. This meant that was little pressure on sellers to keep shipping costs down. This is an example of the 'limited attention' bias. In this context, what is needed is for the industry to embrace the spirit as well as the letter of the new requirements for transparency and explanation.[96]

I explore in Chapter 6 the extent to which that embrace was self- or co-regulatory, as well as the spirit and letter of the guidance to consumers. It should be added that if you do not tell consumers what net neutrality is, they will not be very effective in identifying breaches! Cooper and Brown cite the reaction of consumers who had been educated about IAP traffic management: 'Do they do that? Oh the bastards! I can't believe it! Why do they need to do that?'[97] Behavioural nudging might be effective if net neutrality became as political as it is in the US. There is no sign of that outcome when the UK government refuses to even use the term, as we will see in Chapter 6.

[95] Richards (2010).

[96] *Ibid.*

[97] Cooper and Brown (2015), p. 12.

Limits of competition law in net neutrality analysis

I now summarise the conclusions one can apply from considered examination of competition law in this field (to explain all in detail would require an update to Maniadaki's book). They are as follows:

1. Dominance: it is very difficult to assess long-run competition to local monopoly (incumbent wholesale network) or duopoly (with cable), MNO cartels or Tier 1 Internet peering cases (examined next in Chapter 3). These markets betray behaviours more suited to utilities with natural monopoly characteristics than perfectly competitive markets. MNOs in Europe were assigned by government and remain a regulated cartel.
2. Collective Dominance: there are very specific layers to the problems in the Internet's ecology, with many net neutrality breaches by smaller local monopolies, including wired and wireless data caps, and anti-competitive throttling of Skype. It is clear that the entire class of IAPs have incentives to discriminate, rather than simply the largest such operators.
3. State aid: regulation for neutrality is vital here. Services of General Economic Interest (SGEIs) include rural fibre optic projects, rural networks and dominant former monopolists. All carry severe anti-competitive subsidy problems for regulators.
4. Merger Control: as Ungerer and Vestager identified,[98] mergers provide the best opportunity to induce structural reform of these markets, by divesting networks from dominant vertically and horizontally integrated actors, monitoring content company acquisition, and so on.
5. Evidence base: this is an extremely hard issue to analyse for competition law. Technical know-how, long-run Schumpeterian competition and network effects are all important. A major issue is that telecoms has 'performed' better than other privatised utilities, yet this improvement is about technology innovation (Moore's and Metcalfe's Laws),[99] not competition at network level. Consider VoIP, WhatsApp and fibre capacity, and the examples of Hong Kong and Japan, where RIO costs are extremely low and not telco cost-related.
6. Economic Analysis of Innovation: there are information problems, emerging dominance, 'over-the top' platform players and utility analysis of universal service to consider.
7. Institutional design mechanisms: Ungerer claims that 'essential facilities' competition was the basis for telecom liberalisation, yet in practice he acknowledges there has been little other than retail competition. Merger cases resulted in divestment of cable networks, but the many hundreds of

[98] Ungerer (2013); Speech by Margrethe Vestager: Competition in telecom markets.
[99] Clark and Claffy (2015).

market reviews established that retail/long distance/international markets had liberalised, but wholesale/local networks had not. Structural rather than functional separation is becoming the favourite remedy, as Vestager indicates.

8. Human rights and personal data: there is analysis emerging of 'predatory privacy policies', and of competition barriers to platform competition. The problems of traffic monitoring and freedom of expression are considered in Chapter 5.

The limits of competition policy in the Digital Age are clear. These include 'nudge as fudge',[100] when behavioural remedies are ineffective for consumers who have such limited knowledge. Technical knowledge of net neutrality policy and measurement is akin to making the decision on how close to live to Fukushima. There must be a role for regulation.

Competition law claims there is no problem when Facebook swallows WhatsApp and Instagram, as economists from Chicago tell us there is 'no durable monopoly' and that the EC has failed to even define a market.[101] Former Commissioner Almunia set a very dangerous precedent here, as informal discussions about Google's dominance started in November 2009 when Foundem complained.[102] It has been many years without his 'faster more informal' competition procedure producing a result.

As will be explored in Chapter 5, the IAPs' privacy-invasive activities have become subject to much scrutiny, yet the social media behemoths have escaped more or less scot-free, reflecting the very special place in public policy of the access provider. This is explained not merely by the evident monopoly power of the IAP, but as a feature of its position as the utility provider of a public need. Economists need to incorporate into their models that Internet access reflects the public policy choices about access to what is increasingly declared a human right: access to the Internet. That may sound grandiloquent, self-serving or overblown, but it is an emerging primary means of understanding the net neutrality debate. Without comprehending that users view Internet access as a more utilitarian and therefore profound service than that of social networks or Internet search, it is impossible to understand the net neutrality debate.

Maniadaki's claims regarding the need for behavioural regulation as a remedy to digital services issues looks set to become a watchword for competition and communications regulators.[103] The originator of net neutrality, Lawrence

[100] Yeung (2012).
[101] C(2011) 7279, Case No COMP/M.6281 Microsoft-Skype, with the EC failing to rule on market definition in para. 29 –consumers and paras 51, 55, 57, 62/1 – enterprises.
[102] Marsden (2013c).
[103] Maniadaki (2015).

Lessig, wrote on behavioural issues, normative issues and human rights in the digital economy prior to the Microsoft litigation in the US.[104] European competition law seems on the cusp of discovering those lessons. That is why this issue is so important to competition lawyers, and a useful guide to competition law's limitations for communications lawyers.

[104] Lessig (1998).

3

Noam's Arc and the zettaflood: towards Specialised Services?

> Start-ups need special services more than anyone in order to have a chance of keeping up with large Internet providers. Google and co. can afford server parks all around the world … small companies cannot … If they want to bring services to market which require guaranteed good transmission quality, it is precisely these companies that need special services …
> By our reckoning, they would pay a couple of percent for this in the form of revenue-sharing. This would be a fair contribution for the use of the infrastructure. And it ensures more competition on the Internet.
>
> Timotheus Höttges, Chief Executive, Deutsche Telekom, 28 October 2015[1]

Three data floods? Video, productivity and innovative services, SpS

Eli Noam predicted and regretted the death of common carriage in 1994, as cable economics overcame telecoms principles.[2] Lack of trust on the Internet, combined with a lack of innovation in the QoS offered in the core network over the entire commercial period of the Internet since NSFNet was privatised in 1995 meant that development was focused almost entirely in the application layer, with P2P programmes such as low-grade VoIP and file sharing as well as the WWW designed during this period. However, 'carrier-grade' voice, data and video transmission was restricted to commercial VPNs that could guarantee trust, with premium content attempting to replicate the same using CDNs such as Akamai, or the IAPs' own local loop offerings deployed within the user's own network. Höttges appeared to be

[1] Höttges (2015), quoted in McCarthy (2015).
[2] Noam (1994). See also Steinberg (1996).

celebrating his Specialised Service (SpS) alternative to the Open Internet in 2015.

Network neutrality is only the latest phase of an eternal argument over control of communications media. The Internet was held out by early legal and technical analysts to be special, due to its decentred construction, which separated it from earlier 'technologies of freedom' including radio and the telegraph. It is important to recognise the E2E principle governing Internet architecture. The Internet had never been subject to regulation beyond that needed for interoperability and competition, building on the Computer I and II inquiries by the FCC, and the design principle of E2E. That principle itself was bypassed by the need for greater trust and reliability in the emerging broadband network by the late 1990s, particularly as spam email led to viruses, botnets and other risks. E2E has gradually given way to trust-to-trust mechanisms, in which it is receipt of the message by one party's trusted agent which replaces the receipt by final receiver. This agent is almost always the IAP, and it is regulation of this party which is at stake in net neutrality. IAPs also can remove other potentially illegal materials on behalf of governments and copyright holders, to name the two most active censors on the Internet, as well as prioritise packets for their own benefit. As a result, the E2E principle would be threatened were it not already moribund. Legal scholars still suggest freedom to innovate can be squared with absolute design prohibitions, despite over a decade of multi-billion-dollar protocol development by the IAP community resulting in the ability to control traffic coming onto their networks, and wholesale rationing of end user traffic.

Pioneering network engineer Crowcroft makes three net neutrality policy points: the Internet was never intended to be neutral; there has been virtually no innovation within the network for 30 years; 'network-neutrality has in fact stifled evolution in the network layer'[3]. Network congestion and lack of bandwidth at peak times is a feature of the Internet: it has always existed. That is why video over the Internet was, until the late 1990s, simply unfeasible. It is why VoIP has patchy quality, and why engineers have been trying to create better QoS. E2E is a two-edged sword, with advantages of openness and a dumb network, and disadvantages of congestion, jitter and ultimately a slowing rate of progress for high-end applications such as high definition video.

In 2015 the battleground has moved on, and in this chapter I focus on three areas. The first is the 'zettaflood' of video over the Internet, now the dominant form of consumer Internet traffic according to all reliable traffic measurement surveys (P2P is the previous decade's problem). The future of Internet traffic management is video, but that also means public service broadcasting in the European context. Therefore, this chapter considers online video

[3] Crowcroft (2015).

and the European public policy challenges it faces[4] The second issue is the use of consumer Internet connections by many workers, especially in high-technology-associated industries – the 'consumer' is not easily separated from the homeworker. This is especially critical when considering that productivity growth in the struggling Western economies is enabled by ICT (information communication technology). Without adequate unthrottled Internet connections, this productivity and therefore any measurable growth in those economies may be jeopardised. A closely associated innovation is in data security and back-up, encryption, the 'cloud' and the 'Internet of Things', industrial applications known in Europe as Industry 4.0; they are innovations in Internet use that hold much promise for the future. The final element is Internet interconnection itself and SpS competition with CDNs. How networks transport traffic, which has become highly controversial as CDNs have been denied peering and forced into expensive transit arrangements in the United States, leading the FCC to regulate those arrangements. Neutrality for end users can also be directly impacted by those arrangements.

Exploding facts and Internet traffic growth

Before exploring why and how discrimination can affect users, it is important to slaughter the zettaflood myth: Internet data traffic is growing at historically low levels. The claim by IAPs wishing to traffic-manage the Internet is that Internet traffic growth is a zettaflood which is unmanageable by traditional means of expansion of bandwidth, and that therefore their practices are reasonable.

In order to properly research this claim, regulators need access to IAP traffic measurement data. There are several possible means of accessing data at Internet Exchange Points, but much data is private either because it is between two peers who do not use an Internet Exchange Point, or because it is carried by a CDN.[5] No government regulator has produced any reliable data, and carriers' and CDNs' own data is subject to commercial confidentiality (for instance, Google's proprietary CDN). In this Chapter I explain that HDTV and UHDTV will challenge even faster speed networks. Delays can also make the Internet unreliable for video gaming or VoIP.[6] Regulators engage with measurement companies to analyse real consumer traffic, while Akamai and Cisco issue quarterly 'state of the Internet' traffic aggregation studies. The UK and US regulators and the European Commission employed SamKnows to conduct a wide-ranging measurement trial, while Akamai and Cisco issue quarterly 'state

[4] See van Eijk (2011a, 2011b).

[5] Faratin *et al.* (2008).

[6] See Ferguson (2013).

of the Internet' traffic aggregation studies.[7] Research into the reality of the consumer broadband experience is much needed.

Internet traffic is dependent on local access, which is provided over either wireless means, copper telephone lines or more efficiently over fibre optic cable. The upgrading of consumer Internet connections from copper to fibre broadband is a gradual process, with urban areas and new build/multi-occupier households faster and cheaper to upgrade than rural areas and older as well as single-dwelling properties. This partially explains the rapid deployment of fibre in capital cities such as Stockholm and Paris, as well as Tokyo, Hong Kong, Taipei and Seoul.[8] Even in these early adopter nations, the deployment of fibre outside urban areas, and especially in areas with no cable networks, is patchy. Countries with high cable build, such as South Korea, the Netherlands, Germany and the United States, achieve urban and suburban roll-out rapidly. The telecommunications companies, which own the copper lines (and in most countries the largest mobile provider), provide fibre increasingly close to the end user, originally in telephone exchanges in the local town or suburb, then in roadside cabinets (fibre to the cabinet, FTTC), then fibre to the street (FTTS) via manhole covers,[9] then on remote nodes (FTTrN) on telephone poles at the end of the street (though the latter options are not yet available in the United Kingdom).[10] Some larger office and residential buildings, especially in urban East Asia, even have fibre to the building or basement (FTTB). Local network topography can explain bottlenecks close to the user, but the aggregate of Internet traffic cannot be measured here, only QoE, as we will see.

Router manufacturer and network designer Cisco estimates Western European Internet traffic grew only 18% compound annual growth rate and mobile 45% (and dropping with Wifi hand-off) in 2014,[11] but the European Commission seems almost irrationally fixated on exaggerating 'exploding' growth. In 2013 Western European fixed Internet traffic was estimated to grow at only 17% compound annual growth rate in 2012–17 and mobile at 50% or lower (the latter number is inherently unreliable as mobile was only 0.15% of overall Internet traffic in 2012 and networks jealously guard actual data use).[12] Both are historically low figures, suggesting the opposite of a 'data explosion'. Price Waterhouse claimed that in 2014 mobile data would be 58% of the total Internet traffic costs to end users, yet it was measured in 2012 at only 0.16% of the total data and in 2014 at 4% by Cisco. In all, 1 in 600 bytes were transported across mobile devices and/or networks in 2012.

[7] European Commission, Quality of Broadband Services in the EU, March 2012.
[8] See Marsden (2010), pp. 56–58.
[9] Thöny (2014).
[10] Jackson (2014).
[11] Cisco (2015).
[12] Cisco (2012).

Politicians continue to claim growth is 'exploding', despite the low growth suggested by the figures, which is a worrying divergence from evidence-based policy making. In June 2013 in a European Parliament meeting to discuss net neutrality, Commissioner Kroes stated:

> The fact is, the online data explosion means networks are getting congested. ISPs need to invest in network capacity to meet rising demand. But, at peak times, traffic management will continue to play a role: it can be for legitimate and objective reasons; like separating time-critical traffic from the less urgent.[13]

This astonished me, as I sat next to her and was next to speak in the discussion.[14] My exact words were:

> We now work out that the Commissioner is opposed to real net neutrality ... There is no data explosion on the European Internet so we should not be making policy based on a fallacious assumption ... evidence-based policy making should be based on actual evidence and the evidence does not support that idea.[15]

Internet data is not growing explosively, though the very small proportion that is mobile data is extremely expensive, and growing faster from an extremely small base. Evidence-based policy making is sorely needed in this area, and I return to the issue of independent measurement research in the concluding chapter.

Why this political obsession with explosions that do not exist? Critics continue to explain that telecoms companies have continually underestimated the value of peer-to-peer communications, which is their core value proposition, and have overestimated the 'content is king' argument that IPTV and other forms of content will substitute for decreasing telephone call revenues.[16] Odlyzko explained the dangers of disproportionate usage caps designed to prevent even the existence of rival video competitors to telco affiliates in 2012.[17] Burstein previously stated his belief that current caps are designed to prevent 'over-the-top' (OTT) video to be delivered via broadband, competing with the triple-play offers of IAPs which want subscribers to pay for a telephone line, broadband service and cable- or Internet-delivered video programming.[18] OTT video would compete with the last of these services, and degrading or capping the broadband service can protect the incumbent's video service. Burstein estimates the backhaul costs to IAPs as under $1/month,

[13] SPEECH 13/498.

[14] The hashtag for the event was #NNinEP and can be viewed at https://twitter.com/search?q=%23NNinEP&src=typd – I was not the only observer staggered by her speech.

[15] Marsden, Chris (2013b).

[16] Odlyzko (2012), and his other papers and presentations over twenty years on the subject at www.dtc.umn.edu/~odlyzko/talks/index.html.

[17] Odlyzko *et al.* (2012).

[18] Burstein (2011).

whereas Ofcom estimated the costs of backhaul for BBC's iPlayer video catch-up service to UK IAPs as being in the order of £4–£5 a month.[19] Prices have fallen rapidly with increases in transmission efficiency in that period (Moore's Law alone will have decreased prices by 75 per cent over five years). Much more research is needed into backhaul costs and other constraints on unlimited data offers.

Odlyzko has studied this area more deeply than any other analyst, and explains it as resulting from collective delusion as well as innumeracy. He explains:

> Innumeracy is especially dangerous in situations such as the telecom bubble, where the quantities under discussion are huge, with prefixes such as tera-, peta-, and exa- and refer to photons and electrons, objects that are not very tangible. That may be one reason the Internet bubble fooled people so much more than the Railway Mania of the 1840s, which dealt with far more tangible passenger transport.[20]

Added to this ignorance on the part of politicians is the wilful misleading by the industry, which he examines in the case of both the financial crash of 2008 and the dot-com crash of 2000:

> The FCC, the Department of Commerce, the press, and the VCs [venture capitalists] and entrepreneurs and investors who believed in the myth were all like Wile E. Coyote, happily running effortlessly through the air, unaware they were in a freefall, very far from the hard ground beneath. Their oblivious attitude was enabled by a lot of hot air, emanating principally from WorldCom and its UUNet branch.[21]

For full disclosure, I should declare that I was the Regulatory Director for WorldCom UK in 2001–02,[22] joining the coyote in free fall.

This lack of Internet evidence-based policy making is not restricted to net neutrality. European copyright scholar Hugenholtz makes clear that copyright policy is equally unworthy of expert analysis:

> the Commission's obscuration of the IViR [University of Amsterdam] studies and its failure to confront the critical arguments made therein seem to reveal an intention to mislead the Council and the Parliament, as well as the citizens of the European Union. In doing so the Commission reinforces the suspicion, already widely held by the public at large, that its policies are less the product of a rational decision-making process than of lobbying by stakeholders.[23]

[19] Ofcom, BBC new on-demand video proposals: Market Impact Assessment, 2006.

[20] Odlyzko (2010).

[21] *Ibid.*

[22] Marsden (2002).

[23] Hugenholtz (2008).

Alongside the 'exploding Internet' myth is the Balkanisation myth, that the Internet is breaking apart. This metaphor can be traced back as far as the net neutrality debate itself, to 1995.[24] Johnson and Post's classic article on the dangers of regulation cites the same fear.[25] Yet in 2016, such Balkanisation appears to stem as much from IAPs erecting new walled gardens against net neutrality as from government censorship directly. European governments who pass laws allowing such fast lanes and/or walled gardens are setting a poor example if they wish to avoid fragmentation and restrictions on free expression and flow of information across borders.

UHD video over the Internet

According to packet inspection company Sandvine, NetFlix streamed video in standard, high definition (HD) and ultra-high definition (UHD) accounts for 35% of North American Internet traffic in 2015.[26] The major video suppliers used over 60% of consumer bandwidth with NetFlix (34.7%), followed by YouTube (16.88%), Amazon Video (2.94%), iTunes (2.62%) and Hulu (2.48%). By contrast web traffic HTTP was 6%, Facebook (2.51%) with increasing amounts of video, and the most popular file-sharing platform BitTorrent only 4.35%, having been 7% in 2014. Video thus dominates Internet traffic for consumers, with over 70% of peak-time traffic being streamed video and audio, double the proportion in 2010.

Though public service broadcasting has slightly declined with the exponential increase in channels in the last 20 years, and near ubiquity of multi-channel households, the larger European nations still have over 50% of citizen video share devoted to live public service television. The distribution of these services on the Internet has been described to me by broadcasters privately as 'eating 15% of distribution budgets for only 3% of total viewing'. No one knows how many consumers have high enough minimum speeds to receive UHDTV signals, though it is generally agreed that about 16Mbps is the basic requirement in peak periods. Measurements by companies such as Akamai produce lower speeds than regulators' tests using SamKnows, possibly as a result of the need to work to a minimum to deliver encoded video at the rate a user's connection can accept. Furthermore, Rayburn explains that 'there's no direct correlation

[24] Standards Balkanization of the WWW – November 1995 W3C meeting by David Siegel, available at www.w3.org/Style/Welcome.html. The actual presentation has sadly disappeared but the notes exist … as does an uncomplimentary profile: www.soundbitten.com/siegel.html (Accessed 30 August 2016).

[25] Johnson and Post (1996). 'Balkanization' was also raised by Jon Auerbach in an article on NSFNet privatisation, government filtering in Germany and AOL walled gardens, for the *Boston Globe*, 1 February 1996, cited in Johnson and Post (1996).

[26] Sandvine (2015).

between unique IP addresses and households'.[27] An obvious example: campus connections maintain far higher speeds than commercial IAPs.

This also requires various QoE improvements, such as reduced latency, packet loss and jitter, as due to increased live streaming, 'inefficiency of TCP has become increasingly apparent with greater network congestion errors due to lower throughput and packet loss'.[28] Akamai explain: 'there's quibbling about exactly how much oomph is needed to faithfully render ultra-high definition images on screens, the early consensus is that 4K will demand downstream throughput of 15–20 megabits per second, minimally,[29] citing Reed Hastings of NetFlix. Akamai is deliberately extremely vague about market share and the total Internet traffic it serves, claiming only '15–30% of the Internet's traffic'.[30] Akamai uses FastTCP, an improved protocol, as well as 'Dynamic Adaptive Streaming over HTTP (MPEG-DASH) [which] promises to further advance the craft by creating the first international standard for adaptive bitrate HTTP-based streaming'. It offers:

> Akamai's highly distributed Intelligent Platform, which spans nearly 150,000 globally distributed media servers that are co-located at network edge points, literally sitting side-by-side with servers maintained by 'last mile' internet service providers ... Akamai, Elemental and Qualcomm demonstrated how 4K content can be encoded with HEVC/H.265 using MPEG-DASH.[31]

They explain that:

> a 90-minute movie encoded using H.264 at 20Mbps can weigh in at close to 14 gigabytes, nearly 5x the size of the same film encoded for 1080p delivery over the internet. Although the emerging H.265 codec promises to squeeze more bits into a smaller file, 4K movies will still pose delivery challenges that adaptive streaming can help to accommodate.[32]

While it is thus technically increasingly feasible to deliver UHDTV, the economics do not make sense using the Open Internet. Rayburn cautions that:

> true 4K streaming can't take place at even 12–15Mbps unless there is a 40% efficiency in encoding going from H.264 to HEVC and the content is 24/30 fps, not 60 fps .... With NetFlix already encoding 4K content at 15.6Mbps today, and with the expertise they have in encoding and the money they spend on

[27] Rayburn (2015a).

[28] Akamai (2015a), p. 3.

[29] *Ibid.*

[30] *Ibid.*, p. 5.

[31] *Ibid.*, adding that 'Akamai's Hybrid HTTP/UDP effort isn't so much about "switching" to UDP as incorporating the transport protocol into a broader hybrid solution that leverages TCP-speaking rule sets such as HTTP Live Streaming (HLS) while allowing for forward error-correction and for congestion relief.'

[32] Akamai (2015b) at p. 9.

> bandwidth, they will get the bitrate lower over time. Some observers think it might go down to 10–12Mbps, but that would only be possible down the road and at 24/30 fps, not 60 fps. If you want 60 fps, it's going to be even higher. But even if we use the 10–12Mbps number, no ISP can sustain it, at scale. So while everyone wants to talk about compression rates, and bitrates, no one is talking about what the last mile can support or how content owners are going to pay to deliver all the additional bits. The bottom line is that for the next few years at least, 4K streaming will be near impossible to deliver at scale, even at 10–12Mbps, via the cloud with guaranteed QoS.[33]

The new codec H.265 is more properly termed High Efficiency Video Coding (HEVC) and was ratified in 2013 as a commercially available standard based on over 5,000 proprietary patents, after a decade of work to succeed the HDTV standard known variously as H.264, MPEG4 or Advanced Video Coding (AVC).[34] A major problem is patents for H.265 which succeeds H.264 (an MPEG4 standard):

> MPEG LA pool own about 500 patents (a number that will grow over time as more patents are granted), while HEVC Advance states that they will have 500 at launch. So while HEVC won't have the same number of patents as H.264 (could be more, could be less), there may still be a lot of IP owners out there in neither camp.[35]

HEVC Advance is the more aggressive on licensing[36] and Rayburn has described their lawyers as rapacious in their demands.[37] This is hardly nudging consumers[38] into HEVC-ready UHDTV sets. Video at such high bitrates may be delayed by availability of content encoded to the correct compression for the appropriate bitrate, but UHDTV is coming to the Internet. As we will see at the end of the chapter, the question is whether it arrives via CDNs such as Akamai, or via IAPs' own proprietary walled gardens and fast lanes.

[33] Rayburn (2015a).

[34] HEVC was jointly developed by the ISO/IEC JTC 1/SC 29/WG 11 Moving Picture Experts Group (MPEG) and ITU-T SG16/Q.6 Video Coding Experts Group (VCEG) as ISO/IEC 23008-2 MPEG-H Part 2 and ITU-T H.265. See Wien (2015).

[35] Ozer (2015).

[36] HEVC Advance (2015).

[37] Rayburn (2015b), stating 'HEVC Advance is quick to say how "fair and reasonable" their terms are, they aren't. The best way to describe their terms would be unreasonable and greedy. MPEG LA, another licensing body for HEVC patents, charges CE manufacturers $0.20 per unit after the first 100,000 units each year (no royalties are payable for the first 100,000 units) up to a current maximum annual amount of $25M. HEVC Advance's rates for TV manufacturers is seven times more expensive than MPEG LA's licensing fees.'

[38] Noting that UHDTV sets already have nudge regulation for their energy efficiency in the European Union. See Cave and Cave (2012).

Table 5 Productivity growth percentage by decade (per hour in US Dollar purchasing power parity)

Nation/region	1974–84	1984–94	1994–2004	2004–14
UK	160	66	56	17
US	123	54	51	37
G7	143	65	50	35
South Korea	264	160	97	56

Source: OECD.stat (2015).

Home-worker Internet and productivity

Productivity in developed nation economies has declined significantly with the rapid offshoring of manufacturing industries to lower cost locations, leaving services as the dominant sector. This weak productivity growth is largely driven by ICT adoption and innovation. Adalet-McGowan *et al.* declare that 'acceleration in productivity growth in the United States from the mid-1990s largely reflected the rapid diffusion of ICT, but these benefits were not necessarily realised in all economies, with Europe in particular falling behind.'[39] The UK fell further behind than any other major economy, with only 17% diffusion in the decade to 2014, compared to 43% in the Euro area.[40]

In a study in 2013 the UK government concluded that consumer broadband has a significant effect on national productivity (through business innovation, international trade and effective time management via teleworking), which demonstrates the economic, environmental and social case for state subsidy to ensure a universal fast broadband connection for every home.[41] It is therefore critical that workers are able to connect to their companies', suppliers' and customers' networks, whether in the office, on the move (via Wifi or less frequently wireless networks) or increasingly at home. They note that:

> The incremental social impacts of superfast broadband are likely to include an increase in time spent consuming video entertainment, and an increase in the use of video communications. For areas with poor current levels of connectivity, improvements in broadband speed will mitigate the extent of adverse impacts on the usability from … increased file sizes for webpages.[42]

While correlation is not causation, an enormous body of work in industrial organisation and micro-economics has proven beyond reasonable doubt that a

[39] Adalet McGowan *et al.* (2015).
[40] *Ibid.*, p. 19, Figure 5.
[41] Department for Culture, Media and Sport (2013a).
[42] *Ibid.*

workforce educated in the use of networked information technology ('digitisation') substantially boosts productivity.[43] A 2013 UK survey reported:

> Around 50% of Britons regularly work from home however, some people are unhappy with their current connection; with a third of workers claiming connectivity problems affected their efficiency to carry out a job. Around 34% of those surveyed said slow broadband speeds affected their work, with 16% saying temporary service loss was the main problem.[44]

Home working via broadband has been recognised as an issue since the technology emerged, with Microsoft (and British Telecom) paying for UK home-working employees to receive broadband at home since 2002.[45] It is therefore common currency that high-quality broadband aids productivity in the modern workforce, and most of that home-working connectivity is via a residential broadband line paid for by the employee directly.

Broadband is developing at vastly different rates in different parts of the developed and developing world, between nations, within nations, in rural and urban areas and even within suburbs.[46] As advertising has become increasingly intrusive in commercial websites, especially for mobile and rural users, the need for faster broadband as well as widespread deployment of 'ad blockers' has become more important simply to load such pages.[47] The major defining issue is the distance between a premises (home or business) and the nearest router attached to the Internet via a fibre optic cable. In some areas this may be located inside the premises, or in the next building, the same street, the neighbourhood or the nearest town. Where this distance is greater than 1–2 kilometres, the speed of the broadband connection will be very significantly reduced. The second issue is how many homes and businesses may have to share that connection – the contention ratio. It is significantly less costly for an IAP to decrease the contention ratio by increasing 'backhaul' switching capacity in the router or from the router to a major fibre route, than it is to build the fibre closer to the end user's premises. These improvements are needed for innovations such as those analysed below.

In 2015 the UK government observed:

> emergence of cloudbased services, ever increasing levels of data consumption, increased mobility and new electronic communications networks technology that is more responsive to user needs. Other changes are clearly on the horizon such as 5G and the rapid expansion of the Internet of Things. We are also witnessing increasing pressures for consolidation within markets as well as convergence

[43] Booz Allen Hamilton and Katz (2012).
[44] Havergal (2013).
[45] Richardson (2002).
[46] Emmott and Harris (2015).
[47] Aisch and Keller (2015).

> of traditional electronic communications services, over the top services and media content (e.g. bundling of service) … commercial deployment of new technologies such as 5G, G.Fast (an ultrafast broadband technology), and more responsive networks (e.g. network function virtualisation and software defined networks).[48]

In response they suggest deregulation where possible and emphasise that 'Other aspects of network performance, such as reliability, capacity, latency and resilience, are becoming just as important as connection speed to the user experience, both for business and personal use.'[49] While this is undoubtedly true, it has been shown that consumers react to headline speeds as a default measure of broadband performance,[50] while actively deploying advertising blocking add-ons to web browsers, which are themselves chosen for speed of loading web pages.

Brown for the International Telecommunications Union states that:

> The terms on which IoT [Internet of Things] service providers can access customers across the public Internet will have a significant impact on their ability to enter new markets. Baseline access could be protected by 'network neutrality' rules from communications regulators in the US, EU, and elsewhere. IoT users with very high bandwidth or reliability requirements may be affected by neutrality rules that limit the ability of telecommunications companies to discriminate between Internet data from different sources. Such rules usually still allow telecommunications providers to offer such customers 'specialised services' with specific speed or reliability guarantees. The terms attached to such services will be a key area of review for competition regulators.'[51]

Specialised Services: the exception to net neutrality

IAPs are creating managed service lanes with guaranteed QoS alongside the public Internet. I discuss both the definitions of SpS and the minimum USO to prevent 'dirt roads' in this section. BEREC offered a 2012 definition, rigorous in separating SpS from the public Internet:

> electronic communications services that are provided and operated within closed electronic communications networks using the Internet Protocol. These networks rely on strict admission control and they are often optimised for specific applications based on extensive use of traffic management in order to ensure adequate service characteristics.'[52]

[48] Department for Culture, Media and Sport (2015b).

[49] *Ibid.*

[50] Arnold *et al.* (2015).

[51] Brown (2015).

[52] BEREC, BoR (12) 131, p. 5

BEREC explained it: 'might be the case that all IAPs present in the access markets are blocking traffic of special P2P applications. That situation might be considered as collective SMP, which is difficult to prove.'[53] It observed that: 'Blocking P2P systems or special applications reduces consumers' choice, restricts their efficient access to capacity-intensive and innovative applications and shields the user from innovation. Thus it reduces the consumer's welfare, statically and dynamically.'[54] It concludes that: 'For a vertically integrated IAP, a positive differentiation in favour of its own content is very similar to a specialised service.'[55] This is an important conclusion, that SpS can in reality form a means of evading net neutrality regulations, while diverting traffic away from the public Internet to a less regulated, premium-priced alternative.

The FCC Open Internet Advisory Committee (OIAC) states: 'The business case to justify the investment in the expansion of fiber optics and improved DSL and cable technology which led to higher broadband speeds was fundamentally predicated upon the assumption that the operator would offer multiple services.'[56] In its *Comast/NBC* merger condition in 2011, the FCC held that Specialised Service means:

> any service provided over the same last-mile facilities used to deliver Broadband Internet Access Service [BIAS] other than (i) BIAS, (ii) services regulated either as telecommunications services under Title II of the Communications Act or as MVPD services under Title VI of the Communications Act, or (iii) Comcast's existing VoIP telephony service.[57]

The FCC 2010 Order offers a definition of:

> services that share capacity with broadband Internet access service over providers' last-mile facilities, and may develop and offer other such services in the future. These 'specialised services', such as some broadband providers' existing facilities-based VoIP and Internet Protocol-video offerings, differ from broadband Internet access service and may drive additional private investment in broadband networks and provide end users valued services, supplementing the benefits of the open Internet.[58]

In the US, Comcast was accused of failing to conform to its obligations not to favour its own specialised IPTV service in 2012–13,[59] breaching 2011 merger

[53] BEREC, BoR (12) 132, para. 277.
[54] *Ibid.*, para. 279.
[55] *Ibid.*, para. 307.
[56] FCC, Open Internet Advisory Committee Annual Report, 2013, p. 68.
[57] FCC, In the Matter of Applications of Comcast Corporation, General Electric Company and NBC Universal, Inc. For Consent to Assign Licenses and Transfer Control of Licensees, 2011.
[58] FCC, Report and Order Preserving the Open Internet, 2010, para. 112.
[59] See Lee (2012).

consent terms.[60] Highly controversially, the FCC chose not to open an enforcement action, possibly because it was soon to reopen the Open Internet Docket again in 2014.[61]

The FCC OIAC explains:

> A high threshold or cap may represent an additional factor that shapes the ability of an edge provider to supply its service or conduct business with a user. If an ISP imposes a data cap or other form of UBP [usage-based pricing], this could affect user demand for the edge provider's service, which, in turn, may shape the ability of the edge provider to market and deliver its service. This is especially so if the ISP offers specialised services that compete with the edge provider, and for which a cap or other UBP does not apply.[62]

They continue:

> There is a rationale for separately provisioning between the specialised and non-specialised services, usually to achieve some engineering or market objective, such as improve the quality of service (e.g., reduce user perceptions of delay). In addition, one service often has a set of regulatory requirements associated with it, and one often does not.[63]

The conclusion is:

> a specialised service should not take away a customer's capacity to access the Internet. Since statistical multiplexing among services is standard practice among network operators, the isolation will not be absolute in most cases. However, if a specialised service substantially degrades the BIAS [Broadband Internet Access Service] service, or inhibits the growth in BIAS capacity over time, by drawing capacity away from the capacity used by the BIAS, this would warrant consideration by the FCC to further understand the implications for the consumer and the possible competitive services running on the BIAS service.[64]

As the FCC Open Internet Advisory Committee admits in suggesting technology neutrality is observed where possible: 'There are painful edge-conditions to this principle, which we acknowledge.'[65]

I favour FRAND rules for SpS, not a complete ban. As with all telecoms licensing conditions, net neutrality depends on the physical capacity available, and it may be that de facto exclusivity results in some services for a limited

[60] See Public Knowledge (2013): 'The Commission must show that it has the conviction to actually enforce merger conditions – not merely to impose them.'

[61] Frieden (2015c).

[62] FCC Open Internet Advisory Committee (2013b), p. 18.

[63] *Ibid.*, p. 32.

[64] *Ibid.*, p. 68.

[65] *Ibid.*, p. 70.

time period as capacity upgrades are developed.[66] Interoperability requirements can form a basis for action where an IAP blocks an application.[67] Recall that dominance is neither a necessary nor a sufficient condition for abuse of the termination monopoly to take place, especially under conditions of misleading advertising and consumer ignorance of abuses perpetrated by their IAP. I also support the 2012 BEREC definition which would define (almost) all IP services not physically separated into the normal net neutrality rules.

SpS and the European ASQ research controversy

Almost two decades of multi-billion-dollar protocol development by the IAP community has resulted in the ability to control traffic coming onto their networks, especially via the 3GPP and Cisco work on MPLS (Multiprotocol Label Switching).[68] The most explicitly influential of these corporate-sponsored projects on net neutrality law was the ETICS project, which aimed to create a blueprint for SpS and was funded at a cost of €8m by the European collaborative research project within the ICT theme of the 7th Framework Programme of the European Union programme.[69] Its work was cited in COM(2013) 627:[70]

> This strategy explicitly recognises ASQ agreements as a potential new source of growth and innovation in Europe, but which must operate alongside a well-functioning best-effort internet access service. By studying existing limitations and proposing both new business models and a flexible architecture able to adapt to a maturing interconnection market based on QoS, ETICS has established the basis for developing network interconnections.[71]

The ETICS definition of Assured Service Quality (ASQ) was written into COM(2013) 627 Article 2.12:

[66] See FCC, Report and Order, In the Matter of Preserving the Open Internet, 2009, and Andersen *et al.* (2010).

[67] See Marsden (2010), p. 1.

[68] Lin, Davie and Baker (1996); Rekhter *et al.* (1997). Arguably this developed from the failure to deploy multicast to solve congestion: see Deering (1989).

[69] The partners were six ISPs (BT, Deutsche Telekom, Orange, Telefónica, Telenor, Primetel), the largest networks company in Europe (Alcatel-Lucent Bell Labs France) and several specialist academic research institutes (Athens University of Economics and Business, Telecommunications Research Center Vienna (FTW), Institut Telecom, Politecnico Di Milano, University of Versailles Saint Quentin, Technion and University of Stuttgart. See www.laquadrature.net/files/ETICS_final_publishable_summary.pdf (Accessed 16 September 2016).

[70] Annex II of COM(2013) 627.

[71] See European Commission, Blueprint for next-generation quality-enabled network interconnection, 3 June 2014.

> 'assured service quality (ASQ) connectivity product' means a product that is made available at the Internet protocol (IP) exchange, which enables customers to set up an IP communication link between a point of interconnection and one or several fixed network termination points, and enables defined levels of [E2E] network performance for the provision of specific services to end users on the basis of the delivery of a specified guaranteed [QoS], based on specified parameters.[72]

This promotion of the ETICS standards by the Commission in June 2014 post-dated claims by the Commissioner's spokesperson Ryan Heath that she had never heard of the project,[73] even though its work is cited as central to SpS. This led one of Europe's foremost telecoms journalists to state that:

> This is not obscure stuff from your point of view, Ryan. It's all about the very technology and economics behind the 'specialised services' you were so keen to promote just a few weeks ago in Connected Continent.[74]

The ETICS partners unsurprisingly concluded from their study that the market should pursue ASQ without a European law regulating net neutrality:

> ETICS partners found that coherent conclusions from various studies on the impact of traffic management on net neutrality remain elusive. Safeguards exist thanks to regulation and market monitoring by consumer associations, which should prevent any abusive use of ASQs. The project investigated best practices and business policy rules and concluded that it would be wise to allow consumers and the market to decide on the relevance and value of QoS management in a competitive environment.[75]

ETICS was explicit in its aim:

> increased market value will be split between well-established traditional CDNs (Akamai, Limelight and Level 3) and ETICS' players. Assuming ETICS will serve only the market corresponding to the very unsatisfied customers who will increase video demand with ETICS ASQ launch, the lower bound would then be equal to $68.4 million. The upper limit will consider that ETICS could either develop a 'proprietary' CDN solution or reduce CDN's relevance by creating an [ASQ] pipe, possibly cannibalizing part of the market for traditional CDN providers.[76]

[72] The ASQ definition, also in Annex II of COM(2013) 627 is taken from the ETICS project (2010–12), available at www.laquadrature.net/files/ETICS_final_publishable_summary.pdf (Accessed 16 September 2016).

[73] Heath (2013), commenting on Zimmerman (2013).

[74] Marsden, Chris (2014c).

[75] See European Commission, Blueprint for next-generation quality-enabled network interconnection, 3 June 2014.

[76] ETICS (2012), which also includes regulatory implications in particular.

Software-defined networking could make such telco plans highly problematic and this story of differing standards for QoE is just beginning.[77]

How does this work in practice? It raises heroic regulatory enforcement issues, and as we will see at the end of Chapter 4, BEREC is charged to aid NRAs to do this. Research for the German regulator, a reluctant late convert to net neutrality, has shown that detection will be far from trivial and exposes the fact that SpS is designed to create quality differences between different services: to create a superhighway alongside a 'dirt road' Internet.[78] There is substantial controversy regarding definition of SpS, data caps on public Internet (or 'BIAS' as the FCC calls it), and the limits of public net neutrality rules. This is already apparent in the US, and will be a central feature of the European net neutrality debate.

Universal service to avoid Internet 'dirt roads'

Traffic management techniques affect not only high-speed, high-money content, but by extension all other content too. You can only build a high-speed lane on a motorway by creating inequality, and often those 'improvement works' slow down everyone currently using the roads. The Internet may be different in that regulators and users may tolerate much more discrimination in the interests of innovation. To make this decision on an informed basis, it is in the public interest to investigate transparently both net neutrality 'lite' (the slow lanes) and net neutrality 'heavy': Specialised Services (rules which allow higher speed content). For instance, in the absence of regulatory oversight, IAPs could use DPI to block all encrypted content altogether, if they decide it is not to the benefit of IAPs, copyright holders, parents or the government. (IAP blocking is currently widespread in controlling spam email and copyright-infringing material, and for blocking sexually graphic illegal images.)

As higher speed services develop, IAPs may be distinguished from other services by more than price discrimination, and research into consumer-revealed preference is needed. In 2010 Finland offered a guarantee of universal access to 1Mbps broadband for all households on non-discriminatory terms by 2012, a minimum QoS guarantee backed by law, making it the world's first country to offer such minimal neutrality. Other countries may soon follow, and Australia is building a wholesale fibre-to-the-node network, supplemented by rural satellite, which will offer a much higher speed universal service, though conditions of retail access remain to be negotiated. As we explore in the final chapter, the UK consulted on committing to a 10Mbps USO by 2020, and the European Commission on a 1Gbps floor. The possibility of abusive breaches of net

[77] Bubley (2015).

[78] Marcus and Waldburger (2015).

neutrality by monopoly-funded networks at lower speeds is significant, a fate avoided in the US thanks to its insistence on net neutrality for government-subsidised broadband roll-out.[79] The European Union target in the Digital Agenda strategy aims to make 30Mbps+ speeds available to 100 per cent of households by 2020 (with 50 per cent being within reach of a 100Mbps+ service). This will involve at the very least significant further FTTC vectoring investment. The EC has published rules for direct state investment in broadband networks, in which the only net neutrality concern states that there is 'public interest in funding an open and neutral platform on which multiple operators will be able to compete for the provision of services to the end-users', but there is no specific mention of network neutrality.[80]

There will remain a prominent policy question regarding rural and semi-rural access to high-speed broadband, as a copper telephone line with more than 3 km distance to a local exchange will not achieve high speeds with ADSL, ADSL2 and ADSL2+ technologies (the theoretical maximum at 3 km on a standard copper line is 8Mbps with all three technologies, though much faster over shorter line lengths, or with vectoring using G.Fast technology). Therefore, rural households will depend on a combination of satellite and mobile technologies, though the possibility of higher performance remains theoretical. These rural households may therefore prove the most resistant to significant breaches of network neutrality for basic services such as Skype, and empirical research is needed into the types of service held most valuable by these households.

Detecting IAP bad behaviour in breaching net neutrality will in practice be a difficult undertaking. Exceptions permitted by law for 'reasonable traffic management', SPS and other elements (see Chapter 4) create plausible deniability for operators. Where smoking guns appeared, regulators have not responded, as we will see in Chapter 6. Evidence that government collaborated with IAPs in illegal interception (and vice versa as Edward Snowden's revelations clarify) make the detection of violation of a principle that became European law in April 2016 even less likely. That law is the subject of the next chapter.

[79] American Recovery and Reinvestment Act 2009, and FCC, Report on a Rural Broadband Strategy, 2009, pp. 15–17, especially footnotes 62–63.

[80] European Commission, Community Guidelines for the application of State aid rules in relation to rapid deployment of broadband networks.

4

European Open Internet regulation

> Net neutrality ... is a term apparently the [European] Council shuns like the plague. Because that was their first objective: this term had to get out of the directive ... there is no definition of net neutrality ... no assurance of equal treatment, non-discrimination and free movement of traffic.
>
> Sabine Verheyen MEP[1]

This chapter first considers the 2013 Proposal and Trilogue in 2014/15, then the 2015 Regulation's net neutrality aspects, before finally looking at the details of BEREC's implementation of its Guidelines. European law upheld transparency on a mandatory basis, and minimum QoS on a voluntary basis, under provisions in the 2009 electronic communications framework. In May 2011 Commissioner Kroes stated she was in favour of net neutrality 'lite': 'I am ready to prohibit the blocking of lawful services or applications. It's not OK for Skype and other such services to be throttled.'[2] However, she then asked BEREC to undertake preparatory work on behalf of the Commission to examine the nature of the problems and potential solutions. BEREC published a great deal of useful information regarding the techno-economic aspects of the argument in 2011–14 (see Table 1, Chapter 1) including a survey in which mobile operators admitted to blocking VoIP, especially Skype.

In spring 2013 the European Commissioner was finally prepared to act, though, as I detailed in Chapter 3, she proceeded on the basis of net neutrality 'lite' and explicit support for Specialised Services. On 11 September 2013 the European Commission adopted a proposed Regulation that would substantially impact and harmonise net neutrality provision, allowing priority Specialised Services and generally preventing IAPs from blocking or throttling third-party content.[3] The proposal was extensively strengthened from a July

[1] Verheyen (2015).

[2] SPEECH 11/285.

[3] COM(2013) 627.

2013 draft, and its essential items were in part positive and in part negative for net neutrality policy.

Article 23(5) would enforce net neutrality 'lite', thus conforming to the Netherlands and Slovenian laws:

> Within the limits of any contractually agreed data volumes or speeds for Internet access services, [ISPs may not engage in] … blocking, slowing down, degrading or discriminating against specific content, applications or services, or specific classes thereof, except in cases where it is necessary to apply reasonable traffic management measures.

These are defined as:

> transparent, non-discriminatory, proportionate and necessary to:
>
> a) implement a legislative provision or a court order, or prevent or impede serious crimes;
> b) preserve the integrity and security of the network, services provided via this network, and the end-users' terminals;
> c) prevent the transmission of unsolicited communications to end-users who have given their prior consent to such restrictive measures;
> d) minimize the effects of temporary or exceptional network congestion provided that equivalent types of traffic are treated equally.

It continues: 'Reasonable traffic management shall only entail processing of data that is necessary and proportionate to achieve the purposes set out in this paragraph.' Articles 21–24 then explain users' contractual remedies in switching from providers who discriminate unreasonably.

European law is a deliberate process of negotiation between the Commission (which has sole power to initiate legislation and acts as the 'Guardian of the Treaties'), the Parliament (whose Committees consider legislation in detail, propose amendments and advise the full plenary of Parliament on how to vote) and the Council (comprising the 28 Member States' representatives, including many who are the largest shareholders in their dominant IAP). The process is rather tortuous to outsiders, and Table 6 shows the progress of the first net neutrality 'advisory' law that came into force in 2011, and the second that was passed on 27 October 2015.

In particular, note that in each case the law was amended by Parliament on First Reading, leading to a Trilogue with the Commission and Council of Ministers in which the head of the relevant Parliamentary Committee had to negotiate a compromise agreeable to Member State governments, to then be returned for a further vote in Parliament. In 2009, prior to the five-yearly parliamentary elections, the Parliament held its nerve at Second Reading and forced the Council of Ministers to accept a slightly stronger net neutrality 'advisory' opinion than they had wished. This occurred after the elections in 2015, and Parliament caved in. I briefly outline why that happened, before analysing the new law in detail, and BEREC's consultation on its meaning, in the remainder of the chapter.

Table 6 European Union legislation and regulation on the Open Internet

Stage of Legislation/ Regulation	2007 Telecoms Package[a]	2013 Telecoms Single Market[b]
Commission adoption of Proposal	13 November 2007[c]	11 September 2013[d]
Parliament Opinion on First Reading	24 September 2008	3 April 2014
Parliament Opinion on Trilogue Compromise	6 May 2009	27 October 2015
Legal effect	18 June 2011 – Directive to be implemented	Entry into force on 30 April 2016, BEREC guidelines issued August 2016[e]
National laws of interest	Netherlands, Slovenia 2012	To be determined

[a] Available at http://eur-lex.europa.eu/legal-content/EN/HIS/?uri=CELEX:32009L0136&qid=1445890303556.
[b] Available at http://eur-lex.europa.eu/legal-content/EN/HIS/?uri=celex:52013PC0627.
[c] COM(2007) 698.
[d] COM(2013) 627.
[e] Madiega (2015).

Trilogue agreement on net neutrality regulation 2015

Schaake and almost 200 other pro-neutrality European parliamentarians wrote to the Italian EU Presidency on 25 November 2014, worried that 'The initial impressions of your proposals concern us, as they would water down precisely those strong definitions of net neutrality and specialized services that are needed in an EU Digital Single Market.'[4] Given the Italian government's historic relations with Telecom Italia, that is unsurprising. Julia Reda MEP had forewarned of Deutsche Telekom's views in a written question in Parliament in June 2013:

> Deutsche Telekom has announced plans to change its tariffs and impose severe 'throttling' when a certain data usage is reached. At the cheapest rate, this throttle is to apply to usage from 75GB. This will take effect in 2016 … According to newspaper reports, the Commission intends to put forward, as early as this year, a legal recommendation designed to protect unlimited access to all Internet content for consumers. When will that recommendation be published? [5]

[4] Schaake *et al.* (2014).
[5] European Parliament, Parliamentary Questions, E-006146-13, 2013.

The Regulation was published shortly thereafter, in September 2013. After its trashing by Parliament, Council and BEREC as both a power grab for the Commission and a political obituary for outgoing Commissioner Kroes, all that remained by the end of 2014 was mobile roaming and net neutrality. All other issues were parked until the 2016 Telecom Single Market proposal, which was in Regulatory Fitness and Performance Programme (REFIT) evaluation until autumn 2016.[6]

Though it is always dangerous to make predictions (especially about the future), I had assumed that the First Reading strong parliamentary position would collapse in the face of Commission and Council intransigence and the tempting carrot of a proposed end to roaming in 2017.[7] This occurred even more completely than I had predicted:

> Telecom Italia, guardians of ugly monopolists ETNO, was always going to spike the Italian Presidency's guns, so this will probably roll on to the next Telecoms Council chaired by Latvia in early 2015 – though not presumably as late as the 9 June formal Telecoms Council – handily on the Google Calendar for the Latvian Presidency. There will be a UK General Election before that date but as all major UK parties are neoliberals (except in this area possibly the LibDems) don't expect much change.[8]

Ed Vaizey set out the UK position very clearly in a 'Dear Bill' letter to the Chair of the UK Parliament European Scrutiny Committee in May 2014:

> I begin with noting that the outcome of the EP First Reading deal was not as expected i.e. in line with the recommendations put forward by the ITRE Report. This was, in the main, due to the ALDE (liberal) Group within the EP withdrawing its support for the content of the ITRE Report covering this issue after voting for its adoption, and then aligning itself with the positions previously adopted by the Socialists & Democrats and Green Groups by jointly putting forward a series of amendments. It was these amendments that were voted passed during the Plenary vote rather than those in the ITRE Report ... the result is that the EP First Reading now contains a specific definition of 'net neutrality', as well as a more restrictive approach to 'specialised services' and 'traffic management'. This is in direct opposition to HMG's current negotiating stance and underlines the contentious nature of this issue as previously noted in the most recent Commons Committee Report.[9]

What he identified is that the Committee on Industry, Research and Energy (ITRE) report of Malcolm Harbour, an arch corporatist Conservative, had been

[6] Viola (2015).
[7] Baraniuk (2015).
[8] Marsden, Chris (2014b).
[9] Vaizey (2014).

overturned in particular by the Dutch liberals led by Marietje Schaake. BEREC found this an improvement on the poor original EC proposal, but:

> While some of the language in the text adopted by European Parliament draws upon BEREC previous publications on the subject, improving the original Commission's proposals, it does not yet meet these standards. A balanced approach to promoting net neutrality on the Internet in parallel to the provision of specialised services is a difficult challenge. BEREC considers that specialised services should be clearly separated (physically or virtually) from internet access services at the network layer, to ensure that sufficient safeguards prevent degradation of the internet access services. Therefore BEREC welcomes the European Parliament's acknowledgement of this principle.[10]

Unfortunately, the final version of the Regulation moved backwards to a deliberate obfuscation of the distinction between the Internet and private services, leaving definition to NRAs.

Ed Vaizey for the UK government argued that the net neutrality situation:

> may change but taking into account early indications of Member States' views in this area, we cannot rely on a change on the position from one where UK's [*sic*] remains relatively isolated in its opposition. It is worth noting that the issue of net neutrality is one that is covered by the UK and German initiative.[11]

It was the Germans' apparent change of heart for the three-month Trilogue process in early 2015 prior to the Second Reading that may have proven to be the key change, together with the very obvious compromising quality of the Spanish European People's Party MEP del Castillo[12] supposedly defending the parliamentary position from the First Reading. Trilogue is an entirely opaque process, but the German ruling coalition had agreed to argue for the First Reading text in Council. Between their representative, the other Member States, the new Commissioner (Internet newbie, elderly corporatist, Christian Democrat regional politician Gunther Oettinger[13]) and Del Castillo, the defence of the First Reading almost entirely collapsed in 2015. Oettinger opened the trilogue on 4 March, and the following day famously decried Pirate Party lobbying for strong net neutrality by declaring:

> Net neutrality: Here we've got, particularly in Germany, Taliban-like developments. We have the Internet community, the Pirates on the move, it's all about

[10] BEREC, BoR (14) 50.

[11] Vaizey (2014).

[12] The Spanish Partido Popular is the direct democratic descendant of the Fascist dictator Franco's political movement, the Popular Alliance – the old fascists left the party before 1979. It is right wing and corporatist. It led the Spanish government in 1996–2004 and 2011–15. Del Castillo is a member of the board of the Telefónica Foundation.

[13] For his regional political background, see Kosmopolit (2009). For his Internet policy background, see Hirst (2015).

> enforcing perfect uniformity. They talk about 'the evil industry'. It's not about the industry, it's not about the CEO and his salary. If you want to have real time road safety, our lives are at stake, this has to have absolute priority with regards to quality and capacity.[14]

He appeared to believe that drone vehicles needed a continuous low-latency, high-speed connection via the public Internet. (It is impossible to overstate how important the automobile industry is to southern German politicians, especially his home town of Stuttgart, also the home town of Porsche and Mercedes.) When challenged by the Pirate Party MEP Reda, he took three months until the Trilogue had finished before confirming via parliamentary answer that his technical assumptions were not based on a specific example.[15]

Table 7 shows the timing of the Trilogue negotiations and final European Parliament vote.

This was a deal conducted in private in spring 2015. Though the German government and Parliament were in favour of net neutrality in 2014 after German elections and before European Parliament elections, Freedom House summarises the retreat by the German government from its formal coalition agreement in late 2013 ('Granting net neutrality will be one of the aims of the government'[16]) to its position in 2015:

> The ruling coalition has started to endorse classified net traffic in order to privilege certain services and providers. In October 2014, it was revealed that the government may refrain from promoting net neutrality in order to create incentives for private companies to speed up the development of broadband internet in Germany. According to suggestions published by the Federal Ministry of Transportation, certain internet services may acquire paid priority treatment by the networks in order to refinance infrastructure measures in this sector.[17]

This certainly appears to have been the approach pursued by Oettinger and Deutsche Telekom, and indeed the outcome of the Trilogue.

As the European Digital Rights Initiative (a consortium of national civil society representatives) put it:

> The trilogue system is so opaque that, even when the trilogue process produced an agreement in May 2015, there was no press attention. Besides our blogposts and press releases warning about the situation, there was no outcry at the weak, ambiguous text that would put free speech, innovation and the interests of start-ups at risk. The opaque, closed, undemocratic, bureaucratic nature of the trilogue process had successfully killed off any meaningful public debate. With the

[14] Reda (2015). Reda, the only Pirate Party MEP in the 2014–19 Parliament, is from Germany and provides the YouTube video of Oettinger's comments as documented proof of his remarks.

[15] European Parliament, Parliamentary Questions, E-004461/2015, 10 June 2015.

[16] Emert (2013).

[17] Freedom House (2015).

Table 7 European trilogue debates between Council, Commission and Parliament[a]

March 2015	3rd: Council adopts position 4th: Council's mandate to start trilogue negotiations.	5th: Commissioner Oettinger considered net neutrality as a 'Taliban-like issue'.	19th: MEP Julia Reda asked Oettinger what services would be banned by neutrality. Oettinger cannot name any.
1st Trilogue April 2015	17th: Committee of Permanent Representatives in the Council (COREPER) approved the Council text of 15 April (no change).	21st: European Commission's (pseudo) compromise text proposal.	27th: New Council text, leaked by Statewatch (only changes on roaming).
2nd Trilogue May 2015	8th: European Parliament's first compromise text.	17th: Council's 'compromise text' on net neutrality.	22nd: reviewed version of Presidency 'compromise'.
3rd Trilogue May 2015	27th: Latvian Presidency send new Council text to Parliament.	29th: Council produces a new mandate slightly changing Latvian Presidency 27th May text.	
June 2015	12th: Council confirms trading net neutrality for ending roaming charges. Oettinger seemed to agree.	18th: European Parliament's second compromise text.	4th June European Commission writes a 'non-paper'.24th: New Council text, adopted by COREPER.
4th Trilogue July 2015	30th 2–3 a.m.: provisional agreement regarding the articles.	2 July: Commission proposes technical drafting suggestions for the recitals.	3rd: Provisional joint agreement regarding the recitals.
End of Trilogue July 2015	8th: COREPER approves the political agreement, published by the Council.	15th: Committee on Industry, Research and Energy formally adopted the text by vote: 51–10.	Now a vote in Parliament can confirm negotiated outcome.
October 2015	1st Council adopted the trilogue agreement as its first reading position, published by the Council.	27th Parliament at Second Reading votes down amendments that would restore net neutrality, confirming Council text.	Law will be published in Official Journal – probably early December.

[a] See further Askthe EU (2015) Trilogues on the Telecommunications Single Market Regulation www.asktheeu.org/en/request/trilogues_on_the_telecommunicat_2.

Source: EDRi (2015).

> Socialist (S&D), Conservative (EPP/ECR) Groups in the European Parliament (representing 62% of MEPs) having given their approval, it would have taken a miracle to pass amendments in the rubber-stamping Parliament plenary session on 27 October.[18]

The Trilogue outcome removed the definitions of SpS and net neutrality, and were very unclear on zero rating. While the European Parliament and Commission congratulated themselves, the truth was revealed by the governments of the Netherlands and Slovenia, who already had net neutrality and feared that in 2016 they would be obliged to repeal their laws. Slovenia stated that it was particularly the express permission for SpS to which it objected:

> Slovenia fears that the new arrangements will result in a two-layer Internet: a slow 'best effort' service model and a high-speed Internet with guaranteed quality for an additional charge. Slovenia believes that this is the wrong response to the competitive challenges facing the European industry in the global digital market. Also, given the current legal protection of Internet neutrality in Slovenia, we cannot support the final TSM regulation.[19]

The Netherlands made clear its particular opposition was based on 'zero rating' and price discrimination:

> effective net neutrality rules also require discriminatory pricing practices to be clearly prohibited. Such a clear ban on price discrimination is unfortunately not included in the final compromise. The Netherlands will therefore be obliged to withdraw this ban from its national net neutrality rules, even though it was applied effectively. The lack of a clear ban on price discrimination has been a fundamental concern for the Netherlands throughout the negotiations. This fundamental concern is expressed by a vote against the Regulation.[20]

This makes the national objections clear, and the civil society and parliamentary debates in Horten's 2010 work[21] deserve a sequel: now for the legislative detail.

Regulation 2015/2120

After more than two years of gestation, on 27 October 2015 the net neutrality law was approved by the European Parliament.[22] As the new Regulation states, the safeguard of equal and non-discriminatory treatment of Internet traffic became urgent and necessary because 'a significant number of end-users are

[18] McNamee (2015).
[19] Council of the European Union (2015).
[20] *Ibid.*
[21] Horten (2011). See also Horten (2015).
[22] Regulation (EU) 2015/2120.

affected by traffic management practices which block or slow down specific applications or services'.[23] Indeed, the spread of such discriminatory practices had been clearly demonstrated by a joint investigation by BEREC and the European Commission in 2012.[24]

Reactions to the vote were quite heterogeneous, with the European Commissioners cheering the rules as excellent news[25] and several NGOs stressing the existence of worrying loopholes. The reality lies in the Guidelines issued by BEREC in August 2016 to enforce the rules, and the NRAs' individual and collective actions in enforcing that set of guidelines.

To begin with, note that the Regulation is hideously badly written, even by the standards of European law. If a camel is a horse designed by a committee, this was a digital rendition drawn by a joint committee of the Council of Ministers, Commission and Parliament, none of whom had ever seen a horse before, nor knew how it should function. In particular, it introduces two new definitions which have no objective justification, both signifying 'Electronic Communications Service Provider' (ECPS), the Euro-version of an Internet Access Provider. It fails to define VPN, though the term is used in the Regulation. Article 2 sets out these definitions based on previous e-communications law,[26] but adds two new types: PECP (provider of electronic communications to the public) and IAS (Internet Access Service):

> (1) 'provider of electronic communications to the public' means an undertaking providing public communications networks or publicly available electronic communications services;
> (2) 'internet access service' means a publicly available electronic communications service that provides access to the internet, and thereby connectivity to virtually all end points of the internet, irrespective of the network technology and terminal equipment used.

'Virtually all' may have the meaning 'substantially all' or 'most that are technically possible'. Is this all end user points, or content providers, or both? It is at least a close approximation of the FCC term broadband IAS:

> A mass-market retail service by wire or radio that provides the capability to transmit data to and receive data from all or substantially all Internet endpoints ... This term also encompasses any service that the Commission finds to be providing a functional equivalent of the service described in the previous sentence, or that is used to evade the protections set forth in this Part.[27]

[23] Regulation 2015/2120, Recital 3.
[24] Norwegian Communications Authority, BEREC and net neutrality, 2012.
[25] IP/15/5927 and European Commission, MEMO-15-5275 (2015).
[26] Article 2 of Directive 2002/21/EC (Framework Directive).
[27] See FCC, Open Internet Order, 2015, p. 10, para. 25. Used in both the 2010 and 2015 Open Internet Orders.

The assumption must be that creating a new definition is undertaken in order to capture services otherwise excluded by the Framework Directive (2002/21/EC) Article 2:

> (c) 'electronic communications service' means a service normally provided for remuneration which consists wholly or mainly in the conveyance of signals on electronic communications networks, including telecommunications services and transmission services in networks used for broadcasting, but exclude services providing, or exercising editorial control over, content transmitted using electronic communications networks and services; it does not include information society services, as defined in Article 1 of Directive 98/34/EC, which do not consist wholly or mainly in the conveyance of signals on electronic communications networks.

The Regulation states that the Open Internet – not net neutrality – is to be preserved in Articles 3–6, which impose duties on IAPs, NRAs and the Commission itself.

Regulation 2015/2120 recitals

As almost always with European law, it is the explanatory Recitals which provide more information on what the Regulation's language actually means, as a guide to interpretation. We therefore begin with Recitals 1–19. Recitals 1–10 are relatively short and set out the basics of Open Internet rules. Recitals 11–15 explain the exceptions, for reasonable traffic management measures (TM/M), for PECPs, as well as 'content, applications or services' (CAS). To begin:

> (1) This Regulation aims to establish common rules to safeguard equal and non-discriminatory treatment of traffic in the provision of [IAS] and related end-users' rights. It aims to protect end-users and simultaneously to guarantee the continued functioning of the internet ecosystem as an engine of innovation. Reforms in the field of roaming should give end-users the confidence to stay connected when they travel within the Union, and should, over time, become a driver of convergent pricing and other conditions in the Union.
>
> (2) The measures provided for in this Regulation respect the principle of technological neutrality, that is to say they neither impose nor discriminate in favour of the use of a particular type of technology.

The second Recital is necessarily a bare-faced lie as the Regulation is studded with specific mobile-only provisions against an Open Internet, and of course roaming itself can only be provided by mobile networks. However, claims of 'technological neutrality' are part of the EC's Information Society 'tao', especially when not true. They are more a 'motherhood and apple pie' aspiration than a principle, as Reed points out,[28] rather like net neutrality itself.

[28] Reed (2007).

Recital 3 is a garbled synopsis of the dreadful Impact Assessment[29] that accompanied the original COM(2013) 267:

> (3) The internet has developed over the past decades as an open platform for innovation with low access barriers for end-users, providers of [CAS] and providers of [IAS]. The existing regulatory framework aims to promote the ability of end-users to access and distribute information or run applications and services of their choice. However, a significant number of end-users are affected by TM practices which block or slow down specific [CAS]. Those tendencies require common rules at the Union level to ensure the openness of the internet and to avoid fragmentation of the internal market resulting from measures adopted by individual Member States.

Recital 4 tries to explain basic principles without having to use the term 'net neutrality':

> (4) An IAS provides access to the internet, and in principle to all the end-points thereof, irrespective of the network technology and terminal equipment used by end-users. However, for reasons outside the control of providers of IAS, certain end points of the internet may not always be accessible. Therefore, such providers should be deemed to have complied with their obligations related to the provision of an IAS within the meaning of this Regulation when that service provides connectivity to virtually all end points of the internet. Providers of IAS should therefore not restrict connectivity to any accessible end-points of the internet.

Recitals 5–6 explain the freedom to connect equipment to the network together with content, applications and services – the FCC 'Four Freedoms' of 2005. However, copyright, speech-crime and other laws apply, so 'matters therefore remain subject to Union law, or national law that complies with Union law'.

Recital 7 may be termed the 'zero rating'/SpS clause: 'In order to exercise their rights ... end-users should be free to agree with providers of IAS on tariffs for specific data volumes and speeds of the IAS.' These contracts ('freedom') and their implementation by IAS providers:

> should not limit the exercise of those rights and thus circumvent provisions of this Regulation safeguarding open IAS. [NRAs] should be empowered to intervene against agreements or commercial practices which, by reason of their scale, lead to situations where end-users' choice is materially reduced in practice.

Material effect is not defined, and will presumably emerge as a matter of practice within BEREC, as explored in the next chapter. To take an example from Chapter 3: only HD video really congests – so how can anything else have a 'material' effect?

[29] European Commission, SWD (2013) 331.

> To this end, the assessment of agreements and commercial practices should, inter alia, take into account the respective market positions of those providers of IAS, and of the providers of CAS, that are involved.

The use of competition terminology may be interpreted as either allowing a *de minimis* exception for small IAPs, or to permit further analysis of the type indulged in by the French regulator when analysing the Level3/France Telecom and YouTube/Free disputes. Finally, '[NRAs] should be required, as part of their monitoring and enforcement function, to intervene when agreements or commercial practices would result in the undermining of the essence of the end-users' rights.' Powers must therefore be put in place where they are currently inadequate, though in the UK case such powers are already in place.

I have argued for a decade that common carriage requirements of FRAND should be imposed on IAS, including for zero rated services and SpS. The Regulation does not adopt that usage, but instead settles for the weaker 'reasonable', defined as 'transparent, non-discriminatory and proportionate' (RTNDP), a type of restriction on commercial practice that will depend even more heavily on the regulator than the well-recognised and court-enforced FRAND used in patent and other legal matters. The relevant Recitals state:

> (8) When providing IAS, providers of those services should treat all traffic equally, without discrimination, restriction or interference, independently of its sender or receiver, CAS, or terminal equipment. According to general principles of Union law and settled case-law, comparable situations should not be treated differently and different situations should not be treated in the same way unless such treatment is objectively justified.
>
> (9) The objective of reasonable TM is to contribute to an efficient use of network resources and to an optimisation of overall transmission quality responding to the objectively different technical quality of service requirements of specific categories of traffic, and thus of the CAS transmitted. Reasonable TMM applied by providers of IAS should be transparent, non-discriminatory and proportionate, and should not be based on commercial considerations.

Objectively different management of email as compared with phone calls or streamed video is permitted:

> The requirement for TMM to be non-discriminatory does not preclude providers of IAS from implementing, in order to optimise the overall transmission quality, TMM which differentiate between objectively different categories of traffic. Any such differentiation should, in order to optimise overall quality and user experience, be permitted only on the basis of objectively different technical quality of service requirements (for example, in terms of latency, jitter, packet loss, and bandwidth) of the specific categories of traffic, and not on the basis of commercial considerations. Such differentiating measures should be proportionate in relation to the purpose of overall quality optimisation and should treat equivalent traffic equally. Such measures should not be maintained for longer than necessary.

Recital 10 is then a nod to the privacy regulators, who were extremely unimpressed with the Regulation: 'Reasonable TM does not require techniques which monitor the specific content of data traffic transmitted via the IAS.' Recitals 33–35 explain that the legislative Act complies with fundamental rights, subsidiarity and Recital 35 and that the EDPS was consulted, though the date of its Opinion is incorrect: 'delivered an opinion on 24 November 2013' when it was in fact 14 November.[30] The EDPS had actually stated that the legal exceptions to neutrality 'appear overly broad and have a considerable potential to trigger a wide-scale, preventive monitoring of communications content. A surveillance of this kind will not only go contrary to the right to confidentiality of communications, as well as privacy and personal data protection, but furthermore may seriously undermine consumer confidence'.[31] It also robustly criticises the failure to make it a legal pre-contractual requirement to inform consumers if the IAP is using DPI or other 'communication inspection techniques'[32] and that:

> techniques such as Deep Packet Inspection (DPI) presuppose a detailed analysis of the content of information transmitted over the Internet, which may thus reveal substantial and detailed information about users. Should traffic management measures be based on less intrusive communications inspection techniques, such as those that analyse IP headers, they would nevertheless reveal information on the websites visited by endusers, and thus allow inferring the content of their communications. Communications content may include information on individuals' political views, religious beliefs, or health or sex life, and thus its analysis – both directly and via the IP headers – may entail processing of sensitive data within the sense of Article 8(1) of Directive 95/46/EC.[33]

The institutions ignored the EDPS advice in the final Regulation.

Recitals 11–12 are general:

> (11) Any TM practices which go beyond such reasonable TMM, by blocking, slowing down, altering, restricting, interfering with, degrading or discriminating between specific CAS, or specific categories of CAS, should be prohibited, subject to the justified and defined exceptions laid down in this Regulation. Those exceptions should be subject to strict interpretation and to proportionality requirements. Specific CAS, as well as specific categories thereof, should be protected because of the negative impact on end-user choice and innovation of blocking, or of other restrictive measures not falling within the justified exceptions. Rules against altering CAS refer to a modification of the content of the communication, but do not ban

[30] EDPS, Opinion on the Proposal for a Regulation to achieve a Connected Continent, 2013.

[31] *Ibid.*, para. 9.

[32] *Ibid.*, para. 11.

[33] *Ibid.*

non-discriminatory data compression techniques which reduce the size of a data file without any modification of the content. Such compression enables a more efficient use of scarce resources and serves the end-users' interests by reducing data volumes, increasing speed and enhancing the experience of using the content, applications or services concerned.

(12) TMM that go beyond such reasonable TMM may only be applied as necessary and for as long as necessary to comply with the three justified exceptions laid down in this Regulation.

Recitals 13–15 explain the justifications. Recital 13 deals with the requirement imposed by various branches of government to block CAS:

(13) First, situations may arise in which providers of [IAS] are subject to Union legislative acts, or national legislation that complies with Union law (for example, related to the lawfulness of CAS, or to public safety), including criminal law, requiring, for example, blocking of specific CAS.

In addition, situations may arise in which those providers are subject to measures that comply with Union law, implementing or applying Union legislative acts or national legislation, such as measures of general application, court orders, decisions of public authorities vested with relevant powers, or other measures ensuring compliance with such Union legislative acts or national legislation (for example, obligations to comply with court orders or orders by public authorities requiring to block unlawful content).

The requirement to comply with Union law relates, inter alia, to the compliance with the requirements of the Charter of Fundamental Rights of the European Union ('the Charter') in relation to limitations on the exercise of fundamental rights and freedoms.

As provided in Directive 2002/21/EC … any measures liable to restrict those fundamental rights or freedoms are only to be imposed if they are appropriate, proportionate and necessary within a democratic society, and if their implementation is subject to adequate procedural safeguards in conformity with the European Convention for the Protection of Human Rights and Fundamental Freedoms, including its provisions on effective judicial protection and due process.

Recital 14 deals with cyber security incidents:

[TMM] going beyond such reasonable [TMM] might be necessary to protect the integrity and security of the network, for example by preventing cyber-attacks that occur through the spread of malicious software or identity theft of end-users that occurs as a result of spyware.

Recital 15 deals with lack of capacity and temporary congestion:

measures going beyond such reasonable [TMM] might also be necessary to prevent impending network congestion, that is, situations where congestion is about to materialise, and to mitigate the effects of network congestion, where such

> congestion occurs only temporarily or in exceptional circumstances. The principle of proportionality requires that [TMM] based on that exception treat equivalent categories of traffic equally.

It explains the difference between 'temporary exceptional' congestion and unreasonable lack of network capacity:

> Temporary congestion should be understood as referring to specific situations of short duration, where a sudden increase in the number of users in addition to the regular users, or a sudden increase in demand for specific CAS, may overflow the transmission capacity of some elements of the network and make the rest of the network less reactive[34] ... While it may be predictable that such temporary congestion might occur from time to time at certain points in the network – such that it cannot be regarded as exceptional – it might not recur so often or for such extensive periods that a capacity expansion would be economically justified. Exceptional congestion should be understood as referring to unpredictable and unavoidable situations of congestion, both in mobile and fixed networks ...[35]
>
> The need to apply [TMM] going beyond the reasonable [TM] measures in order to prevent or mitigate the effects of temporary or exceptional network congestion should not give providers of [IAS] the possibility to circumvent the general prohibition ... Recurrent and more long-lasting network congestion which is neither exceptional nor temporary should not benefit from that exception but should rather be tackled through expansion of network capacity.

Recital 16 deals with the specific exception of SpS, though it is classified here as 'other' than IAS:

> There is demand on the part of providers of content, applications and services to be able to provide electronic communication services other than [IAS], for which specific levels of quality, that are not assured by [IAS], are necessary. Such specific levels of quality are, for instance, required by some services responding to a public interest or by some new machine-to-machine communications services. Providers of electronic communications to the public, including providers of [IAS], and providers of content, applications and services should therefore be free to offer services which are not [IAS] and which are optimised for specific content, applications or services, or a combination thereof, where the optimisation is necessary in order to meet the requirements of the content, applications or services for a specific level of quality. [NRAs] should verify whether and to what extent

[34] Adding for the benefit of frustrated mobile network users: 'Temporary congestion might occur especially in mobile networks, which are subject to more variable conditions, such as physical obstructions, lower indoor coverage, or a variable number of active users with changing location.'

[35] Continuing with an example: 'Possible causes of those situations include a technical failure such as a service outage due to broken cables or other infrastructure elements, unexpected changes in routing of traffic or large increases in network traffic due to emergency or other situations beyond the control of providers of [IAS]. Such congestion problems are likely to be infrequent but may be severe, and are not necessarily of short duration.'

> such optimisation is objectively necessary to ensure one or more specific and key features of the content, applications or services and to enable a corresponding quality assurance to be given to end-users, rather than simply granting general priority over comparable content, applications or services available via the [IAS] and thereby circumventing the provisions regarding traffic management measures applicable to the [IAS].

The circumstances under which SpS can be offered are detailed in Recital 17:

> In order to avoid the provision of [SpS] having a negative impact on the availability or general quality of [IAS] for end-users, sufficient capacity needs to be ensured. Providers of electronic communications to the public, including providers of [IAS], should, therefore, offer such other services, or conclude corresponding agreements with providers of content, applications or services facilitating such other services, only if the network capacity is sufficient for their provision in addition to any [IAS] provided. The provisions of this Regulation on the safeguarding of open [IAS] should not be circumvented by means of other services usable or offered as a replacement for [IAS]. However, the mere fact that corporate services such as [VPNs] might also give access to the internet should not result in them being considered to be a replacement of the [IAS], provided that the provision of such access to the internet by a provider of electronic communications to the public complies with Article 3(1)–(4) of this Regulation, and therefore cannot be considered to be a circumvention of those provisions. The provision of such services other than [IAS] should not be to the detriment of the availability and general quality of [IAS] for end-users.

It also contains a somewhat unnecessary infringement of technological neutrality to explain that cell towers can get congested in peak periods and that this may be somewhat unavoidable – which could surely have been left to BEREC's better judgement in 2016 Guidelines.

> In mobile networks, traffic volumes in a given radio cell are more difficult to anticipate due to the varying number of active end-users, and for this reason an impact on the quality of [IAS] for end-users might occur in unforeseeable circumstances. In mobile networks, the general quality of [IAS] for end-users should not be deemed to incur a detriment where the aggregate negative impact of services other than [IAS] is unavoidable, minimal and limited to a short duration.
>
> [NRAs] should ensure that providers of electronic communications to the public comply with that requirement. In this respect, [NRAs] should assess the impact on the availability and general quality of [IAS] by analysing, inter alia, quality of service parameters (such as latency, jitter, packet loss), the levels and effects of congestion in the network, actual versus advertised speeds, the performance of internet access services as compared with services other than internet access services, and quality as perceived by end-users.

Recital 19 explains the duties of NRAs in regard to IAS:

> [NRAs] should have monitoring and reporting obligations, and should ensure that providers of electronic communications to the public, including providers of [IAS], comply with their obligations concerning the safeguarding of open internet access. Those include the obligation to ensure sufficient network capacity for the provision of high quality non-discriminatory [IAS], the general quality of which should not incur a detriment by reason of the provision of services other than [IAS], with a specific level of quality.
>
> [NRAs] should also have powers to impose requirements concerning technical characteristics, minimum quality of service requirements and other appropriate measures on all or individual providers of electronic communications to the public if this is necessary to ensure compliance with the provisions of this Regulation on the safeguarding of open [IAS] or to prevent degradation of the general quality of service of [IAS] for end-users.
>
> In doing so, [NRAs] should take utmost account of relevant guidelines from BEREC.

Recital 18 explains in incredibly conflated language that privacy must be maintained in use of traffic management:

> The provisions on safeguarding of open [IAS] should be complemented by effective end-user provisions which address issues particularly linked to [IAS] and enable end-users to make informed choices. Those provisions should apply in addition to the applicable provisions of Directive 2002/22/EC … and Member States should have the possibility to maintain or adopt more far- reaching measures.
>
> Providers of [IAS] should inform end-users in a clear manner how [TM] practices deployed might have an impact on the quality of [IAS], end-users' privacy and the protection of personal data as well as about the possible impact of services other than [IAS] to which they subscribe, on the quality and availability of their respective [IAS].
>
> In order to empower end-users in such situations, providers of [IAS] should therefore inform end-users in the contract of the speed which they are able realistically to deliver. The normally available speed is understood to be the speed that an end-user could expect to receive most of the time when accessing the service. Providers of [IAS] should also inform consumers of available remedies in accordance with national law in the event of non-compliance of performance.

It then explains what NRAs should do to enforce these consumer protection rules:

> Any significant and continuous or regularly recurring difference, where established by a monitoring mechanism certified by the [NRA], between the actual performance of the service and the performance indicated in the contract should be deemed to constitute non-conformity of performance for the purposes of determining the remedies available to the consumer in accordance with national law.

> The methodology should be established in the guidelines of [BEREC] and reviewed and updated as necessary to reflect technology and infrastructure evolution. [NRAs] should enforce compliance with the rules in this Regulation on transparency measures for ensuring open [IAS].

This imposes on BEREC members collectively the need to coordinate monitoring mechanisms and their certification, and to individual NRAs the obligation to inform consumers of the legal redress mechanisms available. Without this, the Regulation has absolutely no teeth.

Regulation 2015/2120 Articles 3–6

Article 3(1) states the Four Freedoms, Article 3(2) that contracts/practices cannot limit those rights and Article 3(3) that traffic must be treated equally:

> 1. End-users shall have the right to access and distribute information and content, use and provide applications and services, and use terminal equipment of their choice, irrespective of the end-user's or provider's location or the location, origin or destination of the information, CAS, via their IAS. This paragraph is without prejudice to Union law, or national law that complies with Union law, related to the lawfulness of the CAS.
> 2. Agreements between PIAS [providers of Internet access service] and end-users on commercial and technical conditions and the characteristics of IAS such as price, data volumes or speed, and any commercial practices conducted by PIAS, shall not limit the exercise of the rights of end-users laid down in paragraph 1.
> 3. PIAS shall treat all traffic equally, when providing IAS, without discrimination, restriction or interference, and irrespective of the sender and receiver, the [CAS], or the terminal equipment used.

Do we read Article 3.3 in light of Article 3.2, when Recital 11 tells us that 'those exceptions should be subject to strict interpretation and to proportionality requirements'?

Immediately, the general RTNDP condition is detailed, and three justified exceptions are identified:

> The first subparagraph shall not prevent providers of IAS from implementing reasonable TMM. In order to be deemed to be reasonable, such measures shall be transparent, non-discriminatory and proportionate, and shall not be based on commercial considerations but on objectively different technical quality of service requirements of specific categories of traffic. Such measures shall not monitor the specific content and shall not be maintained for longer than necessary.
>
> PIAS in particular shall not block, slow down, alter, restrict, interfere with, degrade or discriminate between specific CAS, or specific categories thereof, except as necessary, and only for as long as necessary, in order to:

(a) comply with Union legislative acts, or national legislation that complies with Union law, to which the provider of internet access services is subject, or with measures that comply with Union law giving effect to such Union legislative acts or national legislation, including with orders by courts or public authorities vested with relevant powers;
(b) preserve the integrity and security of the network, of services provided via that network, and of the terminal equipment of end-users;
(c) prevent impending network congestion and mitigate the effects of exceptional or temporary network congestion, provided that equivalent categories of traffic are treated equally.

Article 3(4) is a basic privacy reminder.[36] Article 3(5) sets out the SpS exception:

> 5. PECP, including PIAS, and providers of CAS shall be free to offer services other than IAS which are optimised for specific CAS, or a combination thereof, where the optimisation is necessary in order to meet requirements of the CAS for a specific level of quality. PECP, including PIAS, may offer or facilitate such services only if the network capacity is sufficient to provide them in addition to any IAS provided. Such services shall not be usable or offered as a replacement for IAS, and shall not be to the detriment of the availability or general quality of IAS for end-users.

O'Donoghue and Pascoe argue the Regulation is 'expressed in extremely telegraphic terms ... not exactly a model of clarity and consistency.'[37] Because the exception for SpS in Article 3(5) provides a general right and not a derogating exception (as with other elements of Article 3), they argue the 'concept of necessity under Article 3(5) is not only independent of how that concept would be defined in other contexts under EU law, but is even distinct from how necessity is defined in other aspects of the Regulation'. They also argue that judging which qualities are 'necessary' for SpS is aided by a recent High Court case accepting reduced latency as a vital element in Google Maps service.[38] While this remains to be seen, I agree with their conclusion that 'the concept of "reasonable traffic management measures" is nebulous and is therefore likely to lead to litigation'.[39] They see the best line of attack as 'to challenge various aspects of the Regulation, or at least certain suggested interpretations of it, on the basis that they would be inconsistent with the concept of non-discrimination as a (higher) general principle under EU law'. They also see litigation grounds on:

[36] 'Any TMM may entail processing of personal data only if such processing is necessary and proportionate to achieve the objectives set out in para. 3. Such processing shall be carried out in accordance with Directive 95/46/EC ... TMM shall also comply with Directive 2002/58/EC.'

[37] O'Donoghue and Pascoe (2016).

[38] *Ibid.*, footnote 10, p. 8.

[39] *Ibid.*, p. 11.

> compatibly with fundamental rights provided by the EU Charter, including the freedom to conduct a business under Article 16 and the right to property under Article 17 ... restrictions on the provider's right to operate its business in accordance with its own wishes, and those of its customers, should be more limited than the terms of the Regulation might suggest on its face.[40]

Expect telecoms operators to threaten Ofcom and other NRAs, and if necessary to launch litigation based on these arguments, in 2017 and beyond.

Article 4 on transparency may have the most immediate impact on users, as it enforces proper information on the actual use of the PIAS, and took effect from 29 November 2015:

> 1. PIAS shall ensure that any contract which includes IAS specifies at least the following:
>   (a) information on how TMM applied by that provider could impact on the quality of the IAS, on the privacy of end-users and on the protection of their personal data;
>   (b) a clear and comprehensible explanation as to how any volume limitation, speed and other quality of service parameters may in practice have an impact on IAS, and in particular on the use of CAS;
>   (c) a clear and comprehensible explanation of how any services referred to in Article 3(5) to which the end-user subscribes might in practice have an impact on the IAS provided to that end-user.

This mandates minimal consumer protection by IAPs within minimal consumer protection by regulators. Translation: how may your video SpS hamper the rest of your family using regular Internet video services?

> (d) a clear and comprehensible explanation of the *minimum, normally available, maximum and advertised download and upload speed* of the IAS in the case of fixed networks, or of the *estimated maximum and advertised download and upload speed* of the IAS in the case of mobile networks, and how significant deviations from the respective advertised download and upload speeds could impact the exercise of the end-users' rights laid down in Article 3(1) [emphasis added].

Translation: some real speed data based on postcode needed for fixed users, but mobiles can continue to use laboratory-based maxima for 3G and LTE without regard to real world data. One hopes this interpretation is cynical.

> (e) a clear and comprehensible explanation of the remedies available to the consumer in accordance with national law in the event of any continuous or regularly recurring discrepancy between the actual performance of the IAS regarding speed or other quality of service parameters and the performance indicated in accordance with points (a) to (d).
>
> Providers of IAS shall publish the information referred to in the first subparagraph.

[40] *Ibid.*, p. 13.

There is then a requirement that users can actually enforce measures, but contract termination if the mobile/fixed provider fails to deliver as promised is not specified:

2. PIAS shall put in place transparent, simple and efficient procedures to address complaints of end-users relating to the rights and obligations laid down in Article 3 and paragraph 1 of this Article.
3. The requirements laid down in paragraphs 1 and 2 are in addition to those provided for in Directive 2002/22/EC and shall not prevent Member States from maintaining or introducing additional monitoring, information and transparency requirements, including those concerning the content, form and manner of the information to be published. Those requirements shall comply with this Regulation and the relevant provisions of Directives 2002/21/EC and 2002/22/EC.

4. Any significant discrepancy, continuous or regularly recurring, between the actual performance of the internet access service regarding speed or other quality of service parameters and the performance indicated by the provider of internet access services in accordance with points (a) to (d) of paragraph 1 shall, where the relevant facts are established by a monitoring mechanism certified by the national regulatory authority, be deemed to constitute non-conformity of performance for the purposes of triggering the remedies available to the consumer in accordance with national law. This paragraph shall apply only to contracts concluded or renewed from 29 November 2015.

Minimal harmonisation clauses inside Regulations are very rare, and it will be of more general interest to non-neutrality lawyers to analyse the outcome of this process.

Article 5(1) (Supervision and enforcement) is the teeth of the Regulation:

> NRAs shall closely monitor and ensure compliance with Articles 3 and 4, and shall promote the continued availability of non-discriminatory [IAS] at levels of quality that reflect advances in technology. For those purposes, NRAs may impose requirements concerning technical characteristics, minimum [QoS] requirements and other appropriate and necessary measures on one or more [PECP], including providers of [IAS]. NRAs shall publish reports on an annual basis regarding their monitoring and findings, and provide those reports to the Commission and to BEREC.

This is the most explicit mention of QoS regulatory requirements in the entire Regulation.

2. At the request of the national regulatory authority, providers of electronic communications to the public, including providers of internet access services, shall make available to that national regulatory authority information relevant to the obligations set out in Articles 3 and 4, in particular information

> concerning the management of their network capacity and traffic, as well as justifications for any traffic management measures applied. Those providers shall provide the requested information in accordance with the time-limits and the level of detail required by the national regulatory authority.

Article 5(3) sets out BEREC's deadline: 'By 30 August 2016, in order to contribute to the consistent application of this Regulation, BEREC shall, after consulting stakeholders and in close cooperation with the Commission, issue guidelines for the implementation of the obligations of national regulatory authorities under this Article.' Article 6 deals with penalties: 'Member States shall lay down the rules on penalties applicable to infringements of Articles 3, 4 and 5 and shall take all measures necessary to ensure that they are implemented. The penalties provided for must be effective, proportionate and dissuasive. Member States shall notify the Commission of those rules and measures.' Notification is thus required by the governments on behalf of NRAs.

Finally, Article 9 exposes the chaotic drafting process and its inevitable denouement; the provisions will be reviewed by 2019:

> By 30 April 2019, and every four years thereafter, the Commission shall review Articles 3, 4, 5 and 6 and shall submit a report to the European Parliament and to the Council thereon, accompanied, if necessary, by appropriate proposals with a view to amending this Regulation.

Given this short timetable, it may be that the 2019 review has nothing to report, or that it recommends codifying the BEREC Guidelines, which are discussed in the next section.

Interpretation: good news, bad news and do not mention the N-word

Although many net neutrality elements have been included in the new Regulation, the lack of any explicit mention of the net neutrality principle is notable. Rather than unequivocally affirming the three pillars of net neutrality, i.e. no blocking, no throttling and no paid prioritisation, the EU policymakers enshrined only the first two components into the Regulation, thus tempering neutrality into a less principled vague 'Open Internet'. Who can disagree with an Open Internet? Telecoms companies very certainly disagreed with strict net neutrality.

The good news for users is that Europeans have the 'right to access and distribute information and content, use and provide applications and services, and use terminal equipment of their choice, irrespective of the end-user's or provider's location or the location, origin or destination of the information, content, application or service, via their internet access service' according to Article 3 of the Regulation. Associated with this right is the IAPs' obligation to 'treat all traffic equally' with reasonable traffic management that should be 'transparent, non-discriminatory and proportionate' and, very importantly 'shall not be based on commercial considerations but on objectively different technical quality of service requirements of specific categories of traffic'. This is an important step forward for those Europeans who were lacking basic protections. The EU Regulation will introduce more effective consumer information with realistic assessments of line speed for the individual, rather than laboratory speeds, and mandatory announcements by ISPs of any traffic management practices in terms which are easier to understand – a process previously researched by the Netherlands regulator.[41]

However, some crucial issues remain unclear and the devil is in the details that are considered in the final part of this chapter.

BEREC and the Guidelines

In order to explore the European law further, it is necessary to stitch together the interpretations of the European Commission issued at the time of the Regulation's approval in MEMO-15–5275, with the clarifications and workplan of the European regulators working as BEREC. This is the most speculative element in the monograph, but is intended to add value by attempting to work out how much the new law will affect, for instance, TMPs, SpS and zero rating. The EC hopefully indicated that 'Europe's regulators have a long track record in safeguarding open, competitive markets, and the Commission will work closely with the BEREC to ensure that clear guidance is rapidly provided in this field to complement the Regulation itself'.[42]

BEREC and the EC must also include an element with which they are less familiar: human rights law. Special Rapporteur on Freedom of Expression to the United Nations Frank LaRue had issued a report in 2011 referencing net neutrality in his ongoing work on rights online.[43] The Council of Europe (CoE) multi-stakeholder dialogue 'Network Neutrality and Human Rights' was held on 29–30 May in Strasbourg.[44] This dialogue was a result of the 2010 Declaration of the Committee of Ministers on network neutrality.[45] The CoE is working towards a soft law instrument to guide Member States in the application of net neutrality rules, that support particularly the aspirations of Articles 6, 8, 10 of the Convention. A short outcomes paper of the major points of discussion was communicated to the 47 Member State representatives of the CoE

[41] Sluijs, Schuett and Henze (2010).
[42] European Commission, MEMO-15–5275.
[43] La Rue (2011).
[44] Marsden, Chris (2013e).
[45] Council of Europe (2010).

Steering Committee on Media and Information Society (CDMSI) to consider and propose further action.[46] The European Commissioner had subsequently on 4 June announced to the European Parliament her intention to introduce specific legislation on network neutrality.

The Commission states that:

> These authorities will thus have the power and obligation to examine how the traffic management practices of internet service providers affect the end-users' (consumers and businesses) rights to access and distribute content, applications and services of their choice. They will have to ensure that the quality of the open internet access service reflects advances in technology. [NRAs] will also have to ensure that the availability and quality of the open internet access service is not degraded by traffic discrimination through internet service providers or by the provision of specialised services.[47]

This is obviously putting the onus on NRAs, who will also have to report annually on the state of the 'Open Internet'. The EC helpfully adds:

> What will happen if a service provider will not respect open internet rules? The Regulation will oblige Member States to set rules on the penalties applicable to infringements of the net neutrality provisions. These penalties have to be effective, proportionate and dissuasive. This means that providers infringing the net neutrality rules will face significant pecuniary and administrative sanctions.[48]

The enforcement of the new Regulation formally took effect after its publication in the Official Journal, but in practice began as of 30 August 2016, nine months after that publication. BEREC – which consists of the 28 national regulators, with its secretariat based in Riga, Latvia – will have to issue specific detailed guidelines to its component national regulators, to ensure that IAPs do not elude the new rules with particular regard to congestion management, Specialised Services and zero rating.

And what about impending congestion? It is unclear what kind of traffic management will be possible in order to 'prevent impending network congestion'. This will be considered a legitimate exception to the non-discrimination rule, but it will be very difficult to foresee what may qualify as 'impending congestion' and what could be reasonable measures to 'prevent' or even to define the imminence of 'impending congestion'.

Introduction to BEREC

BEREC is an example of non-transparent club governance, with a well-established but highly opaque networked model, given that there has been a

[46] Belli (2013).

[47] European Commission, MEMO-15-5275.

[48] *Ibid.*

club of European telecoms regulators since the 'unofficial' Independent Regulators Group (IRG) was established in 1997, and the European Regulators Group in 2002. The BEREC role resulted from the political compromise that, instead of the federal European communications regulator longed for by Commissioners since Martin Bangemann in 1997, there should be a coordinatory club of national regulators.[49]

There now 37 members of the IRG, comprising the 28 BEREC members, 4 EFTA (European Free Trade Association)/EEA (European Economic Area) members (Norway, Iceland, Switzerland, Liechtenstein) and 5 EU accession 'candidates' (Albania, Macedonia, Montenegro, Serbia, Turkey).[50] It is thus an extremely well-organised network of NRAs, as befits telecommunications' status as the least illegitimate of the privatised utilities from the 1990s forwards, and supports the notion of quasi-independence ('orchestration' not 'delegation') of the network from governments and the EC in this instance.[51]

BEREC is a 'loose' regulatory network,[52] not a law-making body, and therefore it does not need to follow better regulation procedures such as entirely open consultation, public meetings or mandatory 12-week consultations.[53] While NRAs 'must take utmost account' of BEREC decisions,[54] it is not a legal requirement to follow BEREC guidelines. BEREC has held an annual public stakeholder forum each autumn since 2013 to 'continue to manifest the open dialogue between BEREC and its members on the one hand and the key stakeholders on the other hand ... provide a platform for a transparent exchange of ideas, concepts and also challenges between all parties involved'.[55] This is videoed and archived and a debriefing session is held after each plenary, as is a consultation on the following year's work programme. It is, however, atrociously disengaged digitally, with between 5 and 25 total views on its stakeholder forum videos from October 2015 by end-2015, and only 11 out of 25 videos in its entire history attracting over 100 views, despite the renewed 2015 interest in stakeholder consultation resulting from the net neutrality rules.[56] By contrast, the IRG is an entirely obscure organisation with no public outreach beyond a bland website. As the IRG and BEREC are regulator groups affecting

[49] Boeger and Corkin (2012).

[50] Independent Regulators Group (2015).

[51] Blauberger and Rittberger (2015).

[52] Regulation 1211/2009, Article 4 states that BEREC's Board of Regulators 'shall neither seek nor accept any instruction from any government, from the Commission, or from any other public or private entity'. The same article grants the Commission observer status when participating at BEREC meetings. The Secretariat is funded by the NRAs, demonstrating BEREC's greater independence from the EC than its predecessor ERG (European Regulators Group).

[53] Simpson (2013). See also Kelemen and Tarrant (2011).

[54] Regulation 2015/2120, Recital 19.

[55] BEREC, BoR (15) 213, Section 17, p. 27.

[56] See the BEREC channel on YouTube at www.youtube.com/user/bereceuropaeu.

trillion Euro industries,[57] it is perhaps surprising that there is so little public consumption of their deliberations or a clamour for their greater outreach. But this does reflect the technocratic and commercial nature of their interactions with telecommunications companies, rare interactions with IT and broadcast content providers, and extremely rare interactions with civil society, user groups and consumer representatives.[58] Similar problems afflict NRAs, with the obscurity of their professed consumer function leading to the expert academic board member of the UK body forthrightly expressing his views of its marginalisation: 'Even those who argue that net neutrality is a non-issue tend to point out important issues where Ofcom might have done more were it not in retreat.'[59]

BEREC's Guidelines: process and interpretation

BEREC was charged with ensuring it issues Guidelines by August 2016 for interpretation of the Regulation by NRAs:

> The Telecoms Single Market Regulation includes a duty in Article 5(3) for BEREC to lay down guidelines for the implementation of the obligations of NRAs related to the supervision, enforcement and transparency measures for ensuring open Internet access. These guidelines should contribute to the consistent application of the Regulation, and be produced after consulting stakeholders and in close cooperation with the European Commission.[60]

The deadlines are as follows:

- Entry into force of the Regulation took place on 30 November 2015; the entire Regulation is applicable 30 April 2016 except for certain provisions (mainly on roaming).
- The deadline for Member States to repeal national measures (including self-regulatory measures) which go against Article 3(2) or 3(3) is 31 December 2016, which must be notified to the Commission by 30 April 2016.
- Deadline for publishing BEREC's implementation guidelines under Recital 19 is 30 August 2016. European Commission's report to the European Parliament and the Council reviewing Article 3 (safeguarding of open internet access), Article 4 (transparency measures for ensuring open internet access), Article 5 (supervision and enforcement) and Article 6 (penalties), including proposals for amendments, if necessary, must be delivered by 30 April 2019.
- The Commission will have to issue a report every 4 years as of 30 April 2019.[61]

[57] European Telecommunications Network Operators' Association (2015). Telecommunications and IT comprise markets of about $250 billion and $600 billion respectively by annual consumption in Western Europe.
[58] Tambini (2012).
[59] Tambini (2010).
[60] BEREC, BoR (15) 226.
[61] EDRi (2015).

The BEREC Board Meeting (heads of NRAs) of 10 December 2015 held at Ofcom contained the following agenda item 12:

> 12.1. Feasibility of quality of service (QoS) monitoring in the context of NN Document(s) BoR (15) 207 Draft BEREC Internal Report on the feasibility of QoS monitoring in the context of NN Introduction by NN EWG Co-Chairs (Nkom/Ofcom) Action requested To discuss and approve for internal use.
>
> 12.2. Oral up-date on the state of play of the BEREC Guidelines on NN ... Introduction by NN EWG Co-Chairs (Nkom/Ofcom) Action requested To discuss the proposed way forward. [62]

BEREC met with stakeholders to gather views on 15/16 December 2015 in Brussels with stakeholder organisations, grouped on the first day as 'end users and consumers, and civil society', with the second 'commercial' day for 'ISPs, equipment manufacturers, content and application providers'. It is explained that:

> BEREC will draw on existing BEREC net neutrality publications as well as on the input received at the stakeholder meetings. BEREC will publish draft guidelines for public consultation following the (provisionally 6th) June 2016 Plenary meeting. BEREC will then take account of all comments received before publishing the guidelines by the end of August 2016.

By the time you read these words, those Guidelines will have provided clarity for NRAs and stakeholders on the rules. The process between January and May 2016 appears to be entirely internal discussion, and the time for consultation is only 4–6 weeks in July 2016. BEUC, the European consumer's body, argued at the 15 December 2015 workshop that this is unacceptable. BEREC themselves are caught in that the Plenary must agree to publish draft Guidelines, which means the working group cannot do much more themselves with the deadlines they were given in the Regulation.

BEREC explains its outstanding concern on four topics: traffic management practices; Specialised Services; transparency in Internet access quality; and 'commercial practices', such as zero rating. These four are the most controversial elements and I deal with each in turn. The co-chairs of the Net Neutrality Working Group (NNWG) – note that BEREC was happy to use the term net neutrality whereas the Regulation will not! – in 2016 were Ofcom for the UK and NKom for Norway (which had co-chaired the group since its foundation in 2011). Team leaders were Italy's AGCOM (traffic management practices); Belgium (Specialised Services); Greece (transparency in Internet access quality); and Ofcom for the UK ('commercial practices', such as zero rating).

[62] BEREC, BoR (15) 190.

All the team leaders prepared a questionnaire for the December stakeholder meetings.

Consultation on traffic management practices

AGCOM prepared the following questions for the meeting:

a) What is your understanding or view on the terms 'specific categories of traffic' and 'specific content, applications or services, or specific categories thereof' in Article 3(3) subparas 2 and 3?
b) In your view, how can day-to-day 'reasonable' TM measures performed by ISPs in accordance with Article 3(3) subpara 2, such as TM for 'specific categories of traffic', affect the end user's choice? It would be helpful if you can provide concrete examples.
c) In your view, how can TM measures 'going beyond reasonable' TM performed by ISPs in accordance with Article 3(3) subpara 3, e.g. 'congestion management', affect the end user's choice? It would be helpful if you can provide concrete examples.[63]

AGCOM is thus seeking views on when practices might be illegal and able to be prosecuted.

The Memo is a vital aid to interpretation, until the BEREC Guidelines are issued, and makes some bold claims. For instance, on traffic management it states:

> Article 3(3) enshrines the principle of equal treatment of all traffic, without discrimination, restriction or interference. But it does not exclude the use of reasonable traffic management to optimise overall transmission quality. Such use has to be based on objectively different technical quality requirements and not on commercial considerations, and it must be transparent, non-discriminatory and proportionate.[64]

It adds the old saw that 'for example vital medical information of a patient receiving home care is delayed because the neighbours are downloading a video. Internet is not different from day-to-day car traffic where you need some basic rules to allow everyone to have access and quality of service.'[65] However, these are not rules for emergency services but for paid prioritisation, as will become clear.

The EC claims:

> Reasonable traffic management therefore cannot be used to discriminate against specific categories of content or services such as P2P. For instance, real-time P2P

[63] Regulation 2015/2120.
[64] European Commission, MEMO-15-5275 (2015).
[65] *Ibid.*

streaming services have to be handled in accordance with their objective real-time requirement. Providers will also not be allowed to deprioritise or discriminate against encrypted traffic simply by invoking the pretext that they are not able to read and classify the content of such traffic, which in any event would not be compatible with reasonable traffic management. Any deprioritisation of encrypted traffic on the sole basis of the sender's identity would be a breach of the non-discrimination principle.[66]

This sounds ideal in principle, but there are exceptions:

- to comply with Union or national legislation related to the lawfulness of content or with criminal law, or with measures implementing this legislation such as a decision by public authorities or a court order, for instance if a judge or the police have ordered blocking of specific illegal content;
- to preserve the security and integrity of the network, for instance to prevent misuse of a network and combat viruses, malware or denial of services attacks;
- to minimise network congestion that is temporary or exceptional. This means that operators cannot invoke this exception if their network is frequently congested due to under-investment and capacity scarcity.[67]

The EC claims strenuously that 'These exceptions have to be interpreted strictly and are subject to proportionality criteria on the scope and duration of traffic management measures', that it provides 'a very clear prohibition to use traffic management beyond such reasonable cases … [in] limited and clearly defined circumstances … [and in] a limited number of tightly defined and exhaustive exceptions'. One might think the EC protests too much in this situation, but actions will speak loudly to this claimed power of the Regulation in 2017 and beyond.

Specialised Services

The Belgian NRA questions were very vague:

a) Article 3(5) subpara 1 refers to providing SpS where 'the optimisation is necessary in order to meet requirements … for a specific level of quality'. What could be the reason for implementing or offering SpS? In your view, are SpS necessary for offering existing or new services?
b) Are you aware of a demand for SpS from end users (including business users)? In your opinion, could content and applications provided on the IAS become a kind of SpS? How should this be assessed under the TSM regulation? If they were allowed, would you see demand for, or benefit to, end users from the provision of sub-Internet offers (i.e. offers where the access to Internet is restricted to a limited set of content and applications)? How should think such offers should be assessed under the TSM regulation?

[66] *Ibid.*
[67] *Ibid.*

c) Do you have a view on the impact of the possibility to provide SpS on future innovation and the openness of the Internet? Do you see any issues arising with the provision of SpS to end users?[68]

These are open-ended questions, though perhaps intended to flush out the intention of telcos to redefine SpS as embracing what are in fact regular Internet services, and thus infringing net neutrality.

The European Commission stated that 'paid prioritization in the open internet will be banned', but the provision of guaranteed-quality services, i.e. 'services other than internet access services which are optimised for specific content, applications or services, or a combination thereof', is explicitly allowed. It will be, therefore, the responsibility of national regulators to make sure that such services are provided only when 'the network capacity is sufficient to provide them in addition to any internet access services provided'.[69]

In the absence of a clear separation requirement between open access and guaranteed-quality services, the evaluation of network capacity becomes a critical element to avoid abusive conduct, but no indication on how to assess 'sufficiency' has been provided by the Regulation. If your Internet access speed and monthly allowance does not change until 2020, you have the same conditions as now, which means you are not impeded – but you will not benefit from the 20 per cent average annual growth in both speed and capacity which has been achieved in the past. Is a static connection acceptable as not 'impeding' you?

The EC also states with regard to SpS that these are:

> services like IPTV, high-definition videoconferencing or healthcare services like telesurgery. They use the internet protocol and the same access network but require a significant improvement in quality or the possibility to guarantee some technical requirements to their end-users that cannot be ensured in the best-effort open internet.[70]

Possibility? That will need defining and is nothing like as straightforward as the BEREC definition of 2012: 'specialised services are provided over virtual or physical networks distinct from networks constituting the Internet, but that will typically operate over the same infrastructure.'[71]

BEREC was unusually highly critical of the original 2013 law proposed by the Commission, stating 'BEREC believes the relevant definition does not adequately capture their provision within closed networks and so risks hindering NRAs' capacity to apply'.[72] It then praised the 2014 European Parliament

[68] Regulation 2015/2120.
[69] European Commission, MEMO-15-5275.
[70] *Ibid.*
[71] BEREC, BoR (12) 131.
[72] BEREC, Statement on the publication of a European Commission proposal for a Regulation, 2013.

amendment providing that 'specialised services should be clearly separated (physically or virtually) from internet access services at the network layer, to ensure that sufficient safeguards prevent degradation of the internet access services'.[73]

NKom, which had co-chaired the BEREC work, finally stated its continued concern at what became the final legal wording: 'As specialised services are exempted from net neutrality, it is especially important that the specialised services are clearly separated from the Internet access services, so as to ensure that Internet traffic is not degraded.'[74] Since Sørensen wrote those words in mid-June 2014, the final wording has actually returned to roughly the original Commission position – a deliberately obscure reading of the definition. It is this that BEREC must clarify in 2016.

Graef points out the differing versions of Specialised Services in the European negotiation of the proposed Regulation – an idea that became so contentious and difficult that it is to be left to NRAs and BEREC to make case-by-case decisions on this, whereas in the legislation of the Netherlands and Slovenia such services are to be defined as 'non-Internet': 'services that are not offered via the public internet but through the closed network of the ISP automatically fall outside the scope of the regulatory framework.'[75] By contrast, the Norwegian regulator and BEREC co-chair explained that Norway's co-regulatory rules in 2009 already considered '"if the physical connection is shared with other services, it must be clear how the capacity is allocated between the Internet traffic and the other services" to prevent the entire data pipe being given over to specialised services at the expense of open Internet capacity'.[76]

This process will affect the market for network access.[77] Specialised Services have the potential to support EU consumer cloud services, notably for home workers and small businesses. As Sluijs *et al.* explain in relation to cloud services, 'Priority services and differentiated prices could enable clouds to perform more reliable services.'[78] However, this will take a sizable chunk out of an IAP's bandwidth, which is a scarce resource, especially in mobile broadband.

Transparency in Internet access quality

The Greek questions were very basic:

a) What information would be beneficial for end users so that they are better informed, e.g. regarding traffic management measures, commercial and

[73] BEREC, Views on the European Parliament first reading legislative resolution on the proposal for a Regulation, 2014.

[74] Sørensen (2014a).

[75] Graef (2014).

[76] Sørensen (2014a).

[77] Sluijs, Larouche and Sauter (2011).

[78] *Ibid.*

technical conditions and their impact on Internet access services? How should this information be communicated to them in the contract? (Ref. Article 4(1)).

b) How should ISPs describe and communicate speed of their IAS offers in the case of fixed and mobile networks? How should the different IAS speed parameters (e.g. minimum, maximum, advertised and normally available speeds in the case of fixed networks and estimated maximum and advertised speeds in the case of mobile) be defined in the contract? (Ref. Article 4(1)(d)).

c) How should ISPs describe other parameters of their IAS offers, such as quality of service parameters (typically latency, jitter, packet loss) and quality as perceived by end users? Should these parameters be defined in the contract? If so, how?[79]

The impression given in December 2015 is that the Greeks have neither the intellectual nor financial resources to devote properly to the task, unsurprising given the highly corporate captured nature of the Greek telecoms market and the exceedingly dire state of Greek government finances, which must impact on the regulator's competence. With 28 NRAs in the EU, there was bound to be a weak link in the four, and this appeared to be it. Given the huge amount of work carried out by other regulators on transparency, these questions appear extremely dated by comparison. As Ofcom had carried out so much research in this area, it is perhaps most odd that Ofcom chose to influence the zero-rating debate instead. One might guess that Ofcom could both assist its impoverished neighbour Greece in this area, as well as dictating as much as possible a liberal interpretation for zero rating to be permitted.

BEREC Work Programme Item 11.2 begins this investigation in the context of user experience:

> work stream will – depending on the outcome of the NN QoS Feasibility study – develop a Net Neutrality quality of service (NN QoS) regulatory assessment toolkit for NRAs by building on previous BEREC guidance on monitoring methods and recommending legal considerations to be taken into account when addressing issues of Net Neutrality and traffic management. As proposed in the BEREC NN QoS Guidelines, the monitoring methods will encompass both Internet access service (IAS) as a whole, as well as individual applications using IAS. The former covers measurement of IAS performance characteristics (download and upload speed, latency etc.) actually enjoyed by end-users, while the latter investigates ISPs' traffic management practices.[80]

The ambition in this work stream is that:

> Accuracy and comparability of measurement results will be emphasised. The methodology will also specify how to assess the IAS measurement results when considering nominal speed and performance of specialised services, and

[79] BEREC, Questions for BEREC stakeholder dialogue with representatives of end-users/consumers and civil society, 24 November 2015.

[80] BEREC, BoR (15) 213.

> comparing different IAS offers and different IAPs, while taking into account market development and technological evolution. The toolkit will also cover monitoring methods for detecting quality differentiation of individual application and the use of traffic management investigations will also be covered. The NN QoS assessment toolkit will thereby describe regulatory best practices on IAS quality and Net Neutrality monitoring.[81]

Because net neutrality monitoring is technically challenging, the thinking of BEREC at end-2015 was to have comparable analysis amongst those NRAs who were able and willing to take part (driven no doubt in part by affording the services of SamKnows or future equivalent), building on:

> findings from the 2014 BEREC NN QoS Monitoring Report and 2015 BEREC NN QoS Feasibility Study, which describe a possible future opt-in quality monitoring system, where individual regulators can participate on a voluntary basis. BEREC's work beginning in the second half of 2016 would consist of specifying the system requirements and describing a framework for NRAs to collaborate in the opt-in system. It is important that quality monitoring is considered trustworthy among stakeholders, in particular within an evolving policy area as Net Neutrality, and a harmonised approach broadly supported by regulators could contribute to this.[82]

The UK had already responded by suggesting something more co-regulatory or less harmonised might be sufficient, as had the European cable operators at the October 2015 BEREC stakeholder meeting.[83]

BEREC is committed to using the '2015 BEREC NN QoS Feasibility Study [which] will be the basis for the BoR to take a decision on whether to move forward to specifying an opt-in quality monitoring system (with a separate decision on whether to establish it to be taken in 2017)'. In 2017 it will thus have to deal with this monitoring framework in detail in respect of descriptions of Deliverables:

- Description of the framework for NRAs to collaborate in an opt-in quality monitoring system: Adoption in P1/2017 (subject to a BoR decision)
- Overall system requirement specification for the opt-in quality monitoring system: Adoption in P2/2017 (subject to a BoR decision)
- BEREC regulatory toolkit for NN QoS assessment: Adoption in P1/2018.[84]

This is thus a future work programme in place for the period September 2016 to June 2018, further developing NRA responses after the Guidelines are published on 29 August 2016.

[81] *Ibid.*, Section 11.2, p. 22.

[82] BEREC (2015) Questions for BEREC stakeholder dialogue with representatives of end-users / consumers and civil society, 24 November 2015.

[83] Marsden, Chris (2015b).

[84] BEREC, BoR (15) 213, Section 11.2, p. 23.

'Commercial practices': zero rating

Zero rating (or sponsored data plans) is a commercial practice used by some IAPs, consisting in the imposition of data caps, i.e. limits to monthly data volumes, and the parallel exemption of selected zero-rated applications from such caps. The data consumption of the zero-rated applications may be sponsored by the IAP itself or by the application provider, depending on the specific commercial arrangement. This practice has the concrete potential to create a two-tier Internet, since non-zero-rated applications may suffer a considerable disadvantage compared to the zero-rated ones. These could be rival social networks, video or other media providers, or messaging apps. The new Regulation does not specifically address zero rating, but, as discussed, it does not allow traffic management which is 'based on commercial considerations'. Both proponents and opponents of zero-rating may find arguments in the new Regulation and the debate is far from concluded.

Zero rating is dealt with in detail in Chapter 7. Memo-15–5275 states:

> Zero rating, also called sponsored connectivity, is a commercial practice used by some providers of internet access, especially mobile operators, not to count the data volume of particular applications or services against the user's limited monthly data volume. Commercial agreements and practices, including zero rating, must comply with the other provisions of the Regulation, in particular those on non-discriminatory traffic management. Zero-rating could in some circumstances have harmful effects on competition or access to the market by new innovative services and lead to situations where end-users' choice is materially reduced in practice.[85]

The EC argues that: 'The new rules therefore contain the necessary safeguards to ensure that providers of internet access cannot circumvent the right of every European to access internet content of their choice, and the provisions on non-discriminatory traffic management, through commercial practices like zero-rating.'[86] This obliges NRAs:

> to monitor market developments, and [they] will have both the powers and the obligation to assess such practices and agreements, and to intervene if necessary to stop and to sanction unfair or abusive commercial agreements and practices that may hinder the development of new technologies and of new and innovative services or applications ... the rules are also directly enforceable before national courts.[87]

Genna very strenuously disagrees with that interpretation, arguing that:

[85] European Commission, MEMO-15–5275.
[86] *Ibid.*
[87] *Ibid.*

> This is completely false and misleading! … power of national regulators will be materially weakened because of the ambiguous wording of [A]rticle 3 of the European regulation … read together with recital 7 (a recital, not a binding provision!) of the same regulation … it is absolutely unclear if and to what extent national regulators can intervene in order to prohibit such discriminations. The Dutsch [*sic*] and Slovenian legislations were quite clear … such legislations will need to be repealed.[88]

I tend to agree with Genna, and the Dutch and Slovenian governments, that the EC interpretation is misleading, whether this is deliberate or not.

Ofcom questions were seeking views regarding different forms of data caps:

> a) What is your understanding of the term 'commercial practices' (Ref. Article 3(2))? Do you think there is a demand for 'commercial practices' such as zero-rating, from the end users' point of view?
> b) Article 3 (2) foresees contractual freedom and ISPs' freedom to conduct commercial practices. Could you provide examples when/under which circumstances commercial practices would limit the rights of end users? (Ref. Article 3(2) and recital 7)
> c) What is your understanding or view regarding the monitoring of traffic for the purpose of traffic management (ref. Article 3(3) subpara 2)? What should ISPs be allowed to do in that regard under the TSM regulation?[89]

It appears from the questions that Ofcom has a much more permissive view of zero rating that the EC memo, which is unsurprising given the UK's long-standing hostility to regulation of net neutrality, as I will explore in Chapter 6.

My initial impression from these questions in mid-December 2015 was the following:

1. Are they suggesting encrypted traffic could be throttled?
2. Is 'sub-internet' to include zero rating?
3. The Greeks don't sound fit for purpose on this. Did they really think these are sensible questions – I would answer 'of course' to all of them as would any regulator that cared about transparency.
4. Ofcom is trying to find a 'de minimis' form of zero rating that can be openly expressed as conforming to the Regulation.

I suggest that is more informative questioning than I would expect, especially from Ofcom. It does not mean their final group outcome will be deregulatory. Both NKOM (Norway) and ARCEP (France) are much more in favour of net neutrality than their Greek and British cousins.

[88] Genna (2015).

[89] BEREC (2015) Questions for BEREC stakeholder dialogue with representatives of end-users / consumers and civil society, 24 November 2015.

Note that while the NRAs for the two net neutral nations, Slovenia and the Netherlands, may lobby against net neutrality inside the opaque hallways of the BEREC plenary, they are in fact infamously opposed to enforcing their national legal provisions for net neutrality, as evidenced by the fact that their respective ministry and consumer organisations had to force them to do so in 2014/15.

The EC also controversially add that:

> certain Member States' existing national rules do not need to change if these can be interpreted by regulators and courts consistently with the Regulation, including to protect end-users from commercial practices that are shown to circumvent the rules and materially reduce users' freedom of choice in the specific national circumstances.[90]

This appears to refer to the legal guarantees of non-discrimination in effect in the Netherlands, Slovenia and Finland. It is not quite as clear to national governments that their rules remain compatible with the new Regulation.

Future Work Programme for BEREC

BEREC does go one step further in its 2016 Work Programme:

> IP-Interconnection (e.g. Peering and Transit Conditions, Routing Policies, Colocation and Caching Policies of operators) is relevant both in the debate on Net Neutrality and with regard to the role of OTT players. However, information on IP-Interconnection is often neither publicly available nor within the remit of NRAs. BEREC will exchange available information. An expert workshop in cooperation with the OECD on the role of IP-interconnection in these debates will be held. The timing is chosen around the December [2016] OECD CISP working party meeting to ensure participation of the FCC and other OECD member states.[91]

This will help to clarify the extent to which IP interconnection, a vital issue in high bandwidth traffic such as video delivery, becomes a live net neutrality issue in 2017. I predict at this stage that it will, and explore this in more depth in Chapter 3.

BEREC also mentions (not in the context of net neutrality) that it will deliver an Opinion to the EC on Universal Service in 2016/17 as required, explaining that this 'will ultimately guarantee an adequate "safety net" to European citizens'.[92]

BEREC completes the summary of its work by acknowledging that its network and in particular the net neutrality Guidelines will be of significant interest to non-European regulators:

[90] European Commission, MEMO-15-5275 (2015).

[91] BEREC, BoR (15) 213.

[92] *Ibid.*, Section 12, p. 24.

> Acknowledging the increasing importance of international cooperation on common issues such as inter alia Net Neutrality [BEREC will continue] cooperation with international regulatory authorities such as the FCC as well as regional regulatory networks (EMERG, Regulatel and EaPeReg[93]). As mentioned in the Joint 4–lateral Summit Declaration of 2nd July 2015 the participants committed themselves to undertake all efforts with a view of keeping this high level regulatory dialogue through future joint summits. If this is confirmed, BEREC would join such a 4-lateral summit in 2016.[94]

The quadrilateral summits of American, European, West Asian and African regulators may become an annual fixture of especial importance in coordinating responses to zero rating by mobile operators based in Europe with subsidiaries in developing nations.

BEREC is thus very active in developing not just the Guidelines legally required in Recital 19 of the 2015 Regulation, but also monitoring QoS cooperation and other areas of interest stretching into 2017–18, not just within the EU/EFTA but also with regulators in neighbouring nations, the Americas, and through the OECD with East Asia also. It has long been recognised that regulators in telecoms exhibit advanced regulatory policy transfer through 'soft law' mechanisms and informal coordination, which is now far advanced, and can be compared to that in environmental regulatory governance and other fields of international coordination.

European law conclusion

Regulation 2015/2120 was passed on 27 October 2015.[95] It was first proposed by the European Commission as a minimal regulation in May 2013, passed at First Amendment in the European Parliament with amendments that would ban both zero rating and tightly defined Specialised Services as physically and/or logically separate to the Internet in April 2014,[96] and then revised in the Council of Ministers to more closely resemble the original proposal, agreed in a highly contentious Trilogue with the Commission and Parliamentary Committee Chair in June 2015. It returned to the Parliament for a vote on potential amendments, which failed, meaning the Regulation became law in all 28 Member States.

After the Trilogue in June 2015, EC Vice-President Ansip claimed:

[93] EaPeReg (Eastern Partnership Electronic Communications Regulators Network), REGULATEL (Latin American Forum of Telecommunications Regulators) and EMERG (Euro-Mediterranean Regulators Group). See EaPeReg (2015) detailing the 38 NRAs present from the EU, EFTA, East of Europe, Caucasus, North Africa, Middle East and Latin America.

[94] BEREC, BoR (15) 213, Section 18, p. 29.

[95] Belli and Marsden (2015).

[96] Marsden, Chris (2014a).

> Internet service providers cannot act as gatekeepers to decide what people can, or cannot, access. Equal treatment and non-discrimination of traffic will be set in law … Paid prioritisation will be banned, which means that a start-up's website cannot be slowed down to make way for a larger company prepared to pay extra to get such an advantage.[97]

However, that fails to clarify either zero rating or SpS, to the anger of some parliamentarians who fear their laws will be undermined by the weaker compromise adopted. Can consumers trust the German or Cyprus NRA to get that technical judgment correct? Does BEREC have the competence to help them? That remains to be seen, as BEREC will have to clarify the application of the law. The lack of clarity in the Regulation means that the guidance in the BEREC Guidelines in 2016 will be eagerly awaited on both zero-rated services, notably already regulated in Slovenia, the Netherlands and Norway, and Specialised Services. It may require the revision of the 2012 Dutch/Slovenian and 2014 Finland laws, but will take direct effect more rapidly than the other 25 Member States' national regulatory debate otherwise promised.

There can be no conclusion nor summary to a chapter examining a law as half-baked as this one. The European 'Open Internet' (*sic*) law is a messy compromise, and I already consider the Regulation as a sort of Internet version of the Treaty of Versailles, heralding the next war rather than settling this one. The 2016 Guidelines will be the first opportunity to make sense of the Regulation, and the 2019 review the opportunity to codify some workable definitions. Until then, net neutrality will be a vague promise rather than a guarantee for Internet users in Europe.

[97] Ansip (2015).

5

Three wise monkeys of net neutrality: privacy, liability and interception

> DPI can look into the content of the message sent over the Internet. To use a real-world example, using DPI is akin to a third party opening an envelope sent by surface mail, and reading its contents before it reaches its intended destination ... it is not clear that examination of content is necessary for network management and may constitute an unreasonable invasion of an individual's privacy.
>
> Privacy Commissioner of Canada[1]

IAPs have been acting as the fabled 'three wise monkeys' in relation to Internet content liability since the dawn of the commercial Internet.[2] These intermediaries are not subject to liability for their European customers' content under the Electronic Commerce Directive (EC/2000/31) (ECD) so long as they have no actual or constructive knowledge of that content: if they 'hear no evil, see no evil and speak no evil.'[3] Regulators have also been acting as 'three wise monkeys' in ignoring evidence that net neutrality is being compromised by IAP decisions to block, throttle and otherwise censor users' access to content. Forms of private censorship by intermediaries have been increasing throughout the twenty-first century even as the law continues to declare those intermediaries (mainly IAPs, but increasingly also video-hosting companies such as YouTube, social networks such as Facebook and search providers such as Google) to be 'Three Wise Monkeys'. The liability question may be paraphrased: will the

[1] Privacy Commissioner of Canada (2013b) at paras 13, 32.

[2] For UK law, see *Shetland Times Ltd v. Jonathan Wills and Another*, 1997 FSR (Ct Sess. OH), 24 October 1996; *Godfrey v. Demon Internet Service* [2001] QB 201. For US law, see *Cubby v. CompuServe* (1991) 766 F. Supp 135; *Playboy Enterprises, Inc. v. Frena* (1993), 839 F. Supp. 1552 (M.D. Fla.); Stratton Oakmont Inc v. Prodigy (1995) NY Misc. 23 Media L. Rep. 1794; *American Civil Liberties Union v. Reno* (1997) 21 US 844 of 27 June No. 96–511; and Digital Millennium Copyright Act 1998, s.512(k)(1)(A–B). See generally Marsden (2012b).

[3] Marsden (2010), pp. 105–149.

monkeys continue to be wise conduits for speech, or will governments make them open their eyes and ears, becoming censors of speech and recorders of our every click online?

Governments have been fundamentally challenged in both the European Union and the United States by the Snowden revelations of mass surveillance.[4] Much of this surveillance took place with the secret cooperation of the IAPs, as required by the conditions of their permissions to conduct their business. This chapter is an empirical examination of the use of interception by IAPs, whether required by law enforcement or for the IAP's own purposes, such as for behavioural advertising. It does not consider the wider theoretical foundations of electronic privacy, the wider implications of Snowden's revelations for providers other than IAPs, or the specific legal reforms imposed in response to the potential illegality of state surveillance under programmes such as Tempora.[5] I do not consider in this chapter the wider implications of the cases that led to the annulment of the Data Retention Directive (2006/24/EC)[6] in 2014, nor the US–EU Safe Harbor in 2015[7] and its putative replacement, the Privacy Shield, announced in February 2016. The General Data Protection Regulation (GDPR) was also confirmed in 2016.[8] Books and articles will be written about such issues in the coming months and years, and the net neutrality blog (chrismarsden.blogspot.com) will refer to them as privacy developments become clearer. Books and articles will continue to be written about various national laws retrospectively securing the legality of surveillance, such as the UK Investigatory Powers Bill 2016, which was proceeding through Parliament as I wrote this book. The so-called Right to be Forgotten (or obscure) is also a topic for other scholars.[9] I focus on the IAP legal regime for net neutrality, the surveillance requirements that may infringe net neutrality, and finally the extraordinary cases of behavioural advertising without the permission of users that were carried out by BT, and which resulted in legal action against the UK government for inadequate enforcement of European privacy legislation, namely the E-Privacy Directive 2002/58/EC.[10]

4 COM(2013) 846; COM(2013) 847.

5 Joined cases C-293/12, *Digital Rights Ireland v. Minister for Communications, Marine and Natural Resources and Others* and C-594/12 *Kärntner Landesregierung and Others*.

6 Directive 2006/24/EC (Data Retention Directive). The Data Retention (EC Directive) Regulations 2007, available at www.opsi.gov.uk/si/si2007/20072199.htm (Accessed 5 September 2016), came into force on 1 October 2007. For a summary see Vilasau (2007). For critical interpretation, see Rauhofer (2009).

7 C-362/14, *Maximillian Schrems v. Data Protection Commissioner*, Judgment of Grand Chamber 6 October 2015.

8 Regulation (EU) 2016/679.

9 C-131/12 *Google Spain SL and Google Inc. v. Agencia Española de Protección de Datos (AEPD) and Mario Costeja González*; Powles (2015).

10 Directive 2002/58/EC (E-Privacy Directive).

It is important that governments consider where best the issue is regulated, by a telecoms regulator or by a ministry. The net neutrality privacy problem is not a lack of regulatory tools per se, but potentially a lack of forensic skills to analyse the potential consumer harms that can be created by unjustified or 'unreasonable' discrimination. Regulators can monitor both commercial transactions and traffic shaping by IAPs to detect potentially abusive discrimination. No matter what theoretical legal powers may exist, their usage in practice and forensic gathering of evidence may make the regulatory task very burdensome. The increasing use of behavioural advertising by third parties is also very concerning to privacy regulators, and any cooperation between IAPs and third parties to share such revenue is likely to need the explicit consent of all IAP users, following the precedent of the Phorm case considered below, and European opinions recently issued about behavioural advertisers.[11]

First, I place interception in the regulatory context, to explore neutrality as a form of 'medium law' as well as the policy implications of the Snowden leaks and other recent developments. Second, I analyse the legal, technical, regulatory and policy discussions that have been applied to this form of interception. I discuss global problems that private or 'co-regulated' filtering and censorship cause, whether for private ends, such as copyright enforcement, or public ends, such as restricting freedom of expression, as well as the potential impact on developing countries. The major part of the chapter deals with the BT and Phorm experiments of 2006–08, the UK government's encouragement of such illegal activities and the European Commission's responses that forced amendment to UK e-privacy law in 2012. Finally, I assess the evidence base for future privacy legislation affecting IAPs, the EC 'Platform Regulation' consultation, and the need for regulators to address privacy in net neutrality policy discussion in future.

The GDPR and the Snowden inquiries 2016

European laws are meant to protect citizens' privacy and liberty. Directive 95/46/EC is the main law giving responsibilities to Member States and data protection rights against corporate actors to citizens. This European law sets a high standard for data protection, arguably higher than that in the United States. National data protection agencies have a permanent joint working group (the Article 29 Working Party) and are required to implement the Directive as uniformly as possible; its tasks include members cooperating with each other and

[11] Council of Europe (2010); Recommendation CM/Rec(2010) 13; Article 29 Working Party WP203, p. 45; Article 29 Working Party, Letter addressed to Ms Le Bail, 2011; Article 29 Working Party, Press Release, 15 December 2011. See also Article 29 Working Party, Opinion 16/2011: 'adherence to the EASA/IAB Code on online behavioural advertising and participation in the website www.youronlinechoices.eu does not result in compliance with the current e-Privacy Directive.'

the European Commission in a transparent manner to ensure the development of consistent regulatory practice, contributing to a high level of protection of personal data and privacy, and ensuring that the integrity and security of public communications networks are maintained.[12] The European institutions are also required by law to consider the Opinions issued on prospective legislation by the EDPS, established in 2002. Directive 2002/58/EC (the 'E-Privacy Directive') includes measures intended to prevent spam, which are supplemented by a 2004 Communication[13] on spam.[14] The critical test in both Directive 2002/58/EC and Directive 95/46/EC is that subscribers have to opt for arrangements that may otherwise infringe their personal privacy, and that sensitive data must not be passed to third parties unless so authorised by subscribers and the data is anonymised.

The GDPR amends the E-Privacy Directive and replaces and repeals the Data Protection Directive (DPD).[15] The draft Regulation COM(2012) 11 in particular contained Sections 42–43 relating to transfer of data outside the European Union, insisting on Binding Corporate Rules (BCR) for such transfers to take place subject to enforcement by national data protection agencies (DPAs).[16]

One area in which European regulators have been forced to investigate potential interception, very much against net neutrality principles, is that of illegal surveillance of IAP users perpetrated by agencies in the 'Five Eyes' multinational espionage coalition.[17] Note that though nation-states funded these activities, they were carried out with the more or less willing cooperation of Internet companies including IAPs acting against their own users' interests in net neutrality. Illegal as well as legal interception activity by 'Five Eyes' within European, Latin American and other nations was exposed by the whistleblower Edward Snowden and *The Guardian* newspaper in June to October 2013.[18] 'Five Eyes' (or more formally AUSCANNZUKUS) describes the cooperation between the intelligence (i.e. espionage) agencies of the Anglo-Saxon powers during what in English-speaking countries was called the Cold War between the US/allies and the Warsaw Pact/allies.[19] The United States, United

[12] Directive 2002/21/EC (Framework Directive).

[13] Communications from the Commission, and Resolutions of the Council are not European legislation and therefore non-binding on Member States, but they have important 'signaling' effects on Member States and companies and therefore are termed 'soft law'.

[14] COM(2004) 28.

[15] Directive 95/46/EC. For current progress on the matter, numbered 2012/0011(COD), see www.europarl.europa.eu/oeil/popups/ficheprocedure.do?reference=2012/0011(COD)#tab-0 (Accessed 5 September 2016).

[16] COM(2012) 11 and COM(2012) 09. See also IP/12/46.

[17] Campbell (1999).

[18] Bowden (2013a).

[19] Richelson and Ball (1985).

Kingdom, Canada, Australia and New Zealand are formal partners, though other allies have subsidiary and subsequent agreements that permit some level of intelligence sharing.[20]

Packet-sniffing schemes such as Carnivore, a system implemented by the Federal Bureau of Investigation that was designed to monitor email and electronic communication, have been active since at least 1997; Carnivore had used a customisable packet sniffer that can monitor all of a target user's Internet traffic.[21] A larger-scale operation, called Echelon, was built by various Western governments, and this was investigated by the European Parliament in a report released on 5 September 2001.[22] Surveillance by Intelligence agencies has vastly increased, as exposed thoroughly by Glenn Greenwald and colleagues at *The Guardian* using evidence supplied by former National Security Agency contractor Edward Snowden.[23] Echelon was later replaced by the US programme PRISM[24] and in 2011 UK–US joint operation Tempora (with sub-programmes called 'Mastering the Internet' and 'Global Telecoms Exploitation'), which intercept communications in fibre optic cables destined for transatlantic transmission.[25] Law in this area is rapidly outflanked by the technological capabilities of public and private parties, which has resulted in inquiries in response to the Snowden revelations, notably by the Intelligence and Security Committee:

> Although we have concluded that GCHQ has not circumvented or attempted to circumvent UK law ... We are examining the complex interaction between the Intelligence Services Act, the Human Rights Act and the Regulation of Investigatory Powers Act, and the policies and procedures that underpin them, further. We note that the Interception of Communications Commissioner is also considering this issue.[26]

This press release absolved GCHQ of any illegality, but a proper inquiry led by the Deputy Prime Minister followed in winter 2014. The Interception of Communications Commissioner (ICC) submits Annual Reports to the Prime Minister in the summer after the conclusion of the previous calendar year.[27] Despite the urgency of the public revelations of the potentially illegal use of Tempora programmes by GCHQ in June 2013, the ICC announced that he

[20] The initial formal UK–US agreement was signed in 1947 after the successful conclusion of the Second World War and just prior to the outbreak of the Korean War of 1950–53, with final partner New Zealand joining only in 1980. See AUSCANNZUKUS (2013).

[21] For a legal perspective on private packet sniffing, see Frieden (2008).

[22] European Parliament, Final report on the existence of a global system for the interception of private and commercial communications, 2001.

[23] For examples, see Borger (2013).

[24] Government code name for a data-collection effort known officially by the SIGAD US-984XN.

[25] See Shubber (2013).

[26] Intelligence And Security Committee Of Parliament (2013), paras 6–7.

[27] House of Commons 571 SG/2013/131.

would submit the investigation into the GCHQ interception of communications to the Prime Minister in July 2014:

> [ICC] is required by section 58(4) of RIPA [Regulation of Investigatory Powers Act 2000] to report annually to the Prime Minister. The Prime Minister lays the report before Parliament except for any sensitive parts of it which he decides to exclude under section 58(7).[28]

He further explained that:

> my role is defined in Section 57(2) of RIPA. I am not appointed or authorised to oversee all of the activities of the intelligence agencies, only those specified in Section 57(2) of RIPA. I can confirm that I am currently conducting an investigation into the various recent media reports relating to disclosures about interception attributed to Edward Snowden.[29]

Much of the relevant intercepted data is metadata, which is more useful to intelligence services and behavioural advertisers than is content of communications, which is not machine-readable and therefore examining it in large volumes would be too time-intensive to be useful. Part I Chapter 2 of the Regulation of Investigatory Powers Act 2000 (RIPA) covers the acquisition and disclosure of communications data (rather than the content of the communications). Therefore, some element of scrutiny of public authorities' use of metadata commenced and was reported internally in July 2015. Metadata was also a key concern of the Article 29 Working Party in its study of the reform of European e-privacy law.[30]

The Snowden revelations allege illegal interception of IAP traffic, which supports earlier European Parliament investigations, and has brought complaints about violations of criminal law to the attention of the European human rights court, national parliaments and information commissioners.[31] The interception laws of most nation states do not permit IAPs to allow or condone interception by third parties, let alone foreign state agencies. Brown provides a useful summary of such laws in several states, including the 'Five Eyes' signatories themselves.[32]

Note that the personal data of EU citizens captured by non-EU countries is still subject to European law under the 1995 and 2002 legislation until the new GDPR comes into effect in 2018.[33] In 2016 there were profound investigations

[28] Interception of Communications Commissioner (2013).

[29] Part I Chapter 1 of RIPA provides the statutory authority for the lawful interception that takes place within the UK.

[30] Article 29 Working Party, Opinion 2/2010, p. 7: 'Article 29 Working Party is deeply concerned about the privacy and data protection implications of this increasingly widespread practice.'

[31] Brown (2013a).

[32] Brown (2013b).

[33] Rauhofer and Bowden (2013).

into the invasion of user traffic streams throughout Europe, with investigations in the United Kingdom, Netherlands, Belgium, France, Germany, Luxembourg and many other nations, as well as investigations by the European Parliament. Focus is on foreign spying, but as details emerge of the interception techniques used, more attention is being focused on the specific national criminal violations by agents acting on behalf of 'Five Eyes', notably IAPs.

While this type of interception is not classified as net neutrality violation, given that it is carried out under the orders of government and is thus presumed to be for law enforcement, should it be proven unlawful, it will amount to interception of user personal data for illegal purposes. Regulators may therefore need to issue instructions to IAPs and others not to cooperate with foreign state agencies and others who instruct them to cooperate with data gathering. In extreme circumstances, that could potentially require IAPs not to interconnect with US- and UK-based IAPs.[34] The OECD has also renewed its privacy guidelines, and referred to the need to ensure that Internet policy conforms to fundamental rights of users.[35] At the 2011 Paris meeting in which the latter declaration was made, the Korean delegation requested that the OECD pay attention to the need for more research and coordinated policy towards net neutrality.

What is changing in IAP analysis of traffic?

IAPs have many reasons to manage traffic:

1. It is required for government law enforcement and security purposes.[36]
2. Network providers already provide filters against the more obvious types of 'spam' – unsolicited commercial communications.
3. Network providers cooperate with national security agencies in tracing potential terrorist activities on the Internet.
4. Network providers can trace non-encrypted VoIP (e.g. Skype) and block these packets.
5. Network providers are increasingly adopting Specialised Services for their networks in order to prevent users from overstraining the network at times of peak usage, and charge content owners for value-added high-volume services such as video files.

These new policies allow network providers to block file transfers, or to charge the users a carriage fee for sending large files. This policy is generally termed a

[34] Agence France-Presse (2013).

[35] OECD (2013b). See also OECD (2007); OECD (2008); OECD (2011). OECD (2013b) also referred to European law and the Asia–Pacific Economic Cooperation's Cross-Border Privacy Rules System (APEC CBPR).

[36] See generally Bendrath (2009).

'walled garden' to denote the isolation of content on the network from other content on the wider Internet.

IAP routers (if so equipped) can look inside a data packet to 'see' its content, via DPI. Less powerful routers conduct only shallow inspection that simply establishes the header information – the equivalent of the postal address for the packet. An IAP can use DPI to determine whether a data packet values high-speed transport – as a television stream does in requiring a dedicated broadcast channel – and therefore offer higher-speed dedicated capacity to that content, typically real-time dependent content such as television, movies or telephone calls using VoIP. Most voice calls and video use a dedicated line, a copper telephone line or cable line: they may use SpS in future. That could make a good business for IAPs that wish to offer higher capability via DPI. Not all IAPs will do so, and it is quite possible to manage traffic less obtrusively by using the DiffServ protocol to prioritise traffic streams within the same Internet channel.

IAPs are using 'black boxes' in their networks to look inside the packets that carry communications and to examine their content, in a change to DPI which has very serious regulatory implications. DPI and other techniques that let IAPs prioritise content also allow them to slow down other content, as well as speed up content for those that pay (and for emergency communications and other 'good' packets). Encryption is common in applications and partially successful in overcoming these IAP controls, but even if all users and applications used strong encryption, this would not succeed in overcoming decisions by IAPs simply to route known premium traffic to a 'faster lane', consigning all other traffic to a slower, non-priority lane (a policy explanation simplifying a complex engineering decision).

Waclawsky stated in regard to MPLS, a mobile industry protocol to permit QoS: 'This is the emerging, consensus view: [it] will let broadband industry vendors and operators put a control layer and a cash register over the Internet and creatively charge for it.'[37] Putting a cash register on the Internet will permit much more granular knowledge of what an IAP's customers are downloading and uploading on the Internet. That means that the formerly 'Wise Monkey' IAPs would rapidly become the all-seeing eye, with many more consequences than simply a new revenue stream to build higher-speed lanes – which generally means laying fibre optic cables closer to the household, replacing the old copper lines. IAPs could filter out both annoying and illegal content. For instance, they could 'hear' criminal conversations, such as those between terrorist sympathisers, illegal pornographers, harassers, those planning robberies, those containing libellous commentary, and so on. They could also 'see' illegal downloading of copyrighted material. They would be obliged to 'speak', to cooperate with law enforcement or even copyright industries in these scenarios,

[37] Waclawsky (2005) cited in Marsden (2010).

and this could create even greater difficulties where that speech was legal in one country but illegal where it was received. Examples include English libel law, Australian pornography filtering, Chinese Falun Gong website blocking, Turkish YouTube bans and United States online gambling bans. Net neutrality is therefore less unpopular with smaller IAPs that wish to avoid a legal liability morass, which the E-Commerce Directive and other national IAP non-liability 'safe harbor' laws are expressly designed to prevent.

IAP (and government) practices have been highly deceptive in places, blocking content for specific anti-competitive and non-specific traffic management purposes. Attempts at least to introduce transparency into the debate, as well as the rights of end users, can be achieved via co-regulation. In European telecoms, this is a prevalent but awkward compromise between state and private regulation, with constitutionally uncertain protection for end users and wide latitude for private censorship.

Mobile IAPs claim the same special protections from regulation that their previous incarnations as mobile voice networks claimed, to enable walled gardens to flourish. It is worth noting that many mobile IAPs use IMSI[38] and other parsing methods to track everything users do on the Internet, in developed and developing countries, with increasing levels of technical sophistication.[39] Any net neutrality solution needs to be holistic, considering IAPs' roles in the round. IAPs are a heterogeneous category, ranging from very large network owners such as (UK examples) British Telecom, Vodafone and Virgin Media (owner of the cable TV/telecoms infrastructure), to large retailers such as TalkTalk and Sky, to hundreds if not thousands of much smaller niche business and consumer operators. Smaller operators do not typically deploy such widespread interception capability, with the most privacy-aware, Andrews & Arnold Ltd, stating: 'We have no so called black boxes to covertly monitor traffic and/or pass traffic monitoring to the authorities or anyone else. Obviously the law is such that we may have to add such black boxes, but we would resist as far as possible.'[40] Other IAPs are not so protective of their users' privacy.

DPI is a technique that may be both unreasonable and invasive of user privacy, and it may be that information/privacy commissioners are best placed to investigate such potentially criminal breaches of user rights. Because net neutrality raises a set of new issues for privacy regulators, the necessary skill set needs to be acquired and developed in consultation with other national and international regulators. Currently, it is not a requirement for most IAPs to

[38] International Mobile Subscriber Identity (IMSI) identifies a cellular network user, and is stored as a 64-bit field on the SIM card.

[39] Vallina-Rodriguez with Sundaresan, Kreibich and Paxson (2015). The paper describes privacy violations and header enrichment practices performed by mobile operators (perma-cookies, x-forwarded-for, IMEI, IMSI, etc.).

[40] Andrews & Arnold Ltd (2016).

notify customers when they block encrypted content, such as P2P-distributed applications. The (often spurious) security or anti-piracy reasons given are not within the remit of typical economic telecoms regulators. Where the reasons given by some IAPs for blocking encrypted traffic, which carries malware and other harmful content, are typically the concern of the security services (Interior Ministry or Prime Minister's office), and occasionally the Ministry of Industry, the regulator defers to these senior agencies because it has little technically specific knowledge of data security.[41] More joined-up regulation is needed with urgency in this field. Regulators and politicians are challenged publicly by such problems, particularly given the ubiquity of email, Twitter and social media protests against censorship.

Network neutrality cannot simply be solved by economic analysis of bottlenecks in transport-based industry, or a convergence of regulation between television and the Internet, but as the delivery mechanism for the global Information Society. The Internet's core values of openness and democracy have been established by accident as well as design. 'Medium law' (i.e. mass market content online that formerly used several media) is intimately tied into telecoms law. Security and antiterrorist measures are also driving IAPs towards filtering all incoming traffic. This may change the entire architecture of the Internet, its business model and freedom of speech. It is happening beyond the analysis of the discrete fields of information security, e-commerce law, media law and telecoms law.

There are at least two critical non-IAP-originated factors at play: concern over illegal and inappropriate content (such as child pornography, music protected by copyright and latterly video files being inappropriately shared, and malware including spam); and the security agenda, which aims to enforce QoS to separate 'good' or preferred from 'bad' or discriminated-against packets. There is a legitimate concern that this represents a division between the rich and powerful senders of packets and the lesser content types. These three policy areas – telecoms, content and security regulation – are coming together. Horten states:

> By authorizing blocking practices, the Telecoms Package puts Europe on a path to a closed series of Internets. It puts at risk innovation, trade, and any policy goals to encourage cross-border trade. It puts at risk the EU's Information Society goals. And, it stands to chill democratic speech.[42]

The problems of development and the global Digital Divide are intimately connected to net neutrality. Internet connectivity is still very expensive for most developing countries, despite attempts to ensure local Internet peering points

[41] See Brown, Edwards and Marsden (2006).
[42] Horton (2009).

(exchanges) and new undersea cables, for instance serving East Africa. Mobile access is considered further in Chapter 7. Mueller argued that net neutrality 'must also encompass a positive assertion of the broader social, economic and political value of universal and non-discriminatory access to Internet resources among those connected to the Internet'.[43] He also argued that the tendency of governments in both repressive and traditionally democratic regimes to impose liability on IAPs to censor content for a plethora of reasons argues for a policy of robust non-interference:

> The flip side of a [network neutrality] policy that valorises the right of Internet users to access each other without interference from intermediaries is the belief that network users wronged by other users must hold the wrongdoer responsible – not the intermediary network operator.[44]

That is especially valuable in countries where there is much less discussion of how government deployment of IAPs as censors can endanger user privacy and freedom of expression. Mueller suggested that the net neutrality metaphor could be used to hold all filtering and censorship practices up to the light, as well as other areas of Internet regulation, such as domain name governance. Network neutrality has become an important policy issue that is discussed at the United Nations Internet Governance Forum (IGF). The IGF discussions of net neutrality and other issues substantially increased from 2009 to 2015, as explored in Chapter 8.

Law and technologies to intercept communications

Committees of both the US Congress and the UK Parliament carried out inquiries into behavioural advertising in 2009.[45] Since 2002, Article 15 ECD has also required European Member States not to impose undue restrictions on IAPs,[46] which continually causes Member States to either derogate from the ECD in the interests of crime fighting and anti-terrorism law or simply

[43] Mueller (2007), p. 7.

[44] *Ibid.*, p. 8.

[45] See http://uk.practicallaw.com/4-500-4840?q=&qp=&qo=&qe= (Accessed 16 September 2016) on the US investigation, and APCOMMS (2014) announcing 'Can we keep our hands off the net?' apComms to investigate the role for Government over Internet traffic.

[46] Directive 2000/31/EC, Article 15 states: 'No general obligation to monitor. 1. Member States shall not impose a general obligation on providers, when providing the services covered by Articles 12, 13 and 14, to monitor the information which they transmit or store, nor a general obligation actively to seek facts or circumstances, indicating illegal activity. 2. Member States may establish obligations for information society service providers promptly to inform the competent public authorities of alleged illegal activities undertaken or information provided by recipients of their service or obligations to communicate to the competent authorities, at their request, information enabling the identification of recipients of their service with whom they have storage agreements'.

to ignore the provision altogether. So many features of wire tapping and anti-terrorism law have been passed or amended since 2001 that there would by now be several thousand derogations across the European Member States, given that interception of communications by IAPs on behalf of governments formally requires a notification for derogation from Article 15 for each of the 28 Member States whenever anti-terrorist law is reformed in this area. The definition of the limits on general obligations to monitor – which are relevant for any imposition of, for instance, copyright monitoring on IAPs by Member States – were explained by the European Court of Justice in the 2012 leading case of *SABAM v. Netlog NV*.[47] The Court held that imposing a copyright-filtering system on an IAP would infringe on the prohibition on general obligation to monitor, and stated that it:

> [Para. 48] may also infringe the fundamental rights of that hosting service provider's service users, namely their right to protection of their personal data and their freedom to receive or impart information ... [Para. 49] Indeed, the injunction requiring installation of the contested filtering system would involve the identification, systematic analysis and processing of information connected with the profiles created on the social network by its users [protected personal data].

Note that the proposed reforms of the ECD including Article 15 were abandoned in 2012 by the European Commission in its E-Europe Action Plan. It is well established that no governmental authority or court can impose a general duty to intercept and monitor, because that would infringe privacy rights.

The range of network and information security requirements at European level, which must then be implemented as national law in the European countries, imposes costs on the network. These are in addition to existing costs for spam filtering, protection against distributed denial of service (DDOS) attacks, phishing and other 'malware' that IAPs typically invest in to protect their subscribers from the worst excesses of IP traffic. Security is a growing problem as dependence on broadband (as a key element of the critical information infrastructure) grows and as the Internet moves towards pervasive computing, and the 'Internet of Things'. There is an escalating arms race as criminal behaviour becomes more sophisticated. The objectives and requirements are also changing on both sides: on the attacking side, the evolution from unauthorised access to data corruption, exposure or access denial; on the defending side, the change in data collection, storage, processing locations (centralised or not), data exchange and transfer of liability among buyers, sellers and IAPs. Loss of Internet privacy, openness and E2E connectivity is one potential casualty of security concerns.

[47] Case C-360/10, Reference for a preliminary ruling under Article 267 TFEU from the rechtbank van eerste aanleg te Brussel (Belgium).

IAPs can either throttle users by cutting off their connections at peak times, once they have exceeded monthly quotas, or try looking inside the packets to see whether they are P2P or not. The latter becomes a very dangerous business to engage in because, as we will see, governments are not only encouraging IAPs to look, they are actually subsidising the DPI equipment required to do so – and this sometimes in breach of both European and UK privacy and interception laws (the latter intended to prevent private spying, even if encouraged by government policy). Felten worried that regulators are used to standard bodies and classes of companies, when, for instance, BitTorrent is a protocol, not a company or a single standard.[48] Blocking BitTorrent or P2P more widely will eventually fail because the protocol designers will route around via encryption or other techniques.

Regulating DPI and interception of traffic

Blocking and other forms of traffic shaping are controversial because, under current network management tools, they are blunt tools. For instance, all P2P traffic using a certain protocol may be blocked. P2P can respond by encrypting its traffic or otherwise spoofing, but this creates an 'arms race' much like that found in security software responses to the threat of breaches. Future networks may try to cap P2P more effectively, which can itself lead to an 'arms race' between encrypted P2P content and attempts by IAPs to detect P2P traffic using DPI.[49]

IAPs have limited liability where they act as 'mere conduits' but not where they have constructive or actual knowledge of illegal content. Their traffic is thus something of a Pandora's box – if they look inside using DPI, all liabilities flow to them, from child pornography, to terrorism, to copyright breaches, to libel, to privacy breaches.

Cooper analysed the choices of whether to introduce DPI equipment into IAP networks, restricting traffic as an alternative to increasing capacity, with the consequent decision to invest in DPI and other management servers instead of greater bandwidth.[50] She points out that US cable companies at the time of the Internet Policy Statement in 2005[51] hoped that the burden of proof on 'reasonable' techniques would fall on complainants, with the presumption that

[48] Felten (2008).

[49] In May 2009, uTorrent announced a future protocol change to UDP, which indicates that (a) TCP may not be the main P2P traffic protocol; (b) that will annoy those who use TCP for YouTube etc.; (c) it will annoy ISPs; (d) so more discriminatory actions can be expected. It is the arms race continuing at a higher level, as filtering UDP is a new departure for many consumer ISPs. See www.theregister.co.uk/2008/12/01/richard_bennett_utorrent_udp/ (Accessed 5 September 2016).

[50] Cooper (2013), pp. 109–120.

[51] FCC, Internet Policy Statement 05–151, 2005.

IAPs were acting reasonably. That has not been the case in the US or Canada. It is for the IAP to demonstrate that its use of technologies such as DPI is reasonable, a test that Comcast failed in its deployment of Sandvine DPI. The presumption that DPI may be unreasonable based on Comcast's experience has been profound for US IAPs. Cooper concluded that while marketing directors still encouraged DPI use and were likely to authorise such expenditure in order to better target services at customers, regulatory departments discouraged its use and had the reverse effect on which engineering choices to deploy.[52] More research is needed into the causes for such differences, but it is very likely that a lack of knowledge and education about the criminal offences from breaching data protection law and intercepting traffic amongst marketing departments of IAPs may account in part for their cavalier approach to installing DPI equipment to monitor customers. By contrast, engineers' typical preference in Cooper's study was to increase bandwidth rather than manage traffic more minutely.

It is also notable from Cooper's study that foreign content providers were unable to influence domestic IAPs' traffic management practices, so that MMORPG (massively multiplayer online role playing game) providers such as World of Warcraft were often significantly impeded in delivering their service because of unreasonable traffic management, a problem significantly worse in the UK where DPI and other traffic management techniques were used much more invasively than in the US.[53] Cooper's conclusions have particularly negative outcomes for those free-to-play MMORPGs such as are commonly found in South Korea, as there would be no likelihood that such MMORPG creators could negotiate or even complain successfully when foreign IAPs block their world.

Cooper establishes that different countries' regulators view of litigation and reputation is likely to colour their view of what is 'reasonable' and how strict their interpretation of that provision may be. Thus the US regulators are not scared of litigation or enforcement and so are likely to prosecute cases more strictly, whereas the lawyer-light UK regulator is committed to alternatives to enforcement and is likely to prosecute only as a last resort. UK regulators, notably the Information Commissioner, have shown no willingness to prosecute even in the infamous case of PHORM/BT's illegal DPI trial,[54] whereas the US regulator Federal Trade Commission (FTC) successfully brought strict settlements with multimillion-dollar fines for social networks that misused their subscribers' data, notably Google and Facebook in August 2012.[55] In the Google case, the FTC declared its first:

[52] Cooper (2013), pp. 122–132. Cooper interviewed 70 elite decision-makers in ISPs, regulators and content companies in the period 2011–12, the broadest sample known.

[53] *Ibid.*, pp. 200–204.

[54] Crown Prosecution Service (2011).

[55] Lardinois (2012).

> FTC settlement order [that] has required a company to implement a comprehensive privacy program to protect the privacy of consumers' information. In addition, this is the first time the FTC has alleged violations of the substantive privacy requirements of the U.S.–EU Safe Harbor[56] Framework.[57]

The likelihood of criminal or civil prosecution for breaches of net neutrality and other abuses of trust with IAP users are thus conditioned by the regulators' willingness to actually enforce regulation. In the US this is clearly the case; in Europe very much less so. This is despite the US corporate need to satisfy European regulators that the theoretically weaker US personal data privacy rules can satisfy European law under existing Directive 95/46/EC. As has become apparent in the wake of the 2013 Snowden revelations, US companies are both constantly in breach of the safe harbour themselves for corporate policy reasons, and obliged by US law enforcement and espionage to mistreat personal data of citizens of other countries, including Europeans and Koreans. Belatedly, this became a significant issue in the renegotiation of the 'Safe Harbor' and the GDPR in 2016.[58]

Alleged criminal breaches of UK e-privacy by IAPs

The continued attempts by IAPs to intercept communications on their own networks are by themselves legal under the law of interception. However, they may not allow others to intercept on their behalf or grant to others the right to intercept for their own purposes. UK law is clear on this point. Interception of communication is subject to RIPA Section 2(2):

> For the purposes of this Act, but subject to the following provisions of this section, a person intercepts a communication in the course of its transmission by means of a telecommunication system if, and only if, he –
>
> (a) so modifies or interferes with the system, or its operation,
> (b) so monitors transmissions made by means of the system, or
> (c) so monitors transmissions made by wireless telegraphy to or from apparatus comprised in the system, as to make some or all of the contents of the communication available, while being transmitted, to a person other than the sender or intended recipient of the communication.

One element of intercepting is that making available some or all of the contents of the communication to a person other than the sender or intended recipient

56 European Commission Decision 2000/520/EC. See also www.export.gov/safeharbor (Accessed 5 September 2016).

57 FTC (2011).

58 European Commission, Consultation on the Commission's comprehensive approach on personal data protection in the European Union, 4 November 2012.

is not permitted. Whether or not some of these contents (via the channels) are made available to anyone other than the IAP or a third party, they are available to someone other than the sender/recipient. The UK test is strict and requires both parties (sender and receiver) to consent.

The most controversial of all attempts by UK network owners to intercept users' communications without consent were the experiments conducted by the behavioural advertising company Phorm with the UK's largest IAP BT (and discussions with the next two largest, TalkTalk and Virgin Media[59]). Phorm employs a user-tracking system by which British Telecom and other IAPs intended to target users more effectively than Google. A variant of this technology was first deployed widely in US wireless IAPs.[60] Phorm operated a behavioural advertising system called WebWise, intending to offer its IAP and website clients a more accurate tracking of customers' Internet use, in order to more closely target advertising and other marketing via that data.

Phorm used DPI to take a copy of IAP subscribers' Web browsing, in order to insert targeted advertising. The original Phorm system trials by BT in 2006 and 2007 did not inform users or ask for their permission.[61] The government department responsible for interception of electronic communications was aware of, and tried to provide helpful regulatory guidance on, the trials and the behavioural advertising system. It emerged in April 2009 that the department, when contacted by Phorm in August 2007, had responded by asking 'If we agree this, and this becomes our position, do you think your clients and their prospective partners will be comforted?'[62] It appears that the consultations between the department and Phorm were extensive and amounted to forming a collaborative view of the law, with comments such as 'My personal view accords with yours, that even if it is "interception", which I am doubtful of, it is lawfully authorized under section 3 by virtue of the user's consent obtained in signing up to the IAPs' terms and conditions.'[63] In an email dated 22 January 2008, a Home Office official wrote again to Phorm and said: 'I should be grateful if you would review the attached document, and let me know what you think.'[64] The publication of this history of emails resulted in a debate in the House of Lords in 2009. Baroness Miller stated that:

[59] Williams (2009c).

[60] See www.pcworld.com/article/163740/us_lawmakers_target_deep_packet_inspection_in_privacy_bill.html (Accessed 16 September 2016).

[61] Dubious value is given to such permission in BT internal documents, see http://wikileaks.org/wiki/British_Telecom_Phorm_PageSense_External_Validation_report (Accessed 5 September 2016).

[62] BBC (2009).

[63] Marsden (2010), p. 79.

[64] *Ibid.*

> The fact the Home Office asks the very company they are worried is actually falling outside the laws whether the draft interpretation of the law is correct is completely bizarre.[65]

As a result of the legal controversy that followed when the trials were made public in early 2008, the IAPs and Phorm itself agreed to insert both notification and consent into any future trial or deployment of the technology, and BT did so for its third trial in December 2008. In legal terms, the system is not just contrary to permissions required in European privacy law under the 1995 and 2002 Directives, but also unlawful interception under the exclusively UK RIPA. In March 2008 the Foundation for Information Policy Research (FIPR) wrote to the Information Commissioner arguing that Phorm's system involved illegal interception contrary to RIPA.[66] Citizens' complaints about the use of behavioural advertising by Internet service providers were handled by the UK Information Commissioner's Office (ICO), the UK personal data protection authority and the police forces responsible for investigating cases of unlawful interception of communications. All had failed to adequately investigate the criminal complaints, in part due to the ICO's weak powers to fine aberrant providers. The UK's Information Commissioner ruled that a 'technical' breach of the law occurred in BT's 2006–2007 trials, and had strong reservations about the nature of the explanation provided for participating in BT's 2008 trial, but took no action.

Clayton, a security expert at Cambridge University, presented a report on the system, to which Phorm responded to ensure technical accuracy.[67] Clayton stated: 'Examining the detail makes it crystal clear that our earlier letter came to the right conclusion. Website data is being intercepted. The law of the land forbids this.'[68] The illegality stems not from breaching the Data Protection Act directly, but arises from the fact that the system intercepts Internet traffic. BT appeared to ignore the fact that they can only legalise their activity by getting express permission not just from their customers, but also from the Web hosts whose pages they intercept, and from the third parties who communicate with their customers through Web-based email, forums or social networking sites.

The EC takes the UK to court for net neutrality privacy breach prosecution

In response to UK citizens' complaints that the ICO was failing to prosecute Phorm and BT for breaching the 1995 Directive in not asking consent for the

[65] LINX Public Affairs (2009).
[66] See FIPR (2008).
[67] Clayton (2008).
[68] FIPR (2008).

original trial, the European Commission formally asked the UK government to explain why action had not been taken. The European Commission is tasked with monitoring Member States' implementation of European law, in this case Directive 2002/58/EC, the Electronic Privacy Directive (EPD).[69] The Data Protection Directive (DPD) of 1995 specifies that user consent must be 'freely given, specific and informed', a formula repeated in the EPD.[70] The critical test in both the EPD and the DPD is that subscribers have to opt for arrangements that may otherwise infringe their personal privacy, and that sensitive data must not be passed to third parties unless so authorised by subscribers and the data is anonymised. The EPD requires EU Member States to ensure confidentiality of the communications and related traffic data by prohibiting unlawful interception and surveillance unless the users concerned have consented to this.[71] Article 24 DPD requires Member States to establish appropriate sanctions in case of infringements. Article 28 requires that independent authorities must be charged with supervising implementation. These DPD provisions also apply to confidentiality of communications.

When the UK response received was unsatisfactory, the EC repeated its request for information in stronger terms. When that second response was unsatisfactory, in January 2009 the Commission threatened legal action[72] and launched legal action in an infringement procedure against the UK in April 2009.[73] Commissioner Reding declared:

> I call on the UK authorities to change their national laws ... This should allow the UK to respond more vigorously to new challenges to ePrivacy and personal data protection such as those that have arisen in the Phorm case.[74]

In October 2009 the Commission requested the UK authorities to amend their rules to comply with EU law, due to inadequate national legal implementation in three main areas:

- There was no independent national authority to supervise the interception of some communications, although the establishment of such authority is required under EPD and DPD, in particular to hear complaints regarding interception of communications;
- Existing UK law allowed the interception of communications not only where the relevant internet users have consented to this but also where the person intercepting the communications has 'reasonable grounds for believing' the

[69] Electronic Privacy Directive supplemented by COM(2004) 28.
[70] Directive 95/46/EC, Article 2(h).
[71] Directive 2002/58/EC (E-Privacy Directive), Article 5(1).
[72] Williams (2009b).
[73] See IP/09/570.
[74] IP/09/1626.

consent to intercept has freely been given under RIPA. RIPA obviously predates the EPD. This is contrary to the EPD, which defines consent as being 'freely given, specific and informed indication of a person's wishes' (Recital 17);

- UK laws prohibiting and providing sanctions in the case of unlawful interception were limited to intentional interception only, whereas EU law was wider, requiring Member States to impose penalties for any unlawful interception irrespective of whether it was committed intentionally or not. UK law did not correctly implement confidentiality of electronic communications, and powers to fine in sanctions for breaches by the UK Information Commissioner's Office (the UK personal data protection authority) were inadequate under Article 28 DPD.[75]

European laws designed to protect citizens' privacy and liberty also include the Framework Directive, which lays down the tasks of NRAs. These include cooperating with each other and the Commission in a transparent manner to ensure the development of consistent regulatory practice, contributing to a high level of protection of personal data and privacy and ensuring that the integrity and security of public communications networks are maintained.[76]

The referral of the UK to the European Court of Justice reflected the Commission's view that the UK was breaching its obligations under the DPD and EPD, implemented in the UK through the Data Protection Act 1998 and Privacy and Electronic Communications (EC Directive) Regulations 2003 respectively, stating 'the Commission considers that UK law does not comply with EU rules on consent to interception and on enforcement by supervisory authorities'.[77] The case therefore challenged much of the legitimacy of the UK communications privacy regime and its powers to enforce those rules, notably by the ICO and the police forces.

The European Commission closed the infringement case on 26 January 2012 in recognition that UK national legislation was amended to properly implement EU law on the confidentiality of communications such as email or internet browsing.[78] Following the Commission's 2010 decision to refer the case to the Court of Justice of the European Union (CJEU),[79] the UK amended RIPA, removing references to implied consent if the interceptor had 'reasonable grounds for believing' that consent had been granted. It also established a new sanction against unlawful interception in Section 1A and Schedule A1 of RIPA,[80] administered by the ICC, who has published guidance with

[75] *Ibid.*

[76] Directive 2002/21/EC (Framework Directive), Article 8(4)f.

[77] IP/10/121.

[78] See IP/12/60.

[79] See IP/10/1215.

[80] SI 2011/1340 Regulation of Investigatory Powers (Monetary Penalty Notices and Consents for Interceptions) Regulations 2011.

practical information on how it will exercise these new functions.[81] The maximum monetary penalty that can be imposed by a monetary penalty notice is £50,000 under the amended legislation. The ICC guidance notes at Paragraph 2.15 state that:

> The Commissioner shall consider serving a monetary penalty notice on a person only if, after investigation, he is satisfied that: the person has without lawful authority intercepted a communication; the conduct cannot be explained by an attempt to carry out an interception warrant; and the person has not committed an offence under section 1 of RIPA.

The criminal investigation into the Phorm trials was finally abandoned by the Crown Prosecution Service Complex Casework Centre on 8 April 2011, choreographed to match the precise day on which the legislative reform was announced. It argued there were:

> several public interest factors against prosecution:
>
> - BT and Phorm received considerable legal advice concerning the use of this software and were advised its use was unlikely to be contrary to section 1 of RIPA. The Home Office also provided informal advice that stated the same. Following the second trial, BT received further and conflicting legal advice that led to it halting the covert trials. As there was no evidence to suggest either company acted in bad faith, it could be reasonably argued that any offending was the result of an honest mistake or genuine misunderstanding of the law;
> - Both companies cooperated with the police investigation;
> - The behaviour in question is unlikely to be repeated. After the first two trials, BT conducted a further single, public trial of the technology (in late 2008). Phorm now requests the user's consent;
> - The trial was of limited duration and limited application. The data gathered was anonymised and processed without human intervention and later destroyed;
> - There has already been an investigation by a regulator, the Information Commissioner's Office, which concluded there was 'no evidence to suggest significant detriment to the individuals involved' and took no action;
> - There is no evidence to suggest that anyone affected by the trial suffered any loss or harm as a result;
> - Taking into account all of the above, a court would be likely to impose only a nominal penalty.[82]

Note that factors included assessments by the ICO, which was itself considered by the European Commission to have inadequate powers, including the lack of significant capacity to fine the parties for their illegal behaviour. It is

[81] Interception of Communications Commissioner (2011).
[82] Crown Prosecution Service (2011).

also noteworthy that in 2009–10 the ICO reprimanded the two IAPs that after discussions had decided not to trial Phorm's system, with both TalkTalk and Virgin Media being reprimanded for their interception of subscribers' communications, in experimental applications of anti-net-neutrality blocking of P2P and streaming services (which will likely become illegal under the GDPR). In relation to TalkTalk, the Information Commissioner stated: 'In the light of the public reaction to BT's trial of the proposed Webwise service, I am disappointed to note that this particular trial was not mentioned to my officials during the latest of our liaison meetings.'[83] It may be that it would in any case be illegal to access the type of details accessed by IAPs and Phorm even with subscribers' consent. The UK authorities did not prosecute for interception of confidential communications by IAPs, preferring to issue warnings.

Corporate interception on behalf of government

Academics and legal experts will continue to pore over the legislative and judicial response to mass surveillance by government in the period from 2013, building on the work before Snowden.[84] I focused in this chapter on its implications for private surveillance of the type used to infringe net neutrality. Rauhofer and the late Casper Bowden remind us that:

> Although EU data protection laws are designed to restrict the private actors handling that data from processing it in a way and for purposes that are unlawful, those laws have no effect on public bodies, including law enforcement and security agencies in third countries whose access to that data may be authorized by the laws of their own countries.[85]

Brown explains that:

> Following Edward Snowden's revelations of large-scale Internet surveillance by the US and UK governments, there has been broad discussion of the relative merits of national legal regimes intended to enable necessary and proportionate Internet surveillance by intelligence and law enforcement agencies … One important but under-discussed part of such regimes is a statutory requirement for telecommunications companies to make their networks 'wiretap-ready'.[86]

He provides examples from many countries, including the apparently obscure s.94 of the Telecommunications Act 1984. Critically, he explains that:

[83] Sir Christopher Graham quoted in Beaumont (2010).
[84] Birnhack (2012).
[85] Rauhofer and Bowden (2013).
[86] Brown (2013b).

> It enables much more sweeping surveillance than is possible using judicial or administrative warrants (or Mutual Legal Assistance Treaty requests) targeted at individuals or individual services. And by reducing the marginal cost of surveillance, it encourages greater use of it.

I reproduce the Table 8 from Brown as it affects case studies explored in Chapters 6–7 (accurate as at 2013 prior to the wave of legislation that followed the Snowden revelations). Note that most nations have similar laws, which make it a legal requirement that networks for security purposes permit exactly the type of interception that would be illegal for private purposes under net neutrality and other interception laws.

Brown concludes that the:

> European Court of Human Rights has not previously shied away from dealing with intelligence issues, commenting in Leander *v* Sweden on 'the risk that a system of secret surveillance for the protection of national security poses of undermining or even destroying democracy on the ground of defending it'. It is not inconceivable that the UK's sweeping Internet surveillance activities will be found, as the Court did in S. and Marper with the UK's National DNA Database, to 'constitute … a disproportionate interference' with privacy that 'cannot be regarded as necessary in a democratic society'.[87]

The frustrated censorious politicians who want the 'wise monkeys' to speak are both reviewing platform regulation and the E-Commerce Directive in 2016, though note that their 2010 review produced little change.[88] National laws favour their copyright industries, such as the Digital Economy Act 2010 in the United Kingdom or the HADOPI law in France.

Interception of communications has eventually been prosecuted when carried out by private investigators, such as those employed by Rupert Murdoch's newspapers in the celebrated 'phone hacking' affair which included computer hacking.[89] However, this has little relationship to net neutrality discussions, and Murdoch's part-owned IAP Sky Broadband has not been implicated in interception of its clients' communications. Note that individuals can bring complaints about alleged illegal interception by public authorities to the Investigatory Powers Tribunal, which publishes its most notable rulings on its website.[90] However, complaints about private parties including IAPs cannot be brought to the Tribunal, but instead to

87 Brown (2013b) (footnotes omitted).

88 See COM(2010) 245, COM(2011) 206, COM(2011) 942, COM(2012) 573, European Commission, SWD(2013) 153.

89 See BBC (2013).

90 See www.ipt-uk.com/judgments.asp?view=archive (Accessed 16 September 2016), noting 'That procedure runs no risk of disclosure of any information to any extent, or in any manner, that is contrary to or prejudicial to the matters referred to in section 69(6)(b) of RIPA and [SI 2000/2665 Investigatory Powers Tribunal Rules] rule 6(1).'

Table 8 Statutory requirement for IAP companies to make their networks 'wiretap-ready'

State	Law	Date
Brazil	Federal Law No. 9.296	1996
European Union	Council Resolution to implement similar lawful interception capability measures[a]	1996
France	Posts and Telecommunications Code §D.98-1	1996
Germany	Telecommunications Act §88/§110	1996/2004
India	Information Technology Act: Procedure and Safeguards for Interception, Monitoring, and Decryption of Information Rules	2009
Netherlands	Telecommunications Act §13	1998
UK	Telecommunications Act §94[b]	1984
	Regulation of Investigatory Powers Act §12	2000
US	Communications Assistance to Law Enforcement Act[c]	1994
	FISA Amendments Act §1881a[d]	2008

[a] Council Resolution of 17 January 1995 on the lawful interception of telecommunications.

[b] Words in s.94(1) substituted by Communications Act 2003, (c.21), ss.406, 408, 411, (Sch. 17 para. 70(2)) (with Sch. 18); SI 2003/1900, arts. 1(2), 2, 3(1), Sch. 1, Sch. 2 with Art. 3(2) (as amended by SI 2003/3142 Office of Communications Act 2002 (Commencement No. 3) and Communications Act 2003 (Commencement No. 2) Order 2003, Article 1(3)).

[c] Communications Assistance to Law Enforcement Act of 1994 Pub. L. No. 103–414, 108 Stat. 4279.

[d] Foreign Intelligence Surveillance Act of 1978 Amendments Act of 2008, H.R. 6304, Stat. 2436, Public Law 110–261.

Source: Brown (2013b).

the ICO or the police. This confirms the analysis by Bowden[91] that UK law as implemented fails to protect citizens from interception, whether by government or private company, and heralding the European Parliament 2013 decision to:

> hold a full inquiry into US surveillance programmes, including the bugging of EU premises … It urged them to examine whether those programmes are compatible with EU law. This element of the inquiry could open up a number of sensitive Signals Intelligence relationships between Europe and the US – particularly the close operational partnership enjoyed by the US, Germany and the UK.[92]

[91] Bowden (2013b).

[92] Davies (2013).

It should be noted that criminal law enters the net neutrality debate in the field of counterfeiting and copyright. Civil liability includes potential to pay damages for every copyrighted item copied, for attorney fees for copyright holders pursuing the case, and for exemplary damages for such 'wilful' abuse of copyright. By contrast, until 2012 it was assumed that criminal liability would be limited as 'in exercising its power to render criminal certain forms of copyright infringement, [the United States] has acted with exceeding caution'.[93] However, the proposed extradition to the United States following the January 2012 arrest of Megaupload executives in New Zealand has caused some surprise and uncertainty in the application of criminal law,[94] as it follows a 2005 restatement of enforcement policy.[95] The 'wilful' requirement in criminal law must be proved beyond reasonable doubt.[96] Nevertheless, a more aggressive prosecution of counterfeiting and other 'piracy' (*sic*) websites was signalled in 2011 with the taking down of domain names belonging to suspected overseas 'rogue sites'.[97] The cooperation of several national police forces in the Megaupload case indicates a more general trend towards aggressive policing of counterfeiting. This overtook the controversies in the latter half of 2011 over US Congress and Senate versions of a more aggressive anti-infringement Bill.[98]

As part of the Digital Single Market initiative launched on 6 May 2015,[99] the EC committed to a 'comprehensive assessment on the role of platforms', launched on 24 September 2015. As the official title indicates, the consultation is extremely wide: 'Regulatory environment for platforms, online intermediaries, data and cloud computing and the collaborative economy'. In the discussions to amend the E-Communications Framework, large well-resourced European incumbent IAPs see the opportunity to make common cause with mobile operators (and others), in an alliance to prevent transparency and permit filtering. The regulation of the Internet that is rapidly taking place is being driven – unquestionably – by European politicians for public safety reasons. They are erecting entry barriers with the connivance of the incumbent players, with potentially enormous consequences for free speech, for free competition and for individual expression. This may be the correct policy option for a safer Internet policy (to prevent exposing children to illegal and/or offensive content, and to counter serious criminal activity), though it signals an abrupt change from the Open Internet.

[93] *Dowling v. United States* (1985) 473 US 207, 222.
[94] See Department of Justice (2012).
[95] Goldstone and O'Leary (2005).
[96] See US Attorney (2006).
[97] See *United States v. Domain Names* (defendants in rem), Affidavit in Support of Application for Seizure Warrant Pursuant to 18 USC §§2323, 981, (31 January 2011) (No 18 MAG 262).
[98] See Law Professors (2011).
[99] COM(2015) 192.

It is therefore vital that regulators address the question of the proper approach to net neutrality to prevent harm to the current Internet, as well as begin to address the heavier questions of positive – or tiered – breaches of network neutrality. The rise in the number of people using encrypted email, Facebook, MySpace, Wikipedia, Skype, Instant Messaging and other applications has extended so far into mass participation that it has truly affected society and the economy in all its facets. Small businesses and solo home-based workers depend on this tool as a vital part of their participation in the economy. The promise of virtual worlds and massive online collaboration is to extend this impact even further by 2020.

The EC suggests that:

> 'Online platform' refers to an undertaking operating in two (or multi)-sided markets, which uses the Internet to enable interactions between two or more distinct but interdependent groups of users so as to generate value for at least one of the groups. Certain platforms also qualify as Intermediary service providers. [Examples are]:
>
> - general internet search engines (e.g. Google, Bing),
> - specialised search tools (e.g. Google Shopping, Kelkoo, Twenga, Google Local, TripAdvisor, Yelp),
> - location-based business directories or some maps (e.g. Google or Bing Maps),
> - news aggregators (e.g. Google News),
> - online market places (e.g. Amazon, eBay, Allegro, Booking.com),
> - audio-visual and music platforms (e.g. Deezer, Spotify, NetFlix, Canal play, Apple TV),
> - video sharing platforms (e.g. YouTube, Dailymotion),
> - payment systems (e.g. PayPal, Apple Pay),
> - social networks (e.g. Facebook, Linkedin, Twitter, Tuenti),
> - app stores (e.g. Apple App Store, Google Play) or
> - collaborative economy platforms (e.g. AirBnB, Uber, Taskrabbit, Bla-bla car).
> - Internet access providers fall outside the scope of this definition.[100]

'Platform' is a generic term that is applied to very different business models: examples included are aggregators/services such as NetFlix and Spotify, which provide a service to consumers but do not connect parties to create a market (as Amazon or eBay do). It is unclear at the time of writing whether such a vast consultation will result in a revision to the ECD, national tweaks to laws to protect domestic services from overseas deregulated competitors (as with local taxi or hotel associations in France and Italy) or no response at all.

[100] European Commission, Public consultation on the regulatory environment for platforms, online intermediaries, data and cloud computing and the collaborative economy, 24 September 2015, with background information at European Commission, Background for the public consultation on the regulatory environment for platforms, online intermediaries, data and cloud computing and the collaborative economy, 2015.

Given that the antitrust case taken by the EC against Google remains unresolved in 2016, it is clear that another academic may need to write a new book on platform regulation in 2017 or later. Fortunately, the consultation expressly excludes IAPs, which means speculation in this book can be limited.

Conclusion: regulatory problems in implementing net neutrality

I continue to argue for a prosumer law (consumer- and citizen-orientated) intervention to protect privacy in net neutrality[101]. That depends on passing regulations to prevent unregulated non-transparent controls exerted over traffic via DPI equipment, whether imposed by IAPs for financial advantage or by governments eager to use this new technology to filter, censor and enforce copyright against their citizens. Unravelling the previous non-liability regime in which IAPs could act as 'wise monkeys' risks removing the wisdom and efficiency of that approach in permitting the free flow of information for economic and social advantage. These conclusions support a regulatory regime involving reporting requirements and co-regulation with, as far as is possible, market-based solutions. Regulatory monitoring of potential abuses, including strengthening investigatory capacity and transparency for end users, is a solution that maintains maximum flexibility and policy choice, while ensuring that any abuses can be quickly detected and dealt with appropriately. Solutions may be international as well as local, and international coordination of best practice and knowledge will enable national regulators to keep up with the technology 'arms race'.

We should not entrench a privacy-invasive 'Lex Monopolium' at the expense of an Open Internet, and the choice is not that drastic: innovation and investment can be encouraged by co-regulatory transparency principles, backed up by a regulator with sufficient comprehension and research into the issues, and teeth that are sharp enough, to make a real political commitment to intervene to protect privacy where economic or social interests dictate. Net neutrality certainly provides an excellent platform to create this wider and better-informed discussion, and to prevent harm to the current intermediary role of IAPs as wise monkeys. In the following chapter, I examine the extent to which the UK regulator has followed net neutrality principles in practice.

[101] See Brown and Marsden (2013a).

6

Open Internet self-regulation in the UK

> The term 'Open Internet' is one that the digital minister, Ed Vaizey and I agreed some years ago was a better term for the UK than the American term – net neutrality.
>
> Richard Hooper[1]

Regulation of Internet access is a difficult process, as we saw in the case of the UK in Chapters 2–3. While regulator Ofcom has been groundbreaking in its research into traffic management measurement, it has tried to solve neutrality problems with co-regulation. This process needs a brief definition, and while I have written extensively on the matter, here I use the term defined by Lord Justice Leveson in the 2,500-page report into 'phone hacking'. He states:

> Co-regulation means any form of self-regulation with some sort of external, independent, incentives, oversight or form of backstop [including recognition] of a self-regulatory body by Government, law or a statutory regulator; approval of codes by Government or a statutory regulator; and compulsory membership or funding arrangements.[2]

He adds detail:

> a co-regulatory model can encompass anything that could be done under self-regulation whilst adding an element of compulsion to make effective enforcement possible. On the other hand, it can encompass anything that could be done by a statutory regulator but put relevant decision making in the hands of those closest to the industry, and rigorously separate from Government, to seek to gain the benefits of self-regulation without losing the benefits of statutory backing. This is a model that is much in use in the UK.[3]

[1] Hooper (2015).
[2] Leveson (2012), p. 1739, para. 2.31.
[3] *Ibid.*, p. 1741, para. 2.38.

Ofcom responded with a combination of:

- denial that there is a problem
- an acceptance and willingness to deal with those problems that were shown to have emerged
- regulation of customer switching problems between IAPs
- regulation of video on broadband offered by the public service broadcaster BBC and
- a light touch attempt to persuade IAPs to offer greater transparency to users.

The first point is denial that a problem existed until 2006, even though BT admitted discrimination in 2001.[4] The second is delay: attempting switching/transparency solutions throughout the period since 2006. This achieved some success: changes in General Conditions in 2006–08, supplemented in 2011.[5] Ofcom also pioneered Quality of Experience measurement, using SamKnows to measure performance from 2010. The third element is to degrade and obfuscate the debate, from a regulatory responsibility to enforce net neutrality to an industry agreement to discuss the 'Open Internet'. Government and Ofcom funnelled regulation into the co-regulatory forum, the Broadband Stakeholders Group (BSG), an industry forum funded in part by government, designing a Broadband Speed Code of Conduct over the period 2008–12.

I have previously described this as a 'corporatist conceit', and it was not agreed to by all IAPs until 2015, only weeks before the critical negotiations over Directive 2015/2120. This was at the very least a happy coincidence for the minister, as the UK attempted to show its European partners that it was successful in achieving what it insisted was 'self-regulation'. I argue that it was co-regulation based on the inevitability that, had it failed, Ofcom would have imposed further changes in the General Conditions for IAPs to operate in the UK. I examine these three elements – denial to 2007, delay to 2009, degrading to 2015 – in turn.

In Section 1, I introduce the regulator and explain why it has not taken on any disputes over net neutrality despite substantial user disquiet about throttling and blocking of content. As explained in Chapters 2 and 3, the UK broadband market has relatively low competitive intensity with only one wholesale network for the majority of the population, and has experienced slower rollout to much lower speeds than the United States, South Korea, Norway or the Netherlands. As a result, government and regulator concern has been to encourage investment in faster services by the former monopoly BT. Direct state aid for

[4] McCarthy (2001).

[5] General Conditions replaced specific licences for operators in 2003, and are based on Directive 2002/20/EC, pp. 21–32. Ofcom provides a useful guide: Ofcom, Guidance on individual General Conditions: Links to Guidance on the General Conditions (undated).

the former monopoly amounted to almost £1.5 billion in the period 2013–15. As we saw, problems of congestion have plagued the network since at least 2005, which were brought to the direct attention of Ofcom in person in 2006 by both IAP TalkTalk, which admitted to consumer anger at its throttling of P2P content (see Chapter 3), and the state broadcaster BBC, whose P2P video streaming was throttled by BT (Chapter 2).

The regulator and government response has been to insist that self-regulation could produce greater transparency for end users, and regulation should be limited to the theoretical possibility to switch should blocking and throttling occur. I examine these in turn in Section 1.

I then explain the slow progress towards a Code of Practice for greater transparency under the auspices of the government-industry partnership, BSG. IAPs had been persuaded by Ofcom to sign a code of conduct on advertising broadband speed and congestion in 2008, though this was inadequate and was then replaced by the BSG Code from 2011–13. Continued government and Ofcom light-touch review of this self-regulation was critiqued by amongst others Sir Tim Berners-Lee. BSG chair Richard Hooper was always pleased to announce: 'The term "Open Internet' is one that the digital minister, Ed Vaizey and I agreed some years ago was a better term for the UK than the American term – net neutrality.'[6] At the IGF 2015, the rest of the world was happy to debate net neutrality, with a European declaration proudly stating the net neutrality law that had just been voted through into law.[7] However, in contrast to that historical commitment to an Open Internet, the UK pursued a TMP Transparency Code 2011 and an Open Internet Code of Practice 2012.

Introduction to UK net neutrality policy

UK network neutrality policy is influenced both by its regulator, Ofcom, and its particular and unique network topology. I first introduce the regulator and explain why it has not taken on any disputes over net neutrality despite substantial user disquiet about throttling and blocking of content. British governments have consistently valued broadband deployment in a semi-competitive market above the needs of end users for open content provision on those networks. This dates to the provision of cable and satellite television in the 1980s, and the convergence between television and telephone services in the 1990s.[8] The demise of the pioneering telecom regulator Oftel to be replaced by Ofcom

[6] Hooper (2015).

[7] European Commission, Internet Governance Forum 2015 Joint Declaration.

[8] See the pioneering Oftel, Beyond the telephone, the television and the PC, 1998, especially point 16: 'Oftel recommends that new rules on opening up interfaces and making infrastructure available to third parties on fair, reasonable and non-discriminatory terms are required, so that others can supply services over existing infrastructure while rewarding investment in infrastructure creation.'

took place gradually over eight years from 1995 (the Office of Communications Act 2002 and Communications Act 2003 fully implemented the new regulator), during which the UK pioneering position in telecoms regulation was lost. By 2003 the UK regulatory landscape had produced only 128,000 unbundled local loop lines, far behind European equivalents.

Ofcom has been the regulator of communications since the end of 2003 (Office of Communications Act 2002), merging the formerly separate regulators for broadcasting, telecommunications, spectrum and radio. Section 3(1) of the Communications Act 2003 (2003 Act) states:

> It shall be the principal duty of Ofcom, in carrying out their functions –
>
> (a) to further the interests of citizens in relation to communications matters; and
> (b) to further the interests of consumers in relevant markets, where appropriate by promoting competition.
> (3) In performing their duties under subsection (1), Ofcom must have regard, in all cases, to –
> (a) the principles under which regulatory activities should be transparent, accountable, proportionate, consistent and targeted only at cases in which action is needed; and
> (b) any other principles appearing to Ofcom to represent the best regulatory practice.

There is therefore a bold statement of the dual nature of the individual member of society: first as a citizen whose needs for information and communications services are paramount, and then equally as a consumer, whose greatest benefits may be secured 'where appropriate by promoting competition'. This is far from a neo-classical view of regulation and competition. Ofcom in practice has tended towards the competition perspective with no intervention until harm is proved, wherever possible, and a 'light touch' in regulatory matters, based on the initial principles established by Lord Currie, its first chair, and its chief executive from 2006 to 2014, Ed Richards.

In the hiatus between the decision to create a new 'converged' communications regulator in 2001 and the actual launch of Ofcom, early net neutrality cases were ignored by the outgoing regulator Oftel, whose staff was preoccupied with finding new jobs, and the farce of Local Loop Unbundling (LLU) in the UK to introduce broadband.[9] In 2007 Ofcom assumed that only the US need worry about net neutrality. By 2009 Richards had publicly disowned the expression 'light touch' in the wake of the economic recession and evidence of disastrous light touch regulation in the financial services in London's markets, stating 'I don't like the expression "light touch" regulator. We try to be as unintrusive as we can be,

[9] Marsden (2005), pp. 6–19.

not to intervene unless we have to, but if we have to and there's a public interest in intervening, we are willing to do so swiftly and effectively.'[10] He reflected that:

> The first five years of Ofcom's life were characterized by focussing on the fundamental problem of the previous twenty years – regulation of the enduring bottlenecks, ensuring downstream equivalence and, through it, effective competition. More recently we have also significantly raised our game in ensuring effective consumer empowerment and consumer protection.

These were very real pro-consumer policy rhetorical shifts within Ofcom in 2009. They reflected a claimed significant realignment with the statutory duties, and a reflection on the previous 'light touch' reliance on market forces to provide for consumers/citizens.

Ofcom is, unusually for a sectoral regulator, required to write a justification for any decision occasioning a conflict between its treatment of the citizen and the consumer.[11] The objective for Ofcom under Section 3(2) of the 2003 Act is to make clear that a regulator's role is far wider than simply acting as a competition tribunal between telecoms providers. Section 4 includes in Ofcom's functions the 'desirability of promoting': purposes of public service television broadcasting; competition in relevant markets; facilitating the development and use of effective forms of self-regulation; investment and innovation in relevant markets; availability and use of high-speed data transfer services throughout the UK; consumers and other users. It is also charged (presumably intended to cover rights to reply and balanced and pluralistic voices in the media) with 'the need to secure that the application in the case of television and radio services of standards … is in the manner that best guarantees an appropriate level of freedom of expression'. Ofcom is increasingly resorting to forms of co-regulation to balance these aims.

Administrative development of Ofcom

From 2004, Ofcom and government policies were designed to achieve:

- functional separation BT local loop and other businesses in the Telecoms Strategic Review 2004/05 and BT Undertakings 2006, resulting in lower

[10] See Marsden (2010), pp. 161–162.

[11] The Communications Act 2003 Section 3 states: (8) Where Ofcom resolve a conflict in an important case between their duties under paragraphs (a) and (b) of subsection (1), they must publish a statement setting out – (a) the nature of the conflict; (b) the manner in which they have decided to resolve it; and (c) the reasons for their decision to resolve it in that manner. (9) Where Ofcom are required to publish a statement under subsection (8), they must – (a) publish it as soon as possible after making their decision but not while they would (apart from a statutory requirement to publish) be subject to an obligation not to publish a matter that needs to be included in the statement; and (b) so publish it in such manner as they consider appropriate for bringing it to the attention of the persons who, in Ofcom's opinion, are likely to be affected by the matters to which the decision relates.

local loop access prices for BT competitors, hastening BT roll-out of ADSL and then VDSL.

- a 'Transparency and Switching' solution to any consumer issues.

Because the inventor of the World Wide Web, Sir Tim Berners-Lee is British, the greatest UK contribution to the consumer Internet experience is associated with his work. In consequence, much public and especially political debate on net neutrality is reactive to Berners-Lee's pronouncements. In 2006 Berners-Lee explained: 'There have been suggestions that we don't need legislation because we haven't had it. These are nonsense, because in fact we have had net neutrality in the past – it is only recently that real explicit threats have occurred.'[12] Berners-Lee was particularly adamant that he does not wish to see the prohibition of QoS because that is precisely the claim made by some US net neutrality advocates – and opposed by the network engineering community.

The chief executive of TalkTalk revealed at Ofcom's October 2006 international conference that he was receiving death threats from online gamers after their connections were throttled during evening peak hours:

> We shape traffic to restrict P2P users. I get hate mail at home from people when that means we restrict their ability to play games. I've got two people that have said they're going to kill me as a result of not allowing them to play certain games. From our point of view, it's not about security, it's about trying to figure out what type of traffic it is.[13]

Though expressed partly in jest, in response to a question to the panel from me, Ofcom revealed that ISPs were receiving complaints when they deliberately throttled users. This revelation coincided with a study commissioned by Ofcom, and widely publicised by it, into traffic management for video on the Internet, warning that mobile and IPTV would become IAP-owned or affiliated walled gardens without regulatory intervention to ensure openness.[14] British Telecom, the former UK monopoly (whose unbundled lines TalkTalk retails), was throttling the BBC's iPlayer service regularly from 2006.[15] The BBC is also a key player in UK net neutrality discussions, being both regulated by Ofcom for its commercial services and itself a major streaming media distributor via its iPlayer service. Network discrimination had arrived, if not net neutrality regulation. Despite this and other evidence, Ofcom confined itself to measuring IAP broadband performance, making it easier for consumers to switch to rival providers.[16]

[12] Berners-Lee (2006).
[13] Dunstone (2006).
[14] Marsden *et al.* (2006).
[15] On these developments, see Marsden (2010), Chapters 2 and 6.
[16] Kiedrowski (2007).

The initial Ofcom position was that there was no evidence of discrimination, no complaint from other IAPs, and therefore no need to examine the problem. In its 2006 response to the Content Online consultation, Ofcom defined net neutrality as permitting no QoS on the network, itself terming this an 'extreme' position:

> Net neutrality implies there is absolutely minimal differentiation. It is therefore at the extreme end of this continuum of different approaches.[17]

This meant that institutionally Ofcom had decided there was little evidence of a problem, by taking that extreme definition. It explained that competition will settle the issue for IAPs: 'if a single operator without SMP were to introduce charging for the delivery of third party content services, or to block specific services, consumers would be able to move supplier.'[18] This is despite mounting evidence that IAPs were non-transparent in their blocking and throttling, and that IAPs were preventing consumers from switching by refusing to release their MAC (Migration Authorisation Code) numbers. Ofcom admitted:

> If this [easy switching] is not the case, then there may be a role for regulatory intervention to protect consumer interests. However, any intervention would be best focussed on addressing the lack of consumer information, consumer empowerment or migrations processes.[19]

That became exactly its policy throughout 2007–08. Evidence was building that BitTorrent throttling was preventing downloads of updates to hugely popular multi-player online game *World of Warcraft* and that subscribers were unable to switch between providers in response. Ofcom's Kiedrowski stated:

> We could apply Article 5.1 of the Access and Interconnection Directive, which allows NRAs to impose ex ante obligations on operators to ensure E2E connectivity, without the need to find SMP. We could use our powers derived from the European Framework which enable us to require suppliers (even non SMP suppliers) to comply with various general conditions in order to take part in the market to address particular issues, for example, in relation to information transparency.[20]

In 2005 it became clear that many IAPs were not cooperating in letting unhappy customers switch away to better providers, whether for throttling or (more likely) other issues. This came into sharp focus in the period before the collapse of abusive IAP E7even UK Ltd on 3 July 2006,[21] but it had been an

[17] Ofcom, Response to the European Commission Consultation on Content Online in the Single Market, 2006.

[18] *Ibid.*

[19] *Ibid.*

[20] Kiedrowski (2007).

[21] Williams (2006).

issue during the crisis at IAP Bulldog in summer 2005.[22] Ofcom took regulatory action to ensure customers could migrate ('switch' or 'churn' in telecoms terminology) between IAPs by ensuring portability of MACs.[23] In December 2006, after public consultation, Ofcom imposed a new General Condition, GC22, governing the obligations of broadband providers to switching customers. GC22 entered into force on 14 February 2007. Ofcom stated: 'Under GC22, all broadband providers must use the MAC Broadband Migrations Process ("the MAC process") if they receive a migration request from an end-user, customer or another provider.' An 18-month enforcement action ensured, which Ofcom stated would force IAPs to comply with customer requests. The length of the action shows how difficult it proved to achieve the basic requirement that customers be able to switch provider – if they had managed to work out why their service was so poor in the first place.

Ofcom engaged in some strenuous IAP arm-twisting to arrive at a mediated self-regulatory position whereby IAPs did agree to provide the information on broadband speeds (if not necessarily service) in their Code released on 5 June 2008. Ofcom could use its powers under the general conditions to require all – even non-dominant – suppliers to provide better consumer information about their products, if self-regulation continued to be inadequate. Ofcom warned IAPs that a failure to continue to improve transparency could result in a regulatory action under the General Conditions of their authority to offer services. Point 39 of the Code states:

> Where ISPs apply traffic-management and shaping policies, they should publish on their website, in a clear and easily accessible form, information on the restrictions applied. This should include the types of applications, services and protocols that are affected and specific information on peak traffic periods.[24]

Ofcom warned IAPs that 'Ofcom also intends to monitor compliance with the Code through a number of methods including, but not limited to, carrying out regular mystery shopping exercises by Ofcom itself or its agents.'[25] The Code states that IAPs are required to:

- provide consumers at the point of sale with an accurate estimate of the maximum speed that their line can support;
- explain clearly and simply how technical factors may slow down speeds and give help and advice to consumers to improve the situation at home;

[22] Richardson (2005).

[23] Ofcom, CW/00946/02/07 Own-Initiative Enforcement Programme to Give Effect to General Condition 22), 2007.

[24] Ofcom, Voluntary Code of Practice: Broadband Speeds, 2008.

[25] *Ibid.*

- offer an alternative package (if there is one) without any penalties, if the actual speed is a lot lower than the original estimate; and
- explain fair usage policies clearly and alert consumers when they have been breached.

Ofcom's approach remained committed to light touch self-regulation where possible:

> Within the UK, the need for specific regulation is likely to be lower, with ISPs and VoIP providers working together through industry bodies to agree a self-regulatory approach to providing consumers with transparency on whether service prioritisation or quality of service charging is being applied.[26]

Richards warned that 'The shibboleth of net neutrality should not be allowed to become an obstacle or a distraction to investment in NGNs [Next Generation Networks] in the UK.'[27] After 2007, domestic policy towards broadband was driven first by the work of the government-funded quasi-independent industry group BSG[28] and then a series of policy discussions held in 2008, called the 'Convergence Thinktank.'[29] The BSG recommended to government:

> Good solutions need to be found that align the interests of operators with upstream content and service providers and end consumers whilst mitigating concerns about blocking or degrading third party applications and services.[30]

The Thinktank was to inform the Digital Britain report,[31] drawn up by former chief executive of Ofcom (2003–06) (Lord) Stephen Carter, who was appointed communications minister in 2008. It was written as a draft report in January 2009, before the final report was presented to the Cabinet on 16 June 2009. Concurrently, the Ciao Review to the UK Treasury of September 2008 was designed to examine how far deregulation and market forces could produce Next Generation Access (NGA) solutions.[32] Carter appointed a list of business advisors on the Digital Britain report. IAPs are given carte blanche to breach net neutrality under Action 2:

> ISPs can take action to manage the flow of data – the traffic – on their networks to retain levels of service to users or for other reasons.

[26] Ofcom, Response to the European Commission Consultation on Content Online in the Single Market, 2006, para. 3.72.

[27] Richards (2008).

[28] See www.broadbanduk.org/ (Accessed 16 September 2016).

[29] See http://webarchive.nationalarchives.gov.uk/20121204113822/http://www.culture.gov.uk/Convergence/index.html (Accessed 16 September 2016).

[30] Broadband Stakeholder Group (2009).

[31] Department for Culture Media and Sport (2009), Chapter 4: Creative Industries in the Digital World.

[32] HM Treasury (2009).

> The concept of so-called 'net neutrality' requires those managing a network to refrain from taking action to manage traffic on that network. It also prevents giving to the delivery of any one service preference over the delivery of others. Net neutrality is sometimes cited by various parties in defence of internet freedom, innovation and consumer choice. The debate over possible legislation in pursuit of this goal has been stronger in the US than in the UK.
>
> Ofcom has in the past acknowledged the claims in the debate but have [*sic*] also acknowledged that ISPs might in future wish to offer guaranteed service levels to content providers in exchange for increased fees. In turn this could lead to differentiation of offers and promote investment in higher-speed access networks.
>
> Net neutrality regulation might prevent this sort of innovation. Ofcom has stated that provided consumers are properly informed, such new business models could be an important part of the investment case for NGA, provided consumers are properly informed.
>
> On the same basis, the Government has yet to see a case for legislation in favour of net neutrality.
>
> In consequence, unless Ofcom find network operators or ISPs to have Significant Market Power and justify intervention on competition grounds, traffic management will not be prevented.

Some trade press journalists picked up on the decision to substitute net discrimination for any actual government support for roll-out:

> In the UK, net neutrality was stillborn as an issue, but Carter was happy today to give its corpse a kick. As well as advocating tiered content delivery, he backed 'traffic management'; the somewhat euphemistic industry term for *BitTorrent* throttling.[33]

The final Digital Britain report instructs Ofcom to tell IAPs to throttle connections of users suspected of illegal file sharing:

> Ofcom will be placed under a duty to take steps aimed at reducing online copyright infringement. Specifically they will be required to place obligations on ISPs to require them: to notify alleged infringers of rights (subject to reasonable levels of proof from rights-holders) that their conduct is unlawful; and to collect anonymised information on serious repeat infringers (derived from their notification activities), to be made available to rights-holders together with personal details on receipt of a court order.[34]

Ofcom would also be given powers to block infringers' connections without a court order, subject to three notifications of infringement. This power would be triggered if the notification process has not been successful after a year in reducing infringement by 70 per cent of the number of people notified. This

[33] Williams (2009a).

[34] Department for Culture Media and Sport (2009), paras 24 and 28.

provision has never been implemented, and warning letters were to be delayed until after the 2015 General Election to the British Parliament. Whether Ofcom will reach the 'throttling point' under even the 2015–2020 government must be open to severe doubt.

Government net neutrality policy 2010

The European laws which became effective in Member States in May 2011[35] stated that Member States may take action to ensure particular content is not discriminated against directly (by blocking or slowing it), or indirectly (by speeding up services only for content affiliated with the IAP). The reality is that the 2009 Declaration on Net Neutrality, helpful though it is in clarifying the legal situation, relies heavily on implementation at national level and proactive monitoring by the Commission itself. Nevertheless, it laid out the principle of openness and net neutrality. The Member States largely opposed the Declaration, and it was only appended as a sop to the European Parliament, which had taken up the consumerist and democratic cause of neutrality, much to the annoyance of the telecoms technical community and economists who insisted it was solving a problem which did not exist.

The devastating long recession of 2008–12 was in 2010 impacting government, after a two-year interregnum under Gordon Brown's 'dead man walking' government, which ended with the historic hung Parliament and coalition of May 2010. The Conservatives, the largest coalition party, had proposed abolishing Ofcom or at least severely reducing its policy-making powers, returning many such policy functions to where they properly belong in government ministries. Thus the super-regulator's wings were to be clipped, along with those of its parent department, by 25 per cent.[36] Ofcom survived thanks to an involuntary decision to reduce headcount by over 150, about 25 per cent, and generally to reduce costs by 23 per cent. Between May 2010 and the end of that year, pre-redundancy letters were sent to 400 employees, or the majority of Ofcom staff. Those who took redundancy included key net neutrality policymakers. As a result, policy focus returned to the ministry, which itself was in upheaval following the Murdoch phone-hacking revelations and an indiscrete conversation by industry minister Vincent Cable, which forced broadband/telecoms functions to move from the Department

[35] Directive 2009/136/EC (Citizens Rights Directive) and Directive 2009/140/EC (Better Regulation Directive).

[36] HM Treasury (2010): 'Over the course of the Spending Review period, the Department for Business, Innovation and Skills (BIS) will reduce its resource budget by 25 per cent. Taking into account anticipated receipts, the cut to capital spending by 2014–15 will be 44 per cent. The Department's Administration budget will be reduced by 40 per cent.'

for Business to the Department for Culture in early 2011.[37] UK policy making on issues such as net neutrality was in a state of suspended animation as key actors faced institutional crises. Alongside these Ofcom and ministry upheavals (exacerbated by the new ministerial home of broadband being in general crisis dealing with BBC governance and the London Olympics 2012), the Communications Consumer Panel was reduced from a semi-detached, semi-autonomous consumer voice to a rump of a few non-experts as a result of similar funding cuts.

When ministries and regulators were taking austerity-induced 25 per cent cuts in staff, it is hardly surprising that they had no regard to reforming and strengthening a perceived 'luxury' policy such as net neutrality. Ed Richards' July 2010 speech was certainly anti-interventionist on openness and interoperability:

> Here in the UK, there have been no formal complaints about anti-competitive discrimination, although there have been a number of modest disagreements between content/service providers and ISPs/mobile operators. It is in this vein that we do not currently see a compelling reason for preventing, ex ante, all forms of discrimination using our sector-specific regulatory powers. But if genuine problems of anti-competitive practices in relation to traffic management emerged, we would of course have the ability to intervene applying our full range of ex post competition powers as appropriate.[38]

Formal complaints were classified as those made by an IAP about another IAP, not those by content providers or consumers. The death threats to TalkTalk's chairman, which he discussed at the 2006 conference at Ofcom, were therefore 'informal'. Note, too, the discussion of 'modest disagreements' as referring to the BT and other IAP blocking of BBC streamed video in 2006–08. The solution *ex post* is competition – which as we saw in Chapter 2 takes effect for Ofcom, based on the various disputes litigated by IAPs, long after the patient has died. Richards continues:

> This allows us to take a measured approach, allowing certain practices – such as permitting operators and ISPs to set differentiated [QoS] – which may prove beneficial to consumers, but which could be caught by a blanket prohibition, whilst at the same time being able to take effective action to curb any genuinely anti-competitive practices that may emerge.
>
> Although the evidence at this stage suggests a blanket prohibition is undesirable, our initial stance in this debate is that consumer transparency must be guaranteed wherever traffic management occurs ... Developing some basic principles around transparency, and ensuring that operators and ISPs comply with these principles, is consistent with our broader functions and duties as a sector regulator.[39]

[37] Marsden (2014a), p. 11.
[38] Richards (2010).
[39] *Ibid.*

The analysis of behavioural economics, 'nudging' and transparency ends the speech, as analysed in Chapter 2. From 2010 to 2015 Ofcom would work to nudge IAPs towards self-regulation and transparency.

Consultation was needed on the implementation of net neutrality by the deadline of 2011.[40] The European Commission closed its consultation period on 30 September 2010. The club of national regulators, BEREC, met on 30 September 2010 to discuss their response to the European Commission on net neutrality.[41] Ofcom's 'so-called net neutrality' consultation closed on 9 September 2010.[42] UK government minister for communications Ed Vaizey spoke on 17 November 2010, stating the government's opposition to regulated net neutrality.[43] He argued that high speed broadband required 'massive investment, and it may also mean networks and the traffic that flows over them are increasingly managed'. He stated that 'ISPs should be allowed to manage their networks' and do already, and Ofcom's 'transparency and switching' solution should be continued: 'it is important that consumers know exactly what service they are buying and have the ability to switch services should the nature of their service change.' He – or his speechwriters – also argued that they must permit possible 'evolution of a two sided market where consumers and content providers could choose to pay for differing levels of quality of service.' He stated:

> This Government is no fan of regulation and we should only intervene when it is clearly necessary ... Quality of service guidelines [in Directive 2009/136/EC] are back stop powers. Competition in the market, combined with transparency, the ability to switch ... should render such intervention unnecessary.

He concluded:

> there is no need for intervention. There is broad agreement on the need for traffic management; and there is broad agreement that there is not yet evidence of any impact either on competition or consumers from traffic management.

UK implementation of the 2009 Directives in 2011 followed this approach of ensuring bare minimum powers for Ofcom to impose QoS guidelines, which in any case the minister made plain they were not to impose.

Unsurprisingly, Sir Tim Berners-Lee (universally known as TBL) was in vehement disagreement with this extremely negative position towards net neutrality[44] – as were gamers, BBC iPlayer viewers, Skype users and other consumers

[40] See Cabinet Office (2010): 'In line with the Government's determination to tackle Britain's deficit, the Cabinet Office has announced today that it will reduce its core resource budget by 35% in real terms, from £280m in 2010–11 to £200m by 2014–15.'

[41] BEREC, BoR (10) 42.

[42] Ofcom, Traffic Management and net neutrality, 2010a.

[43] Vaizey (2010a).

[44] Berners-Lee (2010).

who had seen content throttled for a decade. TBL stated that net neutrality was under threat, and that threatened the openness of the World Wide Web and human online communication. Vaizey phoned TBL at home the Sunday after his anti-net neutrality speech to claim that he was in favour of net neutrality and agreed with TBL. TBL politely suggested that they must agree to differ. This was then reported in the national newspapers,[45] not least because Vaizey's office continued to try to claim no disagreement between the pioneer of Open Internet content standards and a minister who was happy to see differentiated charging and pricing for Internet content.

Three days later, on 25 November, Vaizey set out his considered view to Parliament:

> There is not yet any evidence that discriminatory practices are emerging, or that there is a problem with regards to how ISPs or networks manage the traffic that flows over them (something they all engage in for technical reasons to deliver the best possible service to consumers) ... A contributing factor to the success of the internet has been the lack of legislative restraints that have been placed on it. It is important that we give the market the opportunity to self-regulate. Ofcom will closely monitor how the market develops and if it develops in an anti-competitive way they will intervene.[46]

The difference in emphasis between TBL and Vaizey continued through the first half of 2011. TBL repeated his views somewhat more specifically and publicly after the government convened a private stakeholders meeting on net neutrality in March 2011: 'Best practices should also include the neutrality of the net.'[47] Note that the European Commission was at this point refusing to actively intervene in favour of net neutrality despite its 2009 Declaration, and its formal statement on the issue in April 2011 could have been written by Ofcom: 'transparency and ease of switching are key elements for consumers when choosing or changing internet service provider but they may not be adequate tools to deal with generalised restrictions of lawful services or applications.'[48]

Ofcom's approach to net neutrality 2011

'Not neutrality' became the task Vaizey had set for the neo-corporatist institution beholden to both him and the industry, the BSG.[49] In 2011 Ofcom declared that it would maintain its position that net neutrality could be regulated by

[45] Arthur (2010).
[46] Vaizey (2010b).
[47] Arthur (2011).
[48] COM(2011) 222, p. 9.
[49] Marsden, Chris (2015a).

self-regulation via a commitment to greater transparency by major consumer IAPs, added to its earlier reforms of the process by which consumers can change IAPs at the end of their contract, known as 'switching'.[50] The idea – if intellectually honest rather than ideologically driven by desire not to intervene – rests upon a notion that broadband consumers can ascertain the quality of their IAP's service for their time-critical applications, and make a decision to switch provider based on that knowledge. It is therefore a two-step leap of faith by Ofcom in consumer activism, first that such knowledge can be gained by a reasonable proportion of the populace, and second that they will use that knowledge as an aid to choosing a new provider. I detail this in the following section.

The clearest statement of Ofcom's transparency/switching mantra was issued on 21 November 2011.[51] The document is an extraordinary mixture of relevant analysis and pleas for industry to do its job for it. In particular, it is littered with verbs such as 'expect', 'encourage' and 'review'. I make comments on the relevant passages below.[52]

> [1.10] … if a service does not provide full access to the internet, we would not expect[53] it to be marketed as internet access. [1.11] It is possible that providers may seek to market a restricted service as 'internet access' by caveating this with a description of the restrictions they have put in place. Consideration[54] needs to be given as to whether this practice is acceptable.

This appears a strong commitment but it is unenforced. The reliance on self-regulation rather than any actual Ofcom action continues throughout this aspirational rather than regulatory document:

> [1.14] We do not describe what more detailed information might be provided, over and above the desired outcomes set out above. We note, however, that the self-regulatory model recently proposed by major ISPs provides a good foundation[55]. [1.17] We will monitor progress, and keep under review the possibility of intervening more formally in order to ensure that there is sufficient transparency as to the use of traffic management by network operators.

The document then goes on to dissemble by denying the existence of a problem which had been the subject of press attention for a decade!

> 1.22 From the perspective of protecting the citizens' interest alone it will be important to be vigilant in relation to the core connectivity of the 'best-efforts'

[50] Ofcom, Ofcom's approach to network neutrality, 2011.

[51] *Ibid.*

[52] See Marsden, Chris (2011).

[53] 'Expect' is not regulatory language, but aspirational language.

[54] 'Consideration'? By Ofcom? It is not clear from the document what this language means in regulatory terms.

[55] Ofcom goes on to explain that the BSG implementation work is embryonic.

> open internet and the access to information and services which it provides. It is important to note however that we see no concerns in this regard in the UK at present.[56]

A more concerned approach is signalled in relation to the 'dirt road' scenario in which the Open Internet receives far less investment than Specialised Services:

> 1.27 ... If the quality of service provided by 'best-efforts' internet access were to fall to too low a level, then it may place at risk the levels of innovation that have brought such substantial benefits during the internet's relatively short life so far. This would clearly be a significant concern.[57] [1.29] Any use of a minimum quality of service would need to be considered carefully, balancing the benefits of such an intervention against the associated risks. We are not, at present, aware of any actual concerns[58] which would merit carrying out such an assessment. However, given the importance of 'best-efforts' access to the open internet for innovation, we will keep this issue under review.

Ofcom then entirely admits its lack of concern over blocking and its competition-based anti-intervention strategy:

> 1.32 We do not have a general objection to models of competition where vertically integrated operators do not provide open access to their networks, provided that there is genuine competition and rivalry among the firms. In such circumstances, we do not necessarily regard the blocking of services provided by competing providers, or discrimination against competing services, as being anti-competitive. We do however have a specific concern in the context of the discussion in this document that restricted access to the internet could have a stifling effect on innovation. [1.33] Our stance as a regulator is therefore that any blocking of alternative services by providers of internet access is highly undesirable ... we expect such traffic management practices to be applied in a manner which is consistent within broad categories of traffic. Where providers of internet access apply traffic management in a manner that discriminates against specific alternative services, our view is that this could have a similar impact to outright blocking.

In conclusion at paragraph 1.34, Ofcom will not take any formal action to stop blocking, in the medium term: 'Our current view is that we should be able to rely on the operation of market forces to address the issues of blocking and discrimination.' It is a wait-and-see strategy, as it has been for most of the past decade.

[56] What about BT blocking BBC services? This is apparently not enough for Ofcom.
[57] But no 'dirt road' test is provided, notably for mobile.
[58] Note that actual concerns had been flagged up for five years by this point.

Policy developments after 2012

In October 2012, in a rare public[59] speech made in Seoul, Ofcom's chief executive Ed Richards stated Ofcom's continued belief in 'Transparency and Switching': 'We have undertaken a strategic review of consumer switching in the UK ... We will complete this process in 2013.'[60] On transparency, he flagged up the BSG Code, then made a strong non-statement about blocking:

> We are very clear that any blocking of alternative services by providers of internet access is highly undesirable. We recognise that some forms of traffic management may be necessary in order to manage congestion ... any regulatory intervention in this area (for example the imposition of a minimum quality of service) must be based on careful consideration. The present UK market is highly dynamic, and we would like to see market forces address the issues of blocking and discrimination in the first instance.[61]

He reiterated this in his next published speech 11 months later: 'our work on traffic management is now focused on whether awareness needs to improve and, if so, how.'[62] He added that switching was still not working effectively six years after the previous round of reforms: 'to transform the switching experience in the consumer interest and to support a more competitive market'.[63] By this point, his search for switching solutions instead of consumer protection was becoming ludicrous: one can imagine a similar speech being made in 2023 expressing how with even better information and switching, consumers can protect themselves finally from having Skype throttled.

The UK Ofcom Draft Annual Plan 2012–13 had a small section on traffic management which was bland and uninformative,[64] but promised that Ofcom would 'undertake research on the provision of "best-efforts" Internet access'. In January 2013 Ofcom published its draft Workplan 2013/14 for consultation. Five mentions of net neutrality were made in the workplan draft – with a response to COM(2013) 627 as needed (the consultation was published in September 2013), and analysis in the wider Ofcom Infrastructure Report to be published in November 2013, plus a general commitment to helping government policy formation as required.[65] In April the technology press reported on Ofcom's Annual Plan 2013/14: 'Ofcom has [in its Annual Plan] decided that

[59] Note that Ofcom published only this speech in a 17-month period in 2012–13, an extraordinary period of public policy silence. See http://media.ofcom.org.uk/speeches/.

[60] Richards (2012).

[61] *Ibid.*

[62] Richards (2013).

[63] *Ibid.*

[64] Ofcom, Draft Annual Plan 2012/13, 2012, paras 5.40–5.42.

[65] Marsden, Chris (2013a).

treating all packets of internet traffic as equals without discriminating against particular protocols and services – trendily known as net neutrality – is a non-issue in the UK.'[66]

In July 2013 the government issued the UK Communications 'consultation paper', which was the centrepiece of the coalition government's communications policy strategy.[67] Pages 13–14 confirm that it intended solving net neutrality via transparency and switching:

> Internet traffic management: Cisco predicts that internet traffic will reach 1.4zettabytes (a zettabyte is equal to a trillion gigabytes) a year in 2017. As the amount of data exchanged increases we want to ensure that consumers are aware at the point of sale of the internet traffic management policies that their internet service provider or mobile network operator has in place. For example, some mobile network operators block the use of apps like 'Skype', although this may be reflected in cheaper contracts. We think it should be for consumers to decide what best meets their needs, so, in the first instance, we have asked Ofcom to work with internet service providers to encourage them to make their traffic management policies more transparent on a voluntary basis – the challenge for industry will be to do this in a way that is clear and understandable to consumers. Where there is evidence of consumers not being made sufficiently aware, Ofcom will act to require operators to make their traffic management policies more transparent. We believe in the principle of an open internet and will keep this area under review.

There is a longer regurgitated Ofcom policy line (at page 41) which astonishingly suggests that traffic management could prioritise Skype services – when presumably every incentive exists to do the opposite, that is discriminate to prioritise VoIP that is not a competitor to the IAP: 'firms may prioritise time-sensitive services like video streaming or voice calls over the internet – such as the type of services offered by Skype – over other content which is not as sensitive to time delay'.

The UK Code of Practice, which sets standards for transparency, was negotiated by industry via the government's forum. Ofcom put severe pressure short of actual regulation on major UK consumer IAPs to join the BSG Traffic Management Transparency Code 2011, to which all had finally agreed by 2012.[68] The government-funded BSG produced a Code of Conduct, launched in March 2011,[69] in which Vaizey indicated that TBL would play an oversight role.[70] It is entirely unclear what this represents for TBL, and the industry's July 2012 enhanced Code was at first not supported by a large minority of the

[66] Ray (2013).
[67] Department for Culture Media and Sport (2013b).
[68] This committed them to issuing Key Fact Indicators (KFIs) which were updated on the BSG site in June 2013. See Broadband Stakeholder Group (2013b).
[69] Broadband Stakeholder Group (2013b).
[70] Vaizey (2011).

IAPs. In 2013 a renewed Code was published, which appears to have general consensus of support. It contains a process by which consumers and content providers can complain about traffic management practices which breach the Code. The BSG Open Internet Code of July 2012[71] has interesting language that marks IAPs' attempt to show they listened to Ofcom's November 2011 policy statement:

> The signatories to this code therefore believe that it is right that *Ofcom take ownership of this issue* and also believe that the new proposed process will be a useful input to Ofcom as it continues its work in monitoring the nature and impact of traffic management practices in the market and the effective co-existence of managed services and best efforts internet access.
>
> It is clear that the voluntary commitments being made in this code *closely relate to ongoing monitoring work Ofcom has said that it will conduct.* Signatories to this code are happy to discuss with Ofcom how its future work plans regarding open internet issues could *support or input into a review* of these voluntary commitments [emphasis added].

The idea that content providers might lodge an unresolved complaint with the BSG (see Annex 1 to the Code) instead of going direct to Ofcom or the EC was frankly ludicrous: exactly zero such farcical reports were forwarded from the BSG to Ofcom in 2013. It is too early to assess the effect of the new Code of Practice on the market and society as yet.

Though the Code was updated in May 2013 from its initial version,[72] a new version of that Code was suggested by Ofcom in September 2013 consumer research, with more user-friendly terms. An example of user-friendly Key Fact Indicators (KFIs) is that a 100GB cap on monthly use will equate to about 25 high definition feature films, as explained by the Apple iTunes Store, a standard of transparency not yet deemed necessary by the BSG.[73] It is notable that the Apple warning includes express instruction that low-speed Internet connections are unsuitable for downloading films.

In 2013 Ofcom commissioned market research company Kanders to conduct a multi-stage survey of consumers to find out if net neutrality was important to them, though without using the term 'net neutrality'. In a 63-page report, the term only appeared to sum up Ofcom's 2011 statement dismissing net neutrality.[74] The term 'Open Internet', the FCC/EC preferred

[71] Marsden, Chris (2012).

[72] Broadband Stakeholder Group (2013c), with May 2013 version of Code available at www.broadbanduk.org/wp-content/uploads/2013/08/Voluntary-industry-code-of-practice-on-traffic-management-transparency-on-broadband-services-updated-version-May-2013.pdf (Accessed 16 September 2016).

[73] Apple (2012).

[74] Ofcom, Consumer research into the transparency of traffic management information provided by ISPs, 2013. The report of 7.4MB took 2 minutes to download on Virgin Broadband (based on BT Wholesale not fibre-coaxial hybrid) on Wifi at 10 p.m. on 9 September 2013.

alternative, was also not used. Only once was the term 'throttled' used, to illustrate the most invasive possible practices. However, it emerged that consumers rapidly became concerned once they understood the practices their IAP could indulge in to their detriment: 'once the term and processes of traffic management were explained to them 35% of these respondents felt that they may have been affected by these processes.'[75] Unsurprisingly, 'It was clear that most were not aware of the underlying processes supporting the internet or how it operates.'[76] If government, Ofcom and the industry publicises the implications of 'traffic management', consumer concern rises from 1 per cent to 35 per cent. This reveals a massive failure to educate the public on media literacy.

The research concluded that 'consumers state a preference for information being provided in online formats and by 3rd party independent sources'. Ofcom decided that this represented an initiative to act on gainer-led switching for broadband, which will similarly take years to implement.

As the previous section establishes, there are two camps in net neutrality debates in the UK as elsewhere in the world: the anti-regulatory and pro-regulatory. In 2012 the UK official consumer interest group Consumer Focus released a report into consumer understanding of information on traffic management and stated:

> we conclude that increased transparency alone is unlikely to safeguard effectively the open internet and prevent discriminatory restrictions online. Our research finds consumers are not aware of traffic management practices, and even if they find information on traffic management restrictions on providers' websites they cannot digest the meaning of unfamiliar terms such as P2P or VoIP.[77]

They suggest:

> The findings of our research demonstrate the need to extend the existing regulatory framework by additional non-blocking and non-discriminatory principles that would clarify which type of traffic management is legitimate. If a self-regulatory or co-regulatory solution is a preferred option it must have a robust compliance and enforcement mechanisms monitored by Ofcom. In addition broadband providers need to do more to raise awareness of traffic management through improved marketing of information to customers. This is the only way to ensure consumers can use the broadband connection of their choice to access the internet and any legal online content and applications they wish, free of negative discrimination, and to protect the innovation.[78]

[75] *Ibid.*, p. 31.
[76] *Ibid.*, p. 35.
[77] Consumer Focus (2012).
[78] *Ibid.*

This proves the need for neutral consumer champions to find out what IAPs are doing and stop it if it harms their Internet connection – especially when it is blocking or throttling rival applications such as VoIP or media streams. The same of course applies to IAPs routinely providing a slower connection than advertised, which was the subject of updated 2013 advice given by the Advertising Standards Authority, the self-regulator of text and Internet advertising.[79] It is rather like research on smart metering: if the consumer lacks the technical knowledge, they also do not know what is going wrong. A more appropriate question would be to ask the remote workers if it is wrong for their IAP to throttle Skype during the working day. That might raise a flicker of recognition.

Ofcom explained its 2013 holding pattern: 'As part of our monitoring program we expect to include an update on IAP's traffic management policies in the 2013 edition of the Communications Infrastructure Report. Ofcom will also support BEREC to assess market and regulatory developments at the EU level.'[80]

BSG Code update 2015

BSG Chair Hooper stated in November 2014:

> This Code of Practice plus competition obviates the need for statutory regulation which is being backed by some other European countries despite the UK approach being taken up around the world – most recently by Switzerland. Incidentally, the digital minister Ed Vaizey and I agreed that we should not use the American term 'net neutrality'. In the UK we call it the Open Internet.[81]

It should be noted that the BSG is not of the multi-stakeholder kind – it consists of big corporates with no prosumer representatives at all.

In 2015 BSG could finally claim UK IAPs had signed the voluntary Code. Not only did it take EE, Vodafone and Virgin almost three years to sign up, but the alleged 'consumer champion' Jo Connell, Chair of the Communications Consumer Panel, sounds like a mouthpiece for government, stating:

> The Code usefully supports open access to the internet and builds on previous commitments by ISPs to provide transparent information to consumers about their traffic management policies. We are delighted that EE, Virgin and Vodafone have now agreed to become signatories. The Code has gained significant interest

[79] There have been 69 rulings against misleading broadband speed/quality advertising in under five years: see www.asa.org.uk/Rulings/Adjudications.aspx?SearchTerms=broadband#results. For the most recent example, see Advertising Standards Authority (2013).

[80] Ofcom, Consumer research into the transparency of traffic management information provided by ISPs, 2013.

[81] Broadband Stakeholder Group (2014).

> internationally as a positive example of industry responding to a developing consumer need.[82]

The announcement of a successful funded review by BSG after five years, on 17 November 2015, was that the UK industry was pleased to award itself high marks for its self-implementation not of neutrality but of 'Open Internet', with no 'official' complaints. WIK conducted the actual review and it is the WIK footnotes and page numbers to which the analysis below refers.[83] Note from Hooper's comments what the client BSG thought of net neutrality regulation as opposed to self-congratulation:

> The UK approach, which has proved to be eminently successful, has required no statutory regulation [but] our approach has been undermined in some ways with the Connected Continent Regulation passed by the EU in the summer. This regulation is directly applicable in the UK. Unfortunately, they have pursued a more prescriptive approach than is necessary or desirable in the UK and potentially hinders the ability of network providers to provide innovative services.[84]

WIK heard claims that the value of self-regulation was proved, in that dialogue between content providers and IAPs was just as important as the Codes themselves. The Open Internet Forum (OIF) has no website, is membership only and has no transparency. There has been no 'official complaint' to the OIF from industry players.[85] This may be unsurprising given that the Code is not enforced.[86] Effectiveness of the Code is only measured by market share: 'more than 90%' of users are served by IAPs that have signed the Code, though no exact figure is available. These IAPs claim that the 'overwhelming majority' of users enjoy full access, that is very little blocking, and that blocking is mainly of spam. No statistics were available on consumer awareness, nor do signatories or Ofcom show any willingness to test how much awareness there may be. This indicates that the Code's signatories, BSG and Ofcom have created a 'Potemkin' self-regulator,[87] with no actual

[82] Broadband Stakeholder Group (2015).

[83] WIK (2015).

[84] Hooper (2015).

[85] WIK (2015), p. 18.

[86] Discussion in the OIF took place about an IAP-blocking VoIP. ITSPA was unhappy as VoIP was degraded and not improved by new router standard implementation, as the IAP argued. Information from Huw Saunders, Director, Network Infrastructure, Ofcom, personal communication. See https://indico.uknof.org.uk/conferenceOtherViews.py?view=standard&confId=33 (Accessed 7 September 2016).

[87] Marsden (2011), p. 60, footnote 75: 'In the original "Potemkin" villages, General Potemkin (or Potyomkin) infamously created facades of villages in 1787 to present an image of prosperity to Empress Catherine II of Russia, in which there was no substance to the buildings, a myth for which a website of equally contested veracity provides discussion.'

powers or consumer awareness.[88] There were no recorded consumer complaints.

The Code needs to be reformed in 2016 in order for the UK to comply with European law, and a process was announced that should culminate in autumn 2016. An obvious problem lies in the IAP KFIs which throttle 'unreasonable' traffic management during all hours when people are home from school/work. A further clear issue is that of zero-rated and Specialised Services. Finally, adult content filters will need legislation in order to continue to be legal, especially for crude blocking such as that exercised by TalkTalk, or opt-out blocking as exercised by Sky.

There is no published research into whether consumers knew about KFIs, though Ofcom in 2013 recommended Code improvements.[89] The only organic change to the Code in 2013 was public Wifi use.

Net neutrality law 2016

In 2014 Ed Vaizey informed Parliament that an actual net neutrality law may be agreed, though in the final analysis it was less strict than he and UK IAPs feared:

> We will continue to work closely with industry and Ofcom to ensure that our input into negotiations is as influential as possible. Should the European institutions decide that Regulation is the only way forward and UK is unable to gain wider support for its self-regulatory stance, we should prepare to ensure that any adopted text is as workable as possible given the current state of the UK market and the existing self-regulatory approach.[90]

UK 'transparency and switching' policy has been maintained for almost a decade, supported by the European Commission's wait-and-see approach up until 2013. It is now supplanted by the European Regulation which must be implemented in 2016.[91] In July 2015 Vaizey informed Parliament that:

With regard to net neutrality, the Minister says that:

- overall, the text agreed is principles-based and service and technology-neutral, will ensure an open internet across Europe where all legal traffic is treated equally and end the unfair blocking of rival services;
- it thus fully meets the wider criteria in the Government's negotiating position;

[88] WIK (2015), footnote 111, p. 36 explains that the 'Code neither addresses remedies nor penalties' – it is toothless.

[89] *Ibid.*, p. 33, footnotes 99–100.

[90] Vaizey (2014).

[91] COM(2013) 627.

- with regard to the impact on the UK domestic regime, the work of the Internet Watch Foundation (IWF) and the current voluntary parental control filters regime can continue without further intervention, by implementing the necessary legislation in the UK;
- his Department is now taking this forward, on the basis of a December 2016 deadline for implementation that provides 'ample time' to complete this process;
- bearing in mind how it evolved from Commission and European Parliament proposals that were largely unpalatable for the UK, this outcome is 'largely positive overall' and 'well within a range of outcomes that would have been acceptable to HMG'; and

  he can therefore, on balance, support it.[92]

The UK domestic obsession with its co-regulatory model for child pornography removal from the Internet has been documented elsewhere,[93] but it is clear from Minister Vaizey's reports that it is the single matter that was raised as a red line in negotiation, rather than the principle of a neutral Open Internet. This may be seen as a uniquely British position – censorship but without law – with which the Latvian Presidency of the European Council had to grapple before finally telling the British to stop their ridiculous charade, as is clear from Vaizey's note and my interviews. There will therefore be a new law more formally authorising the IWF as a co-regulatory body in 2016, to which one can anticipate some co-regulatory recognition for the Open Internet Code of Practice. Ed Vaizey stuck to his script on net neutrality self-regulation, even producing it as evidence of effective self-regulation before Parliament in March 2016.[94] Yet we have seen there is no evidence to demonstrate its effectiveness, and the forensic examination by Alissa Cooper to show that it is not.

In November 2015 the Department for Culture, Media and Sport was very vague on whether there will be a need for co-regulatory legislative provision that can be shared with Internet pornography filter amendments that will be needed in late 2016. The Prime Minister announced on 28 October 2015: 'I can tell the House that we will legislate to put our agreement with internet companies on this issue into the law of the land so that our children will be protected.'[95]

[92] European Scrutiny Committee (2015).

[93] Marsden (2011), pp. 166–178.

[94] House of Commons 571, Business, Innovation and Skills Committee (2016).

[95] Cameron (2015).

Evaluation of UK regulation and future development

UK network neutrality policy is influenced both by its regulator, Ofcom, and its particular and unique network topology. Such has been the dismissal of net neutrality in the UK as compared to its source of telecom and regulatory inspiration, the United States, that it has led to substantial scholarship on the differences in stakeholder mapping.[96] For our purposes, what matters is the degree of cajoling of IAPs and others into a self-regulatory mechanism backed by Ofcom's powers to enforce their General Conditions (licences) to provide public communications and/or to bring a competition case against dominant players. This was a vanishingly slight risk in the political circumstances pertaining to 2015, as we shall see. My analysis of Ofcom's arm-twisting was that it became so extreme as to approximate to co-regulation, especially by 2015 when all IAPs had finally signed a 'voluntary' Code of Practice on Traffic Management.

In 2016 the difference between co- and self-regulation becomes academic, as Ofcom will have to bring that Code into a form of formal co-regulation via statutory regulation – not the least reason why the intransigent 'refusenik' IAPs finally signed up to the (in)voluntary Code in 2015. But this is not the end of the story: the Code is currently a 'Potemkin' regulator with no enforcement or dispute resolution practice to support its principles. It is in 2016 the equivalent of the Association for Television on Demand (ATVOD in 2009).[97] UK communications experts will be aware that ATVOD was reformed into a co-regulator under the Digital Economy Act 2010, then abolished in 2015 with its functions subsumed into Ofcom.[98] I would be very surprised if it takes as long as five years from 2017 for the net neutrality 'self-regulator' obligations to become a part of Ofcom's remit.

Both as a practical and normative matter, it is extremely unfortunate that the NRAs charged with implementing net neutrality have such a poor record (in general) of upholding state obligations under Article 10 of the European Convention on Human Rights, and Article 19 of the International Covenant on Civil and Political Rights. For instance, Ofcom has about 300 responsibilities under its empowering legislation, the Communications Act 2003 (as amended, hence the estimate of its powers). It is only in Section 3(4)(g) that you locate any responsibility for freedom of expression, and that is mealy-mouthed: 'that

[96] Cooper and Powell (2011).

[97] Marsden (2011), pp. 147, 222.

[98] Ofcom, Future regulation of on-demand programme services, 2015: 'From 1 January 2016, Ofcom will be sole regulator (other than in relation to advertising) for on-demand programme services ("ODPS") under Part 4A of the Communications Act 2003 (the "Act").'

best guarantees an appropriate level of freedom of expression'.[99] Reference to implementation of network neutrality by BEREC needs to include a coda that agency officials implementing net neutrality must be trained in the latest case law of the European Court of Human Rights, and Declarations and Guidelines of the Council of Ministers,[100] in order to fully comprehend what freedom of expression means in their duty to implement net neutrality and the 'Open Internet'.

[99] Communications Act 2003, Section 3(4)(g): see www.legislation.gov.uk/ukpga/2003/21/section/3.

[100] Declaration of the Committee of Ministers on network neutrality adopted 29/9/2010: 1094th meeting of the Ministers' Deputies, a soft law instrument to guide member states in the application of net neutrality rules: aspirations of Articles 6/8/10 of the Convention. Declaration on freedom of communication on the Internet, Adopted on 28 May 2003 at the 840th meeting of the Ministers' Deputies. See also Council of Europe (2013) 29–30 May, Multi-stakeholder dialogue on network neutrality communicated to the Council of Europe Steering Committee on Media and Information Society (CDMSI).

7

Implementing mobile net neutrality

> Internet.org is misleadingly marketed as providing access to the full Internet, when in fact it only provides access to a limited number of Internet-connected services that are approved by Facebook and local ISPs. In its present conception, Internet.org thereby violates the principles of net neutrality, threatening freedom of expression, equality of opportunity, security, privacy and innovation.
>
> Access Now[1]

This chapter critically examines the relatively few examples of regulatory implementation of network neutrality enforcement at national level outside the EEA examples of Norway, Slovenia and the Netherlands explored in Chapter 2. I draw on co-regulatory and self-regulatory theories of implementation and capture, and interdisciplinary studies into the real-world effect of regulatory threats to TMP. This involved appropriate fieldwork to assess the true scope of institutional policy transfer.[2]

Zero rating

The developed countries have recently legislated for or regulated for 'net neutrality', the principle that IAPs should not discriminate between different applications, services and content accessed by their users.[3] This victory for net neutrality proponents came after 20 years of attempted discrimination between content streams within the walled gardens of both fixed and mobile IAPs, such as AOL in the 1990s, and Vodafone Live/360 in 2002–11, which was intended to challenge the Apple App Store and Android/Google Play.[4] Alongside their walled gardens, these IAPs enforced monthly data caps preventing

[1] Access Now (2015).

[2] See further Marsden (2016a).

[3] Wu (2003b, 2007).

[4] Wray (2009).

their customers having unlimited use of the Internet. Fixed-line walled gardens failed in view of the easy access to the Open Internet at increasingly low cost offered by broadband access. Continued attempts to maintain walled gardens since 2006 have focussed on both 'negative' and 'positive' net neutrality. I explain both in turn.

Negative neutrality is the blocking and throttling of content that threatens the business model of the IAP. This can be relatively benign when it is spam email and viruses that are blocked. It can also be self-serving and anti-competitive when it is unjustified and unreasonable restrictions on user's preferred content that is affected – for instance P2P file sharing or video streaming. It is this 'negative' net neutrality which is the target of most legislation in the area, based on the generic regulatory principle of 'first, do no harm', in this case eliminating the harms caused by unreasonable negative blocking or discrimination. Cases in the US such as *Madison River* and *Comcast* were about blocking, and is it this that rouses much consumer anger and political action.[5]

Zero rating is 'positive' net neutrality violation involving not blocking, but treating some content better than general Internet traffic. As cable TV provides high definition and standard video and television channels at high fees in a separate logical pathway to the general Internet traffic on its cable, some telecoms companies hope to partition their Internet traffic to replicate this business model. Several IAPs attempted this practice over lengthy periods, notably by excluding television channels from monthly data caps for users, positively discriminating in favour of their affiliated content and against other video providers (such as YouTube).

'Walled gardens' thus reappear with much more specialised walls – restrictions that affect only certain non-affiliated types of Internet traffic, such as social networks or video. This exclusion of preferred content from data caps is described as 'zero rating' because all that downloading costs precisely zero in terms of counting towards monthly bills.[6] Note that many fixed IAPs have virtually unlimited data use as part of their offer, made possible because maximum speeds and user profiles mean that the cumulative download burden does not overstrain the network.

Zero rating is only possible when users take an IAP subscription which has a data cap, which is generally at a much lower limit when imposed by mobile rather than fixed IAPs. Unlimited data plans mean users can download as much data as needed using the Open Internet pipe, whose speed is restricted only by the Internet itself, or the type of CDN used to supply media.[7] When a cap

[5] Marsden (2013b).

[6] Marsden (2010) pp. 38–39, 96; Odlyzko *et al.* (2012); Maillé and Tuffin (2014), pp. 89–90; Eisenach (2015).

[7] See BBC (2014).

applies to a monthly subscription (such as 1 Gigabyte a month[8]), that limits the amount of content that a user will choose to access. If data is as expensive as it can be in developing countries, any content can prove too expensive to access for the average user. Offering certain content on a 'zero-rated' basis means that content will not be included in the data cap – which is particularly useful if that content is streamed video, audio or an application used regularly, such as social network Facebook or messaging app WhatsApp. That content may be locally stored, relieving congestion in the network, as a result of partnership with the IAP, justifying in network engineering cost terms the decision to reduce the apparent end user cost, if not eliminating it completely.

Research into comparative net neutrality law in the Global South has recently been carried out by several NGOs and is well reported in the specialist media.[9] Odlyzko notes that the zero-rating debate exists in one Asian country, but does not explore this in depth, while Marsden discussed monthly caps before zero rating had become commonly identified.[10] Net neutrality dates to the 1990s, as does zero rating, even if the term of art was coined much later.[11] There are ten times more mobile (5.6 billion) than fixed line connections (572 million) in developing countries, whereas the developed world ratio is 3:1. There are five times more mobile broadband subscriptions in the developing world with 2.37 billion to only 429 million fixed subscriptions (developed world 1.09 billion mobile to 365 million fixed at a ratio of 3:1). Seventy per cent of Internet users totalling over 2 billion people are outside the EU/US.

Social networking using Web2.0 software expanded from a very low base at the start of the smartphone era. Brown and Marsden explain that 'Facebook grew from nothing in 2004 to become the second most popular destination Web site in the world by 2012.'[12] It grew to 100 million users by autumn 2008, surpassed 1 billion monthly active users (MAUs) in 2012 and 1.5 billion in 2015, which included 210 million in the US/Canada and 112 million in its second largest market, India, in autumn 2014. Facebook was floated on the stock market in 2012, warning investors in its prospectus that 'There is no guarantee that popular mobile devices will continue to feature Facebook, or that mobile device users will continue to use Facebook rather than competing products.'[13]

[8] The typical UK 2015 limit was 2GB/month, see http://kenstechtips.com/index.php/what-does-500mb-or-1gb-internet-actually-mean-explaining-mobile-data-limits#Download_Limits_in_the_UK (Accessed 9 September 2016).

[9] Rossini and Moore (2015); Marques *et al.* (2015). Additionally, many regulatory documents are available in Spanish, Portuguese and English on regulator websites. The consultation process for net neutrality regulation was very well publicised in Brazil, while Chile's 2010 law was well noted but little researched in academia outside Latin America.

[10] Odlyzko *et al.*(2012), p. 15; Marsden (2010) citing Fierce Wireless (2011).

[11] Lemley and Lessig (1999), Marsden (1999).

[12] Brown and Marsden (2013a), p. xii.

[13] Cited in Brown and Marsden (2013a), p. 123.

A particular business model for this practice is that of dominant social network Facebook, which from 2009 introduced Facebook Zero with mobile IAP partners, and in 2015 introduced a wider walled garden called 'Internet.org' (which despite its name is an Intranet for 30–40 affiliates), which was rapidly renamed Free Basics in late September 2015.[14] In May 2015 opposition to the highly exclusive and non-transparent Internet.Org had led to content owners abandoning their previously negotiated tenancies, and mobile IAPs dropping the service.[15] Free Basics has less powerful gatekeeper functions than Internet.Org and more content is permitted, with officially only technical grounds for refusal, but it is still only governed by a contract with Facebook, which Facebook can change unilaterally.

Politicians and telecoms executives who now claim to be in favour of net neutrality are in fact conceding that blocking and throttling users is no longer acceptable to politicians and therefore regulators. They largely only concede 'negative' net neutrality. 'Positive' net neutrality is a much more contested topic, and where download limits apply or ill-defined Specialised Services carry the zero-rated content, this concept of zero rating will be heavily contested. That is more the case with mobile than fixed networks, and more the case with developing nations' mobile IAPs than developed.

Case studies

The book thus far has relied to a large extent on the experiences of developed nations, especially the United States, EU and United Kingdom.[16] I focus here on four case studies, beginning with the earliest effective regulation in Chile. This chapter summarises each nation's development of net neutrality, and focuses on its implementation of regulation against zero rating since 2014. The methodology used both literature review and empirical interviews in the course of 2015.

Chile

Chile has the world's earliest net neutrality law (from 18 August 2010)[17] and an implementation of regulation permitting zero rating from 2014. Law 20.453 includes a provision which adds Article 24(h–j) to Law 18.168 'General de Telecomunicaciones'. Article 24H expressly forbids IAP practices that 'arbitrarily

[14] Galpaya (2015).

[15] Marsden (2016b). Arguably the main reason Facebook Zero was rebranded and relaunched as part of a less exclusive but noxiously titled 'Internet.Org', then as Free Basics, was to access data-poor mass markets in India, Brazil and Nigeria.

[16] Marsden (2014a).

[17] Chile, Ley 20.453 de 18 de agosto 2010, which is implemented by Decree 368 of 15 December 2010.

distinguish content, applications or services based on the source or ownership thereof'. This would be relied upon by those opposed to zero rating. The original law required IAPs to self-report on any violations, resulting in infringement only for failure to report. Cerda reports that there were 'allegations of negligent supervision of the law by public authority' in failing to enforce consumer rights.[18]

In Chile,[19] all four mobile IAPs (Claro, Entel, Telefónica and VTR) were notified to cease zero rating in 2014.[20] The regulator's (sub-secretary of communications: Subtel) conclusion was misreported in the developed nations' media as banning all zero rating from 1 June 2014, when it applied to social networks, notably Facebook and therefore Internet.Org.

Subtel stated: 'las empresas que entregan algunas redes sociales gratis, lo que hacen es privilegiar el uso de estos servicios, mediante el acceso a una Internet bloqueada, excluyendo las redes sociales privilegiadas' – social networking apps received positive discrimination ('privilegiadas') when included in the zero-rated offer. The Chilean situation is complicated by Wikipedia Zero announcing on 22 September 2014 that it had negotiated an exemption from the rules, on the basis that it is neither a social network nor a commercial offer.[21] As carriers have not asked Subtel to confirm this exemption, and Wikimedia does not have standing (as a non-carrier) to request that official explanation, the evidence for this is Wikimedia's version of the exchange and its continued zero-rated offer in Chile.

In fact, Claro (a subsidiary of Mexican operator America Movil, also active in Brazil, Columbia and other Latin American nations) was permitted by the Chilean regulator to continue zero rating as long as it formed part of a wider data plan that customers could choose.[22] This was because data plans were included in the new zero-rating offer, removing the part of the complaint relating to 'cuando los usuarios salen a través de un enlace externo, las empresas piden pagar' – that non-zero-rated websites have to pay for users to exit zero rating onto the wider Internet. Zero rating would have to stop when users exhausted their data plan each month, in order that they were not left with only zero-rated content, which would be very explicit discrimination.

[18] Cerda (2013).

[19] Roa and Mariano (2015).

[20] In Chile, a total of 40 cases may sound substantial, but 25 were in the first two years, and 29 fully relate to those four major ISPs. Most were for infringement of transparency rules or network self-measurement. Zero rating was considered by many observers as the first true test in 2014.

[21] Welinder and Schloeder (2014).

[22] The draft Direction of May 2014 apparently banned all zero rating, but the final decision of August 2014 permitted those plans offered only in addition to a data plan – i.e. where users had purchased wider access to escape the walled garden.

Brazil

Brazil has had zero rating since prior to 2014, when it was a common practice by several mobile ISPs. Like Chile, Brazil has a bicameral constitution with a powerful directly elected executive president. Brazil had discussed net neutrality since the mid-2000s, with its formal advisory committee on Internet governance passing a resolution known as the Decalogue in 2009, which in part stated: 'Filtering or traffic privileges must meet ethical and technical criteria only, excluding any political, commercial, religious and cultural factors or any other form of discrimination or preferential treatment.'[23] This led to a period of public consultation led by the Ministry of Justice in 2009 (29 October–17 December) over a potential new legal framework. In 2011 the Chamber of Deputies (lower house of parliament) began negotiations on a law on privacy and net neutrality led by Deputy Alessandro Molon, which stalled in 2012/13.

In late 2013 the political process was accelerated due to President Roussef's concerns over foreign surveillance of telecoms and Internet traffic (specifically her own communications), resulting in the Senate ratifying the Chamber of Deputies' proposed law in a single month.[24] Law No. 12/965 (the Marco Civil da Internet) was signed by the president at the opening ceremony of the Net Mundial conference in São Paolo in April 2014.[25] The relevant section is Article 9, which states: 'The party responsible for the transmission, switching or routing has the duty to process, on an isonomic [equality before the law] basis, any data packages, regardless of content, origin and destination, service, terminal or application.' According to Article 9(3) ISPs must 'act with proportionality, transparency and isonomy' and 'offer services in non-discriminatory commercial conditions and refrain from anti-competition practices'. The question for regulators implementing zero rating is whether it is proportional, transparent and non-discriminatory.

Unsurprisingly for such a rushed final law, the consequent implementation has proved controversial, not least because it is not clear which of two consultative bodies and the Ministry of Justice should be in charge of the drafting and enforcement of the subsequent rules.[26] Article 9(1) states that it 'shall be regulated in accordance with the private attributions granted to the President … upon consultation with the Internet Steering Committee [CGI] and the National Telecommunications Agency [Anatel]'. In 2015 both the regulator and the Ministry issued consultations, the latter organised together with the CGI in the period 28 January-30 April.[27] The results of the consultation were

[23] CGI (2009).
[24] Wohlers *et al.* (2014).
[25] Law No. 12.965, 23 April 2014 by the Presidency of the Republic, Civil House Legal Affairs Subsection.
[26] Cruz, Marchezan and dos Santos (2015).
[27] Brazil Ministry of Justice (2015) and Chilvarquer (2015).

made public in a Presidential Decree on 11 May 2016, which banned zero rating.[28]

It is unclear whether zero rating or Specialised Services will be effectively regulated at the time of writing, despite the 11 May 2016 Regulation. At the 2015 Summit of the Americas in Panama on 10 April, President Rousseff met Mark Zuckerberg and was photographed with him,[29] he in a suit, she in a Facebook hoodie.[30] Her pronouncements in favour of Facebook's work in Brazil with poorer communities, and by inference Internet.Org, were a public scandal in view of the open consultations then ongoing. However, it is not clear what benefit such public lobbying achieved for Facebook/Internet.Org/Free Basics.

In practice, in 2014 Anatel chose not to regulate zero rating. TIM (the Brazilian subsidiary of Telecom Italia Mobile), in partnership with WhatsApp, released a zero-rating plan that allowed subscribers to use the app in zero rating. Marcelo Bechara, counselor of Anatel, refused to regulate in the absence of specific prohibitions: 'If there is no prioritized traffic, I do not see why it breaks the Marco Civil. This is the free market. It's free business.'[31]

In 2015 Claro abandoned a previous offer that provided zero rating only, and adopted the Chilean approach with free WhatsApp, Facebook and Twitter offered only to users who also subscribed to data plans (pre- or post-pay).[32] Claro CEO Carlos Zenteno had said in April that zero-rating plans were no longer part of the carrier's strategy, as less than 1 per cent of customers used only Facebook or Twitter, and in June added: 'It's an evolution. We realized that it has no purpose only to offer zero-rating access to one site.'[33] Claro argues that zero-rating on top of existing data plans represents a positive discrimination that the consumer chooses. Anatel's decision on this issue will be critical to the future of Brazilian zero rating.

Ramos states that:

> the gap between those who can pay for data caps and those who cannot afford them could lead to a two-tier internet: the 'internet of the rich', or those who are wealthy enough to pay for the unlimited access; and the 'internet of the poor', which would give access only to a few applications that would be affordable to poor people.[34]

In Brazil, such a digital divide has a potent political force, given that the policy of progressive governments, since Cardoso was elected in 1994, has been to

[28] Brazil, Decreto No. 8.771 de 11 de maio de 2016.

[29] Brazilian Government (2015).

[30] Antunes (2015).

[31] Marques *et al.* (2015), pp. 66–67.

[32] Prescott (2015).

[33] *Ibid.*

[34] Ramos (2014).

narrow the inequalities that grew in the military dictatorship and before. Brazil was becoming a less unequal society until its recession, which began in 2015. But as Ramos explains:

> the existence of two different 'internets' could distance the rich from the poor (with application providers creating services aimed for the rich and 'light versions' aimed for the poor). Ultimately, it could lead to a replica of the social apartheid currently perceived in many developing countries, where slums have limited access.[35]

It could lead to a perceived 'gringo net' where only the rich can afford to access the full Internet with its many foreign apps and services. That said, the ISPs in Brazil plead not to be made tools of social engineering, arguing that inequality is a matter for governments not companies, however integral their service to the socio-economic landscape.

Brazil has consulted on net neutrality in two phases, the first running in spring 2015 in which the zero-rating issue emerged as the most significant and commented-upon controversy, the second from 27 January 2016. The second phase resulted in the Ministry of Justice Regulations via Presidential Decree in 2016,[36] and the eventual fate of zero rating remains uncertain. It remains legal in the absence of Anatel action.

India

On 8 February 2016 India banned zero rating.[37] India has a population of 1.25 billion, with a billion mobile users or almost 80 per cent of all citizens, but low data use on smartphones, and only 26 million fixed telephone connections.[38] Only 57 per cent of Indian (and 43 per cent of Brazilian) smartphone users actually use data plans at all, and the average amongst those Indians who do was 80MB a month in 2015 (3–5 per cent of developed national average usage).[39] With a very low fixed Internet subscription rate, most Indian consumers primarily rely on the mobile Internet for data. The regulator is the Telecom Regulatory Authority of India (TRAI), which had consulted on net neutrality in 2006 when the issue first arose, with little public debate.[40] By contrast its spring 2015 consultation produced over a million emails in reply, focused on zero rating.[41]

35 *Ibid.*
36 Brazil, Decreto No. 8.771 de 11 de maio de 2016.
37 India, Prohibition of Discriminatory Tariffs for Data Services Regulations No. 2 of 2016.
38 World Bank (2015).
39 Olsen (2015).
40 TRAI (2006).
41 TRAI (2015).

In India, three zero-rated options were offered in 2015, by both Internet.Org, owned by Facebook using the Reliance network, and Airtel (the largest mobile IAP in India with 226 million customers at April 2015). In summer 2015 an Indian government committee suggested that the locally based Airtel's zero-rated option should be permitted but foreign-controlled Facebook's Internet. Org prohibited.[42] In response to concerns most vociferously raised in India but also in Brazil, the US, and other nations, Facebook made the terms of Internet. Org more transparent in May 2015, effectively opening access in principle to any app developer who could meet its terms.[43] Nevertheless, Facebook's privacy policies continue to apply and it is not possible to use Internet.Org without also being a Facebook user, while Facebook accesses all your tracking behaviour while logged in to any partner sites and can share that with mobile IAPs.

Internet.Org's policies were carefully analysed by the Centre for Internet Studies in India.[44] It was a matter of great priority for Facebook to expand its mobile network partnerships rapidly internationally, especially in India, in the face of a decline in youth MAUs in its home US market from 2013. The prize for Free Basics was to grow the number of subscribers in the Indian market more effectively. Zuckerberg stated:

> [through] Internet.org in India now, there are already more than a million people who now have access to the internet who didn't otherwise ... in terms of DAU (Daily Accessing User) growth, the three largest countries were India, the US and Brazil.[45]

The threat of regulatory action was expressed in July 2015 by the Joint Secretary of the Department of Telecommunications, V. Umashankar:

> [I]f the need arises, the government and the regulator may step in to restore balance to ensure that the internet continues to remain an open and neutral platform for expression and innovation with no [IAP], or for that matter any content or application provider, having the potential or exercising the ability to determine user choice, distort consumer markets or significantly controlling preferences based on either market dominance or gatekeeping roles.[46]

He explained that the Telecoms Committee report delivered in July 2015 proposed *ex ante* regulation: 'a licensee has to file the tariff plan with TRAI prior to the launch. TRAI would examine each such tariff filing carefully to see if it conforms to the principles of net neutrality and that it is not anti-competitive

[42] Indian Department of Telecommunications (2015).
[43] Facebook (2015a).
[44] Jain *et al.* (2015).
[45] Facebook (2015b).
[46] Doval (2015).

by distorting consumer markets.'[47] Should zero-rating have already begun, as with Internet.Org and Airtel, 'penalties will be levied if there is a violation'.[48]

Facebook's partnership with third largest mobile operator Reliance Communications (RCom) to deliver Internet.Org was suspended on 24 December 2015 by Reliance, based on a request from the regulator TRAI.[49] The sequence of events was apparently that RCom informed the regulator on 23 November that it offered Free Basics, to which the regulator replied on 21 December, and asked the carrier not to deploy before submitting the terms and conditions, which included tariff plans. This led Facebook CEO Zuckerberg to interrupt his paternity leave to write an extremely aggressive statement in a major Indian newspaper on 28 December, accusing critics of misrepresenting Facebook's plans.[50] This backfired spectacularly by raising the spectre of economic colonialism, which is a very emotive issue for India, even 70 years after gaining independence from the UK. Guha and Aulakh explain that:

> On December 9, Facebook started a mass campaign on its platform asking users to support Free Basics and urged them to email Trai declaring their support of 'digital equality'. Free Basics was sought to be conflated with digital equality, with Facebook pitching the product as a solution to connect the unconnected billions. [TRAI] had called Facebook's Save Free Basics campaign a 'crudely majoritarian and orchestrated opinion poll'. It also pulled up Facebook for the responses, which the regulator said didn't address any of the questions posed in the consultation paper. On January 1, Trai asked the company to alert its users to send revised responses to the questions on the consultation paper as a vote for Free Basics did not hold up as a valid response.[51]

The Prime Minister, who had been a supporter of Free Basics less than four months earlier, advised Facebook to behave less aggressively: 'government must not allow any platform, no matter how popular, to monopolise any information system in the country as it can have far-reaching social, political and economic ramifications.'[52] This was the clearest indication of political pressure on the regulator to find against Facebook, which it did four days later.

The resulting regulations ban zero rating by both Free Basics via its Indian partner mobile network RCom, and domestic network Airtel's own zero-rated offer. Those offers that subscribers have already received were permitted to continue for six months (to August–September 2016), but any breach of that or any zero-rated (called 'differential pricing' in the Regulations) offer to new subscribers would make the licensed network operator liable to daily fines of

[47] *Ibid.*
[48] *Ibid.*
[49] Economic Times (2015).
[50] Zuckerberg (2015).
[51] Guha and Aulakh (2016).
[52] Mankotia (2016).

50,000 Indian Rupee (about US$700–750). Licensing is permitted and controlled by the Indian Telegraphy Act 1885. Though these fines are low, the context of the regulator's power over other licence conditions makes it unlikely that a network operator would not comply.

India's road to a zero-rating ban has been unusual: the regulator in spring 2015, and the Prime Minister in September 2015, appeared minded to support differential pricing, but the strength of public opinion and lobbying directed by civil society coalition SaveTheInternet.in, compounded by Facebook's culturally insensitive aggressive lobbying, led to a complete reversal within months.[53] Whether that decision leads other (post-colonial or otherwise) regulators into similar bans remains to be seen.

Canada

Canada had a chequered record on net neutrality until 2015, with rules proclaimed by the regulator in 2009 but not enforced until 2015. In 2011 the regulator explicitly supported capacity-based billing (rate caps),[54] which led the main IAPs to stop throttling video and other high bandwidth content as they had admitted to doing since 2008. It then adopted greater enforcement practices for net neutrality in 2014.

In 2008 the dominant incumbent Bell Canada was not ordered to stop throttling smaller IAPs to whom it provided wholesale connectivity, CRTC instead launching a wider inquiry into Internet Traffic Management Practices ('ITMP' was the acronym used).[55] In October 2009 Canada's regulator CRTC announced that it would in future examine infringements of net neutrality on a case-by-case basis,[56] using existing powers under Section 36 of the Telecommunications Act 1993, which states: 'Except where the Commission approves otherwise, a Canadian carrier shall not control the content or influence the meaning or purpose of telecommunications carried by it for the public.'[57] Thus the regulator chose not to act on any individual complaints until 2011. In 2011 Geist then documented failures to investigate, let alone act.[58] A much-heralded 2011 ruling on ITMP and data management caps was little enforced.[59] Until 2013, Canada's regulator claimed the power to regulate net neutrality, but

[53] Srivas (2016).

[54] CRTC 2011–703 Withdrawal of ITMP Letter.

[55] CRTC 2008–108 Telecom Decision, 20 November 2008.

[56] CRTC 2009–657, Telecom Regulatory Policy, 21 October 2009.

[57] Telecommunications Act S.C. 1993, c. 38 Assented to 1993-06-23 at http://laws-lois.justice.gc.ca/eng/acts/T-3.4/page-1.html This replaced the Railways Act 1906, which had been adapted to apply to all federal communications networks.

[58] Geist (2011).

[59] (TRP 2011–703) Telecom Regulatory Policy. CRTC 2011–703, Withdrawal of ITMP Letter.

chose to forebear, claiming no evidence of problems that would justify action by the regulator.[60] Even in mid-2015, research by Geist reveals that CRTC emphasises transparency over fining miscreants where IAPs are shown to have misled consumers over net neutrality violations.[61] The main form of Canada's net neutrality rules is not the ITMP decision itself, but rather provisions in the Telecommunications Act that predate the Internet (section 27(2) (no unjust discrimination) and section 36 (no interference with content).[62]

Jean-Pierre Blais became chair of CRTC in 2012 on the standard five-year term, announcing his arrival with the intention to properly regulate the sector in which the regulator 'has reputational baggage, I want to build it back up'.[63] This was in contrast with his laissez-faire business-friendly predecessors who 'would rubber-stamp almost anything they [the corporates] proposed',[64] including the net neutrality issue. In his first year, Blaise carried out four major interventions: he rejected former incumbent Bell Canada's initial takeover of Astral Media, until conditions were imposed that later revised and approved the merger; he limited mobile phone contract durations to two years; he pressured mobile IAPs into halving international roaming fees with the United States; and he investigated unbundling television channels, leading to a decision to force unbundling in March 2015. Bell Canada's president was also rebuked officially by CRTC for trying to interfere in editorial decisions to ban its TV station's coverage of CRTC, Blais stating 'An informed citizenry cannot be sacrificed for a company's commercial interests … corporate interests may have been placed ahead of fair and balanced news reporting'.[65] The president of Bell Canada was immediately replaced on 9 April 2015.

Note that Bell has cross-media ownership of CTV, Canada's most popular TV channel, and until 2005 also owned the largest circulation newspaper, *The Globe and Mail*. The Bell Canada attempt to purchase Astral Media (the owner of TV channels HBO Canada and The Movie Network) was announced in March 2012, but regulatory clearance was only given when the majority of English-language programming was divested, alongside local programming and unbundling requirements which would dilute any perceived threat to the public interest posed by dominance of Bell's programming in English-speaking Canada (note that Quebec, which has a quarter of Canada's population, is officially Francophone with only 7.7 per cent native Anglophones, mainly around the city of Montreal[66]).

[60] Miller (2012).
[61] Geist (2015).
[62] McTaggart (2008).
[63] CRTC, Statement from Jean-Pierre Blais, 2014.
[64] Ladurantaye (2013).
[65] Sharp (2015).
[66] See Wikipedia (2015 undated).

Zero rating is not common practice, and has not been definitively banned. However, a new CRTC case may lead to a definitive ruling: the Videotron 'Unlimited Music' case.[67] The net neutrality regulatory battle in Canada played out as a broadcasting ownership battle, in which programme unbundling had as an integral part the decision to regulate zero rating in 2015. It was to be expected that net neutrality violations favouring the company's preferred content would also form part of broadcasting regulation, specifically oversight of channel diversity. Unbundling of TV channels would not be appropriate alongside increased bundling of Internet distributed channels. CRTC ruled in February 2015 that Bell had been 'unlawfully' setting a double standard by exempting its $5-a-month Bell Mobile TV app from download limits it places on subscribers to its mobile network, giving it until 25 April to correct its pricing.[68] It also ruled against Quebec rival Videotron. Both are required to change to per-Gigabyte pricing. Bell had argued that the Mobile TV service provides 43 channels, only 12 of which are owned by Bell, the remainder being owned by other Canadian channel operators. The action was based on a 22 November 2013 complaint by student Ben Klass, supported by Telus, who argued that Bell in effect was marking up prices for competing streaming services by as much as 800 per cent.[69]

On losing the action in 2015, Bell immediately filed a lawsuit in the Federal Court of Appeal, whose hearing is pending, arguing that the CRTC was wrong to issue its decision under the authority of the Telecommunications Act, because Bell Mobile TV app is a broadcasting service, but it acts solely as an IAP for other parties' video. Bell says that broadcasting rules should therefore not apply, an argument approximating to that of the IAPs in the US who claim Title II telecoms regulation should not apply to their IAP activities. Moreover, given that Mobile TV was providing Canadian content in competition to over the top player NetFlix, nearly all of whose content is from the United States, the Mobile TV decision may be portrayed as opposed to Canada's national content policies. It illustrates that not all zero-pricing plans may be opposed on the same grounds and potential public interest at stake.

[67] CRTC, Unlimited Music Service, 2016. The two competing complaints filed by rival activists were items 2 and 4: In the Videotron case, the #4 wireless carrier in one province (Quebec) is trying to increase music streaming usage by doing non-monetary deals with a wide variety of players (even Apple is reputed to be joining soon) and zero-rating streaming. Both complainants say that if Videotron sets a precedent, all other carriers will do it too, but Guy Laurence, CEO of the largest carrier (Rogers), said they want to monetise data, not give it away.

[68] CBC News (2015).

[69] Stastna (2013).

Regulating zero rating

The issue of zero rating is highly contentious – a 'bad case' on which to make net neutrality law, as van Eijk describes it. I suggest two regulatory actions to encourage the correct use of zero rating:

1. treating zero rating as a short-term exception to net neutrality, and
2. ensuring any such short-term exception is not exclusive, by subjecting such contracts to FRAND conditions.

These conditions are not dissimilar to the principles by which the Wikimedia Foundation permits Wikipedia Zero to be offered by mobile IAPs, in that it: 'allows other public interest websites to ride onto its own scheme, eschews any exclusive rights or exchange of payment between itself and mobile carriers, and forbids carriers from selling the service as part of a limited bundle'.[70] I consider exceptions, non-exclusivity and FRAND in turn.

Short-term exceptions to net neutrality are likely given the post hoc nature of regulation: regulators lay out ground rules then respond to complaints regarding infringing practices. Difficult marginal cases can require extensive investigation. Such processes can take several months in the case of effective regulators, requiring both technical and economic analysis, a call for evidence, hearings and enforcement notices. In the case of litigious market actors, appeals against decisions can take months, years or longer to reach constitutional courts as the final appeal court. There is nothing in zero rating to suggest it is anything but a straightforward case of discrimination, which should not be subject to such long appeal processes. As explained earlier, walled gardens are nothing new, represent obvious discrimination and have been outlawed by those countries with effective net neutrality regulation. Any attempt to offer a time-limited zero-rated offer as an introduction to mobile data use could be flagged as such and limited by regulation to perhaps 3–6 months. This would be subject to FRAND conditions and regulatory enforcement.

FRAND conditions could be applied to:

1. mobile IAP contracts with Free Basics and other affiliated content providers, including the IAPs' own subsidiaries, and
2. conditions under which the content providers offer access to their own portals;
3. however, if zero rating is not taken up by a significant part of the subscriber base (e.g. 10 per cent of each operator's users), there may be a case for a *de minimis* exception from FRAND/non-exclusivity. It would be difficult to argue in practice that such a small number on a short-term basis distorts innovation significantly.

[70] De Guzman (2014).

The first condition is relatively straightforward to implement in theory but difficult in practice, as it is basically vertical unbundling of the mobile IAP's business unit arrangements. One could also compare it to the regulatory treatment under EU antitrust law of competitors to Microsoft's applications interoperating with their dominant Windows operating system.[71] However, not all regulators are capable of equal treatment of subsidiaries with competitors, especially in the resource-challenged developing world where independence and regulatory commitment are less easily maintained.

An alternative form of FRAND may therefore be to regulate de facto at a regional or global level, in establishing the ground rules for access to the zero-rated platform which mobile IAPs will offer. In this case, the regulated actor is the 'host' platform for those applications that will be offered. If applications to join such a platform offer – such as Free Basics or Wikipedia Zero's offer – are established under FRAND terms that can be examined and monitored independently, then the platform which is established for one developing market may, with few modifications, prove to be that offered in many others.

Mobile operators would like as much content delivered onto their networks as possible, including zero-rated and directly peered CDNs, such as Akamai or Level3. The appeal of Free Basics is the low bandwidth demand of its apps (no graphics, flash video). Some suggest directly peered CDNs should also be zero rated. It should be much cheaper (though not cost-free) to deliver content from a locally peered source. That should be passed on to the consumer, and zero rating is as good a way as any. Actual costs may be nearer zero than full price in any case. Note that without a data package alongside free content, content providers would be obliged to contract with a directly peered CDN – unless the zero-rating offer is very short term (e.g. three months maximum) to let new users 'taste' the edge of the Internet. I argue that FRAND and non-exclusivity should always be applied to zero-rated offers, short term or long.

I argued that zero rating is a relatively minor short-term problem, not technologically but price determinist as I now explain. The majority of 'mobile' data traffic is actually downloaded to devices via Wifi in the home, office or a hotspot location. It is not the cost of mobile data plans that is the dominant price driver, but that of hardware and prevalence of Wifi. There can never be as much Wifi in developing as developed countries, but open Wifi can be accessed relatively widely in countries where Internet policy is not dominated by the copyright maximalist lobby and morality (anti-pornography) cybercrime lobby. Hardware for mobile data is much cheaper than at its introduction a decade or more ago in the developed world, whether that be smartphones, laptops or

[71] Coates (2011), pp. 245–263.

tablets.[72] Combining the huge advances in technology pricing/performance with the prevalence of Wifi hotspots in 2015, it is clear that the environment for rapid adoption of mobile Internet access is far better than for fixed access in 2000. This applies despite the extremely high prices for mobile IAP data, which only forms a small part of the adoptive environment required to access the mobile Internet (arguably, no mobile IAP access is required at all given that schools, cafes, universities and other public areas offer free Wifi). Only 43 per cent of Brazilian smartphone users even use their data plans.[73]

Jurisdiction will be the greatest challenge to any attempt to regulate the platform rather than the mobile IAP offering zero rating. There are three obvious routes to enforcement:

- via the telecoms regulator's enforcement of platform neutrality on the mobile IAP, and therefore into the contractual terms of its agreement with the platform;
- via antitrust as a merger condition for any platform that chooses to expand into this area; or
- by a considered coordinated response by a network of net neutrality enforcement agencies at regional level, such as in BEREC.

The first has resource constraints, except that the better-resourced early mover regulators may establish ground rules that can be 'copy and pasted' by later-acting less-motivated regulators. The second is the type of net neutrality regulation that was adopted in the United States from 2005 onwards as an antitrust 'default' rule against large IAPs that wished to merge. In the global view of such mergers, a net neutrality undertaking for a limited time period was considered by the merger partners to be a small price to pay. The third is also difficult in practice to implement, though larger, well-resourced regulators (e.g. Germany/BEREC) advising their smaller cousins (e.g. Cyprus or Malta) can issue a decision or opinion that will help other regulators to take similar or identical action to enforce neutrality. Given the networks of regulators, consultants, civil society actors, academics and law firms that have exported and shared 'best'practice in telecom regulation since the first liberalisations in the 1980s (in Japan, the US, Sweden and the UK), such networks can be expected to actively engage in spreading such practices internationally.

Conclusion and further research needs

I considered whether zero rating poses a serious challenge to Open Internet use, examined the country case studies that demonstrate its regulation, and

[72] Freischlad (2015) states: 'Even in China, which is a more mature market [than Indonesia] by most measures and smartphone penetration is higher, data usage itself remains low. This tells us either Chinese smartphone users are not interested in using their phones on the go, or they are simply being thrifty.'

[73] Olsen (2015).

suggested areas for further independent research into the effectiveness of net neutrality regulation. I argued that zero rating is a relatively minor if highly controversial short-term problem as compared to Specialised Services, not technologically, but from a price-determinist view, as I now explain. Next to such a pervasive Internet policy problem versus privacy or free speech, is net neutrality an over-inflated sideshow, or a necessary precondition? Examination of national case studies helps to shed light on the extent to which net neutrality proves an essential pre-condition for solving other less technical, more politically accessible communications policy problems. More research is needed in this field as implementation of national and regional net neutrality legislation increases, especially in Europe; but this examination has shown that the roles of regulatory commitment, civil society activism and national political and market conditions are critical to the resolution of hard cases in net neutrality, specifically zero rating.

8

Net neutrality postponed

> This is for everyone #london2012 #oneweb #openingceremony @webfoundation @w3c
>
> Sir Tim Berners-Lee[1]

> Now this is not the end. It is not even the beginning of the end. But it is, perhaps, the end of the beginning.
>
> Winston Churchill[2]

Zero-rate friends and family; specialised-service your enemies

After 20 years we have reached the point of no return: we have net neutrality law in Europe, and in many other countries. It has even been accepted as a principle of the United Nations, as I explore later in this chapter. It is here to stay, however watered down its principles, however complex its enforcement, however unreasonable or overzealous its defenders, or duplicitous its enemies.

On 24 February 2016 I gave a presentation at the closed BEREC workshop on net neutrality, attended by national regulators and the European Commission, alongside Professor Barbara van Schewick, Dr Scott Marcus and Dr Alissa Cooper.[3] While the latter pair of presentations focused on the lack of evidence gathering to prove net neutrality breaches, and the problems created in the technical protocol stack by differing attempts to introduce QoS, my presentation and that of van Schewick focused on the manner in which the BEREC Guidelines needed to clarify Regulation 2015/2120. In particular, we both warned of the danger of operators both favouring their own and

[1] Berners-Lee (2012).

[2] Churchill (1942).

[3] BEREC, BoR (16) 49.

affiliated content providers ('friends and family' as I termed it) using zero rating, especially of video content, and discriminating against rival content by charging extra for service that is barely different in any respect from standard best efforts Internet traffic ('specialised-service your enemies'). The questions from the workshop suggested that there will be several national regulators who are sympathetic to such arguments, as they hold net neutrality as not so much a Friday afternoon job in their remit, as a never-to-be-enforced power. This is hardly surprising when regulators have always resisted new sector-specific regulation, favour consolidation of operators to increase investment in 'superfast' (*sic*) broadband, and have no new resources to meet the multifaceted challenge of net neutrality enforcement. Expect the smaller regulators in the Baltics, Cyprus, Malta, Luxembourg, the Visegrad Four and perhaps Ireland to be the first regulators whose very uncertain regulatory commitment to net neutrality will be tested by zero rating and/or specialised service plans. This double whammy of positive and negative discrimination appears to be emerging in the United States, where non-affiliated services, in particular NetFlix, has been rejected from some zero-rating plans while accepted on others.[4]

The Introduction explained that net neutrality directly regulates the relationship between IAPs and content providers, providing rules about how IAPs may contract with and treat the traffic of those content providers, especially that they may not discriminate against certain providers (either blocking their content or favouring commercial rivals such as IAP affiliates). It does not regulate those content providers directly. There are two types of net neutrality regulation: 'lite' and heavy. The former prevents IAPs from banning or throttling other content, application and service providers; the latter dictates non-discrimination on fast lane broadband, known as Specialised Services. Regulation and its enforcement has been delayed until 2017, given that 2015/16 laws and regulations failed to define those 'lite' and heavy regulations in detail. Chapter 1 explained the beginnings of net neutrality regulation in the US and Europe. Chapter 2 outlined competition policy's purpose and limitations, and regulating telecoms access based on the UK case study, providing insights into how difficult net neutrality regulation will prove in practice. It considered the possibility of platform neutrality or some other form of platform regulation. It assessed the possibilities of behavioural regulation to overcome some of the consumer detriments identified in nascent net neutrality regulation, and the wider use of behavioural 'nudge regulation' in Internet policy. Chapter 3 explained the current debate over access to Specialised Services: fast lanes with higher QoS. Chapter 4 examined the new European law of 2015, with Chapter 5 examining the interaction between those laws

[4] Readers can track these abuses on the book's blog at http://chrismarsden.blogspot.co.uk/.

and interception/privacy. Chapter 6 took a deep dive into UK self- and co-regulation of net neutrality. The majority of 2015 net neutrality regulatory cases related to mobile (or in US parlance 'wireless') net neutrality. Chapter 7 explored some of the wider international problems of regulating the newest manifestation of discrimination: zero rating.

In the final chapter, I consider the various means by which government can regulate net neutrality, focused in four parts: *ex ante* sector-specific and *ex post* generic (competition/consumer protection) regulation; educating the public and encouraging greater transparency; incentives for research and development of new technologies that can overcome (or exacerbate) the problem; and procurement and adoption of technologies to set a standard that the market can follow. The first is addressed with a toolkit for regulation. The second is summarised based on findings in Chapter 2 of the roles of competition law and behavioural regulation. The third is examined using the example of the tools of Internet science, notably the adoption of Responsible Research and Innovation (RRI) by the European Commission for Horizon 2020 projects, based on the Rome Declaration of 2014.[5] This approach is intended to explicitly include human rights to privacy and free expression in the design of Internet technologies. The final part, procurement, is analysed by examining both commitments to Universal Service Obligation (USO) and to an industrial policy of 'Gigabit infrastructure by 2025' set out by the European Commission. If broadband Internet is vital to future socio-economic development, then net-neutral fast infrastructure is an essential part of that process. In conclusion, I consider what net neutrality regulation may teach regulatory research in other areas, the law more generally, and the future of the social sciences as data science and other techniques emerge, making the interdisciplinary challenges of net neutrality regulation of more general application.

A toolkit for neutrality regulation

The case studies have provided a variety of responses to net neutrality violation in practice, with zero rating as the main concern in 2016. I now offer some elements that may be suitable as a toolkit for regulators to respond to net neutrality concerns. It offers several elements:

- how to engage stakeholders;
- how to measure neutrality;
- how to access prior knowledge in technical advice; and
- an example of how to respond to zero-rating offers.

[5] Europa (2014).

Table 9 Net neutrality policy, regulatory basis and major cases by case study

Country	Legislation/regulation	Published	Date enforced
Norway	Guidelines[a]	24/2/2009[b]	Zero rating declaration by NKOM of 2014
Costa Rica	Sala Constitucional De La Corte Suprema De Justicia[c]	13/7/2010	2010 by Supreme Court precedent
Chile	Law 20.453[d]	18/8/2010	Decree 368, 15/12/2010[e]
Netherlands	Telecoms Act 2012	7/6/2012	2014 and Guidelines 15/5/2015[f]
Slovenia	Law on Electronic Communications 2012	20/12/2012	Zero rating 2015
Finland	Information Society Code (917/2014)	17/9/2014	2014
India	Regulations (No. 2 of 2016)	8/2/2016	August: six months after Gazette publication date
Brazil	Law No. 12.965	23/4/2014	Consultation 2015–16, no implementation[g]
Canada	Hearing of 2010	Telecom Act 1993	Zero rating 2015
United States	Open Internet Order under Title II, Communications Act 1934 as amended by Telecoms Act 1996	2015	Zero rating unenforced except by merger condition
UK	Code of Practice 2011	Self-regulatory	Unenforced

[a] Sørensen (2013).
[b] Olsen, T. (2015).
[c] Costa Rica, *Guzmán et al. v. Ministerio De Ambiente, Energía y Telecomunicaciones* (2010).
[d] See Chile, Ley 20.453 de 18 de agosto 2010.
[e] Chile, Decree 368 of 15 December 2010.
[f] Netherlands Department of Economic Affairs, Net Neutrality Guidelines, 2015.
[g] For updates, see 'Ministry of Justice' (2016) available at http://pensando.mj.gov.br/marcocivil/

It is not prescriptive but descriptive, and points out that in all these areas, as well as Specialised Services, there remain serious research gaps in the analysis. These gaps were predictable five years ago but have only slowly been addressed, reflecting the political, economic and forensic uncertainty of net neutrality regulation.

Table 9 summarises the date of introduction of net neutrality policy, its regulatory basis and the major cases dealt with by the regulator.[6]

As seen, no decision has been made in the United Kingdom (and the EU states whose BEREC Guidelines are pending). All other case studies implemented some type of regulation of zero rating, though in the United States and Chile this appears to have exceptions (for music and video streaming in the US, Wikipedia Zero in Chile). The nations with the fastest median Internet access, Netherlands and Norway, also have the strictest net neutrality regulation in practice.

Co-regulation was used extensively in Norway, and to a lesser extent in the UK to form policy. The use of multi-stakeholder forums to consult on policy was made, in addition to parliamentary discussion, in the United States, Brazil and Canada. In the former two countries, and in India in 2015, thousands of replies were received (four million in the US, two million in India) in favour of some form of neutrality. The Netherlands and Slovenia had extensive parliamentary debate about their net neutrality laws. This confirms that, at least in form, the telecoms regulators became best of breed in terms of making consultations widely available and receiving significant numbers of non-traditional responses. The July 2016 BEREC consultation may show Europe to be the exception to this greater participation.[7]

Better regulation requires better evidence of impact and actors. Two outcomes have been presented to the EC in the Code of Practice Agora to improve the evidence base.[8] The first is a formal research/impact analysis task for the EC, rather than just claiming Option Zero if regulation is undertaken by corporate/NGO/standards actors instead of government (i.e. the other two points of the regulatory triangle in our case studies). For example, hardware governance and Border Gateway Protocols (BGP) need to be understood by government in order to formulate useful net neutrality policy even in the absence of formal regulation. How do these emergent areas interrelate with other parts of Internet governance? How does such governance interrelate with regulation, for instance where new actors and institutions are forming new coalitions of interest and epistemic communities? Further research into areas such as cyber-security, jurisdiction and borders, and standard setting is needed urgently to identify the complex international patterns of interdependent regulation of net neutrality's many interlinked technical, corporate and social facets.

[6] The classification of decisions as co-, self- or state regulatory is made on the basis of the analysis in Marsden (2010), pp. 159–180.

[7] EDRi (2016).

[8] Marsden, Chris (2013d).

These new areas also shine a light on the increasing role of non-traditional actors – institutions, the third sector, multi-stakehoderism (MSH), IAPs, social networks and participation of individuals in policy-related activism as evidenced by responses to the net neutrality legislation in the US, India and Brazil. How do we understand the meaning of online activism? What made a difference was stakeholders acting through main political veins (Obama's reference to the four million emails sent to the FCC in 2014 which opened this book, or the two million sent to TRAI in India in 2015), instead of the network make-up of corporate regulatory actors. Dynamic activities are taking place in different places. Recent work by Powell[9] identifies how participation in policy making employs networked dynamics but also created new discourses related to the Internet that countered the ways that governments had attempted to position these regulatory interventions.

In addition to the 'what' and 'who' questions, net neutrality research also reveals 'where' answers: non-conventional venues – international, non-state, code-based, for instance. We are moving towards a more multi-stakeholder environment (and our case studies demonstrate this, with more forms of regulation by market actors), and away from legislative bodies. 'Exotic' actors include prosumers (Anonymous, hackers, Wikileaks and others) and have a strong impact on Internet governance. The recent calls by President Roussef, Chancellor Merkel and others for a 'sovereign cloud' is related to the Snowden revelations, but impacts powerfully on net neutrality. Governments need to commission a research programme to understand these processes, actors and venues.

The net neutrality case studies illustrate 'how' participation can occur in many modes, but they also stress that effective participation in governance is not only a matter of greater numbers of people representing different groups, but is also contingent on the legitimacy of spaces for participation. For code-based governance, this is often linked with expertise, but as the case of net neutrality demonstrates, this legitimacy also emerges in relation to other actors and through the use of the technical solutions. Standard-setting processes can be hijacked to further private interests. In some case studies we see that the more stakeholders there are, the less effectiveness there is. Experts, notably engineers, may migrate to another forum to avoid tedious legitimacy discussions and to conduct 'real work' on QoS. Thus, institutional contexts remain important, and understanding depends on development of new analytic models that:

1. Identify the manner in which governance and legitimacy emerge socio-technically;
2. Employ analysis of power, including the power of policy networks and the significance of discourses as developed by activists, individuals, the media and governments;

[9] Powell (2013).

3. Avoid justifying Potemkin multi-stakeholderism by separating policy domain and policy issues.

Net neutrality demonstrates itself as a Potemkin stakeholder process. Processes it puts in place are *post facto* opportunities for input. Loss of legitimacy of institutions is important, whether due to mission creep or issue linkage, such as between net neutrality, interception and privacy. How do you separate increased participation from decision making in drafting new processes? What do you do when greater participation breaks effectiveness? Proximity allows stakeholders to take each other's interests into account, but a research question that emerges is: Do we take into account the direction of 'travel' in the case studies, e.g. downstream effects? Repulsion and attraction of different multi-stakeholders, such as civil society in the case of net neutrality, is a dynamic process that needs more research.

Measurement

Research is needed to examine both enforcement of transparency in TMP by governments and their agencies, notably through use of SamKnows monitoring (Brazil, US, UK, EU, Canada) and the publication of key metrics, and enforcement by regulators following infringement actions where published. Seven of the national case studies are now using measurement devices in the consumer's home. SamKnows is now active in measuring end user TMPs in contracts with regulators in the US, Brazil, the UK, Canada and the European Union as a whole.[10] This has supplanted self-reporting of violation by the IAPs, and network measurement by downloaded diagnostic tools, as the preferred method of discovering TMPs. Given the lack of clarity in the latter, and the obvious incentive paradox in asking IAPs to self-report violation, this approach appears to be the best fit.

The US regulator is taking action to actively consult on future TMPs that may violate neutrality via its Advisory Opinion approach. Even critics of net neutrality acknowledge that better measurement of end user experience is a vital contributor to forcing IAPs to offer increased transparency to end users.[11] A report for Ofcom published in August 2015 concluded that an approach based on a quality floor (i.e. minimum service quality, possibly based on a new

[10] See SamKnows (2015); they claim to be working with over 30 regulators.

[11] See Geddes (2015), stating that the 'findings obsolete several papers and books on the subject by legal scholars. Their understanding of network performance is unsound, and they have been unintentionally fuelling the conflict as a result. Furthermore, that we can't yet properly measure the services we are offering in a customer-centric way is an industry embarrassment. The technical weakness of these tools is a cause of industry shame. This should be chastening for all of us in the broadband business to do much better.'

Table 10 BITAG and OIAC technical reports 2011–14

BITAG 2011–14[a]	**OIAC 2013**[b]
2014 Interconnection and Traffic Exchange on the Internet	20 August 2013 Economic Impacts of Open Internet Frameworks
2014 VoIP Impairment, Failure, and Restrictions	20 August 2013 Policy Issues in Data Caps and Usage-Based Pricing
2013 Real-time Network Management of Internet Congestion	20 August 2013 AT&T FaceTime Case Study; Openness in the Mobile Broadband Ecosystem
Port Blocking 2013	20 August 2013 Specialized Services: Findings and Conclusions
SNMP DDoS Attacks 2013	20 August 2013 Open Internet Label Study
Large Scale Network Address Translation 2012	17 January 2013 Specialized Services
IPv6 DNS Whitelisting 2011	17 January 2013 Economic Impact Data Cap

[a] See www.bitag.org/ (Accessed 13 September 2016).
[b] See www.fcc.gov/encyclopedia/open-internet-advisory-committee (Accessed 13 September 2016).

Universal Service Obligation) would help app designers and users understand better how SamKnows-type measurement can help them make better choices.[12]

The advanced measurement standards emerging may help regulators and consumers understand how best to enforce net neutrality standards.

Technical advice

Technical elements of net neutrality remain complex in both resource and interpretation for regulators, especially those with fewer human resources and technical experience. It would be helpful if greater clarity on such future approaches were to build on the former role of the Open Internet Advisory Committee (OIAC) of the FCC in 2011–12, and Broadband Internet Technical Advisory Group (BITAG) in the period since. Between OIAC, BITAG and BEREC, many very useful technical and policy reports have been produced since 2011. I highlighted BEREC's contribution earlier in Chapter 4, but in Table 10 I list valuable US contributions.

These reports were all either written by a co-regulatory group, as with OIAC and BITAG (though the latter claims to be formally self-regulatory), or consulted on with many stakeholders. Note that the BEREC site lists several other

[12] Predictable Network Solutions Limited (2015).

draft papers developed in 2011–15. BEREC has consulted very widely on its approach within the various regional regulator groups, including in what might be termed the 'Regulators' regulators' forum in Barcelona on 2–3 July 2015, when no less than ten national regulators explained their approaches to net neutrality.[13] BEREC met with EaPeReg (Eastern Partnership Electronic Communications Regulators Network), REGULATEL (Latin American Forum of Telecommunications Regulators) and EMERG (Euro-Mediterranean Regulators Group) for the high-level Regulator Summit, representing over seventy regulators.[14]

In terms of the value of net neutrality to consumers, regulators in the Netherlands, UK and BEREC[15] all commissioned specialist reports to use focus groups to ascertain consumer ignorance and anger. These are in addition to the SamKnows reports released on an annual basis by regulators.

A Paris conference in 2005 of the most senior IAPs and academics concluded that the only foolproof method of discovering net neutrality violations is when IAPs not only admit to the practice but use it as marketing to sell their services as superior to 'neutral' competitors.[16] This has occurred frequently where net neutrality 'lite' blocking of, for instance, Skype and other IM services has occurred, but may occur less frequently as IAPs respond to the new European law which outlaws this type of blocking. In future it will be even harder to identify violations. Academic research in this field is often industrially and governmentally supported, which has made it very perilous to conduct such research without accusations of bias in policy. Interference in the research agenda by corporate and/or government sponsors of other researchers in the departments in which a researcher works can put collegial pressure on the researcher to curb investigation into malfeasance.[17] Research into detection of violation in Europe has been deliberately blocked by potential funders, whether or not under pressure from corporate interests, while research building the technical and economic case for violation has been richly rewarded.

Those academics conducting research that measures net neutrality have been accused, often incorrectly, of accepting support from content companies, specifically Google until 2010 (Google has become less committed to net neutrality since that point). In this respect, academia is facing that wider issue of identifying sponsors and corporate–government influence over research agendas that has become particularly acute in Europe since 2008 as research council funding has declined and direct corporate and government-sponsored research support has replaced it, yet has been commonplace in the United States throughout

[13] Olsen (2015), slide 6.

[14] BEREC, Outcomes of the BEREC–EMERG–EAPEREG–REGULATEL Summit 3 July, 2015.

[15] BEREC, BoR (15) 90.

[16] Marsden (2010), p. 242, footnote 32.

[17] This applies much more thoroughly in finance, economics and accounting. See Sikka (2013, 2015).

its academic history.[18] Net neutrality is a small regulatory issue area compared to the global financial, pharmaceutical and arms proliferation crises caused by both explicit deregulation and failures to regulate in the twenty-first century.

The European funding gap is likely to become a less severe ethical hurdle as European law finally requires companies and governments to support net neutrality from 2016, and 2015 workshops for both EINS and the SMART Internet Measurement project (which I attended) were permitted to become substantial net neutrality discussions.[19]

The United Nations to the rescue? Free expression and privacy

Law is typically grounded in national policy outcomes, with international law an extension of national norms and processes, notably in the doctrine of extraterritorial application of national laws. Enforcement and policing are typically carried out in cooperation between national jurisdictions, and the most significant international law relating to the Internet remains the 2001 Council of Europe Cybercrime Treaty (to which the United States, Japan and Canada are non-European signatories). International lawyers respond to transnational problems with summitry, followed by an international treaty, paralleled by continued extraterritorial application of national law in cooperation with other jurisdictions: a pattern that is repeated with regard to the Internet, with one exception. The lack of competence and legitimacy for state action in regard to the Internet, and the continuous opposition of the United States government (supported by its complicit partner the European Union) to an international treaty in any but specific cybercrime policing matters, has meant that the frequent pleas from international lawyers, the United Nations system and regional human rights bodies for a treaty to establish international norms for the Internet have been met by deaf ears. In the World Conference on International Telecommunications (WCIT) renegotiation of 2012, varied attempts were made to establish a decision-making body to replace the toothless consultative summitry of the Internet Governance Forum. International lawyers' calls for summitry to lead to binding treaties failed. The lack of policing and enforcement of extraterritorial jurisdiction make calls for a more vigorous international treaty or application of human rights legal norms equally frustrating.

Net neutrality regulation is a blunt telecom regulatory instrument for a multifaceted problem such as Internet access, which also includes such policy issues as privacy and free expression, as well as universal access and many Millennium Development Goals. Belli and Foditsch have written extensively about

[18] Veysey (1965).
[19] See SMART 2010/0030.

the modelling of a universal principle-based network neutrality law, an experiment conducted through the Dynamic Coalition on Net Neutrality in the UN Internet Governance Forum led by Belli since 2012:

> it seems possible to distil some essential elements from the existing net neutrality frameworks, in order to define a common principle base on which interested policymakers or market actors can develop compatible net neutrality frameworks. Indeed, while the Internet is usually seen as an interconnection of electronic networks, it is important to stress that the online environment also determines an interconnection of juridical systems that may benefit from shared policies.[20]

Crowd-sourcing, however led by experts, is a novel form of legislative initiative, with Rasdu *et al.* cautioning that there is greater need for 'leveraging sufficient community interest for substantial input; defining procedures for the collection and screening of inputs; and committing to institutionalizing rules for incorporating feedback.'[21] This issue will not go away quickly.

United Nations policy documents remain generalist and somewhat naive, playing catch-up with telecoms regulation in developed nations. The official UN 'State of Broadband' report appeared to have only a single author, who thanked in preface over forty equipment vendors and telecoms lobbyists but only one person who might be seen as an independent expert. The use of corporate statistics without criticism or any objectivity was immediately condemned by the technical press, who saw the report as captured. It stated:

> Internet companies and Internet content providers need to contribute to investment in broadband infrastructure by debating interconnection issues and revenue/fee sharing with operators and broadband access providers to increase investments in broadband infrastructure and energize the broadband ecosystem.[22]

It does offer support for a FRAND solution to Specialised Services, however: 'Although various strategies for open access exist, it is vital that policymakers ensure that access to new facilities is provided on fair, reasonable and equivalent terms.'[23]

David Kaye, United Nations special rapporteur on freedom of expression, argues that:

> In the longer term, net neutrality policies should be guaranteed wherever Internet infrastructure is being built out. The 13 'Necessary & Proportionate' Principles, which apply human rights to communications surveillance, should also be adopted and implemented as a framework for rights-respecting connectivity.[24]

[20] Belli and Foditsch (2015), p. 297.
[21] Radu, Zingales and Calandro (2015).
[22] Broadband Commission For Digital Development (2015), Section 6.6, p. 75.
[23] *Ibid.* at Section 6.3.
[24] Kaye and Solomon (2015).

In December 2015 he argued for a human rights-oriented connectivity programme to flow from the UN General Assembly debate on WSIS+10 (a ten-year review of the World Summit on the Information Society) and the newly updated Millennium Development Goals ('Global Goals for Sustainable Development' (GGSD) as adopted by the UN General Assembly in September 2015). The GGSD emphasise that access to technology underpins every other 'Global Goal' toward the eradication of extreme poverty. He particularly urged cautious adoption of the multinational connectivity platform pursued by Facebook and endorsed by celebrities, explaining that:

> Mark Zuckerberg and Bono issued a call to 'unite the earth' and, with other global opinion shapers and business leaders, released a Connectivity Declaration to 'connect the world'. The U.S. State Department's Global Connect program makes Internet access a foreign aid priority ... But connectivity alone cannot be global policy. Respect for privacy and the freedom of expression must go hand in glove with the drive to connection.[25]

He argued strongly that the Facebook-sponsored Free Basics project, which offers free access to basic low-bandwidth versions of sponsored websites such as Facebook itself, Wikipedia and local news websites, offers a false equivalence with Open Internet access, warning that government may 'bless deals creating a two-tiered Internet pushed by so-called zero-rated service providers that limits browsing to pre-selected applications and establishes new gatekeepers'[26] such as Facebook. This may be especially pernicious as Free Basics is rolled out in the least developed countries with very low fixed Internet access, and thus greater dependence on low bandwidth mobile connections. Examples are Zambia, Myanmar, Kenya, Peru and Guatemala.

Privacy is an area of clear theoretical distinction between the EU and US, even though in practice smaller European states have highly inadequate regulators, while the US has a strong federal regulator which has imposed fines on a scale far beyond those imposed by its weakling European counterparts.[27] The UK shares the US's ambivalence towards privacy, its government campaigning in the 2015 General Election to leave the 47-member European Convention on Human Rights as a result of media-inspired fears of Article 8 privacy rights.[28] In most developed countries, neutrality developed from privacy concerns, a dynamic which needs further empirical comparative research in the developing nation context.

The calls for a 'Magna Carta for the Internet' in the wake of the Snowden revelations miss the point that since the OECD 1980 Guidelines on the

[25] *Ibid.*

[26] *Ibid.*

[27] Brown and Marsden (2013a), p. 186.

[28] Wagner (2015).

Protection of Privacy and Transborder Flows of Personal Data,[29] and especially since Directive 95/46/EC, there has been such a document in regard to privacy. In the concept of Privacy by Design, the standard use of such guidelines in funded research would prevent some of the most egregious failures to innovate with basic concern for human dignities, which impact users of net neutrality-violating technologies. These are not difficult outcomes to effect, though they require attention to social science and basic regard for regulation.[30]

If international multi-jurisdictional legal doctrine fails to regulate the Internet, this arguably calls for more sophisticated transnational responses, a reality that legal systems and policymakers have embraced cautiously since the mid-1990s. Law is to a great extent the attempt to enshrine controls over technosocial systems; and given the technological and social challenges of controlling the Internet, combined with many lawyers' understandable reluctance to make bad law that leads to ridicule of their technosocial incompetence, new approaches that offer behavioural guidance, rather than prohibition and control, have been attempted. These follow the behavioural science approach of using legal nudges to achieve regulatory aims, as discussed in Chapter 2.

Privacy remains a thorny issue, and is largely unregulated in developing countries. The wider issue of how Internet users of 'free' apps such as Facebook and others are being monetised by advertisers is associated with the net neutrality and zero-rated debates, and in particular the correct policy responses. In countries such as Indonesia where monthly Average Revenue Per User (ARPU) is only $2.20 for calls, texts and data, it is unsurprising that advertising is attractive as a further revenue partnership with zero-rated apps.[31] Freischlad considers:

> Users of zero-rated apps should definitely be aware that aspects of their browsing, downloading, and searching behavior are likely being recorded and analyzed, as both the zero-rated app itself and the sponsor who footed the bill are interested in monetizing this data further. Is there no alternative to sponsored data? It's almost cynical: the most vulnerable people – low income communities just making their first steps on the internet – become easy targets of marketing messages and data mining.[32]

A much more popular service than Facebook (described by Pahwa as 'privacy nightmares'[33]) is Jana Corporation's mCent, a service that lets users use mobile

[29] OECD (2013a).

[30] Not least in the applications for funding from, for instance, the European Research Council or Horizon 2020. A simple initial answer may be to ensure a social scientist with privacy regulatory experience is involved in evaluating each bid with some personal data component. See http://ec.europa.eu/research/participants/portal/desktop/en/experts/index.html, noting that data protection forms part of an illustrative list under research ethics, and http://ec.europa.eu/research/participants/portal/desktop/en/support/legal_notices.html (Accessed 13 September 2016).

[31] Freischlad (2015).

[32] *Ibid.*

[33] Pahwa (2015).

data as a reward if they try a new app – many of which are privacy-invasive. The choice of trading your privacy for basic Internet access is a daily occurrence for the reported 30 million mCent users.[34]

When considered next to such a pervasive Internet policy problem as privacy or free speech, is neutrality an over-inflated sideshow, or a necessary precondition? Examination of national case studies helps to shed light on the extent to which net neutrality proves an essential pre-condition for solving other less technical, more politically accessible communications policy problems.

Interdisciplinary analysis of regulation

Regulatory concepts in multi-stakeholderism, co-regulation, algorithmic regulation and conceptual models of regulation processes are becoming mainstream, as is measuring the effect of multi-stakeholderism. The net neutrality case studies demonstrate the breadth and depth of emerging institutions and actors in regulation and governance, from hardware and BGP to international organisations and multi-stakeholder governance, to bottom-up communities creating innovative open network solutions.[35] Moreover, many ideas to educate politicians about regulating the Internet have been accelerated by the Snowden revelations, causing an intense interest in Internet governance and net neutrality.

Net neutrality is an intensely complex regulatory problem, with implications for a research agenda into both regulatory impact assessment and regulation of support for science and technology. There is continued lack of integration between technical and social sciences in regulatory assessment.[36] In 2001 I wrote: 'The omission of research from nationally-oriented agendas due to funding and resource constraints, is compounded by the disciplinary gulf between social scientists and computer scientists.'[37] In bridging this gulf, more systemic research is needed. *Nature* editors stated:

> If you want science to deliver for society, through commerce, government or philanthropy, you need to support a capacity to understand that society that is as deep as your capacity to understand the science.[38]

That means using social science inputs to help support better regulation and governance of society.[39] Integrating social and technical sciences has been vital both to the innovation engine that the Internet represents and its success. Now that the Internet, and digital information sharing more generally, is becoming

[34] Freischlad (2015).
[35] See Musiani *et al.* (2015).
[36] On impact assessment in general, see Radaelli, Dunlop and Fritsch (2013). On the 2009 Package see Horten (2010).
[37] Marsden (2001, 2004).
[38] Nature (2015).
[39] Flyvbjerg (1998).

the growth engine for the post-industrial economies, these lessons need to be reinforced throughout policy making, notably in both the assessment of regulation and the use of regulation of technology funding.

Governance of research funding can be controversial with respect to fundamental regulatory requirements for new technologies, notably in privacy but also security, interoperability and in other respects. Whereas regulation is often considered reactive, in the field of scientific research it can be proactive, saving vast amounts of time and expense in ensuring that innovation meets basic societal needs in its planning. Evaluation of funding needs connecting to the policy process to ensure regulatory outcomes can be matched to potential innovations in technology. A classic case in point of regulation anticipating and responding to such concerns is the Internet of Things, where the Commission introduced several innovations in the policy process to address regulatory concerns, having been intimately involved in the funding of such components as Radio Frequency Identification (RFID).[40] A rather less successful example might be net neutrality, where the funding of QoS was not accompanied by a commitment to ensuring an Open Internet with fundamental freedoms observed in implementing Specialised Services. As a result, such regulatory requirements have had to be retrofitted into the ongoing deployment of such technologies.[41] Much time and effort can be saved by ensuring that regulatory requirements are addressed at an early stage in such processes, through standardisation and implementation of privacy impact assessments, for instance.

The EC Code of Practice Agora provides an example of a limited but successful umbrella gathering of experts on co-regulation, and may provide a template for a 'foresight' assembly of experts. Certain issues arise with regard to expertise versus advocacy mapped onto policy controversies, for instance on privacy and net neutrality.[42]

Internet industrial policy for gigabit infrastructure?

One of the central governance gaps in information technology policy has been that shared more broadly in techno-economic policy: markets are failing badly, with governments abandoning national strategies in favour of reliance on buccaneering hedge fund-led investment, which has proved neither far-sighted nor strategic for developed economies. While it is more newsworthy to bemoan the

[40] European Commission Decision of 10 August 2010 setting up the Expert Group on the Internet of Things, and European Commission, Conclusions from the Internet of Things public consultation 2013.

[41] See BEREC, BoR (14) 117.

[42] See http://caps-conference.eu/ and more generally https://ec.europa.eu/programmes/horizon2020/en/h2020-section/collective-awareness-platforms-sustainability-and-social-innovation-caps for community action platforms (Accessed 16 September 2016).

fate of the government-supported banking sectors in developed countries or the collapse of the British steel industry in mid-2016, and the polity is obsessed with the failure of the European Union's political-economic vision and the attempt by the British non-political class to leave the European project entirely in favour of neo-colonial ambitions with developing countries, the future of industrial policy lies at a crossroads.

Law can create property rights and use the machinery of government to spur the development of technology via both procurement of private sector expertise and the use of government funding to conduct research and development and primary research. In practice, many innovations were brought to market by a combination of both government procurement and government-funded research. Examples include virtually every technology that we can document from the ancient world and, more especially (given their archaeological prominence) the major civil works and transportation projects from earlier civilisations, notably sanitation, road and harbour building, great libraries, temples and mausolea of gods and emperors, city walls, forts and castles. In particular, military expenditure on innovation has played a well-documented role in technological innovation, and the literature on government expenditure is voluminous. Law played the role of authorising such government expenditure. The Internet uses common carrier public telecommunications systems based on ancient rights of way via what were formerly Roman roads and Victorian railway tracks and telegraph lines, for instance.

There is a particular need for policy making in dealing with disruptive innovation. 'Black swans' have been an issue of great interest to policymakers in the wake of the long recession and Euro crisis since 2008.[43] Turk details the events in 2008–10 that led to the Reflection Group final report: the failed referendum in Ireland in June 2008; the collapse of Lehman Brothers in September 2008; December 2009, when the Treaty of Lisbon came into force; the Greek and Euro crisis in March 2010 that is ongoing. One could add for information policy the various attempts to suggest cyberwar, such as the North Korea–Sony farce of December 2014; Assange's work in Wikileaks since 2008; Snowden's revelations in 2013; and the Brexit referendum result and Trump election in 2016.[44] A pressing need is to strengthen the EC capacity in foresight for technological innovation in the area of Internet policy. Internet Science grew out of three pioneering and highly successful foresight exercises: Towards a Future Internet (TAFI),[45] Reflection Group on the Future of Europe[46] and EIFFEL.[47]

[43] See Taleb (2007) on the concept of 'black swans'.

[44] See some Internet Science outputs on black swam effects: Marsden (2014b, 2015).

[45] Blackman *et al.* (2010).

[46] On final results see Turk (2010).

[47] See http://cordis.europa.eu/fp7/ict/programme/publications1/books/futint/fi36-eiffel_en.html (Accessed 16 September 2016) and Trossen (2010).

While the Internet and technological issues are prominently represented in EU strategic work such as ESPAS,[48] there is a clear need for a much larger-scale foresight exercise identifying the many challenges that are presented by digital social innovation (DSI).[49] A foresight panel would at least be the start of an attempt to identify some of the issues at stake and potential outcomes.[50]

Information technology policy is an aspect of industrial policy, a sector which has gained enormously from government funding for research and development (R&D). Mazzucato has shown how companies such as Apple, Hewlett Packard, Qualcomm, Motorola and Microsoft gained hugely from such investments, most notably at Xerox Parc but also in ongoing programmes. UK companies such as Marconi (until its demise in 2006), Vodafone and even ARM benefitted greatly from UK promotion of their international expansion via favourable tax regimes, tax breaks for research and development, and purchase of their products.[51] For instance: 'ARM uses legitimate tax exemptions and reliefs to minimise its tax liabilities. A large proportion of ARM's products are developed in the UK, where the government offers R&D tax incentives, namely R&D tax credits and the Patent Box, to companies with R&D commitments.'[52] A highly successful and innovative company, ARM licenses its technologies to 105 partner semiconductor chip companies to manufacture its products, selling 12 billion chips in 2014. However, it is feared by Mazzucato and Andy Grove (who led Intel for two decades) that the hollowing out of manufacturing to lower-cost locations (especially China) will lead to mass employment moving away from the developed nations that develop the technologies. Grove cited FoxConn, which manufactures on behalf of Apple and others, employing 1.3 million people mainly in China and its Taiwan base.[53]

Employment outsourcing by technology supported by developing nation workforces has become a major political issue in 2016. De-industrialisation is so advanced, especially in the United Kingdom, that hardware manufacturing at scale has been abandoned to other nations, especially those outside Europe. The European Commissioner responsible is German, and therefore more dedicated to skilled manufacturing than the British rentier class who occupy the Cabinet of the United Kingdom government. On 14 March 2016 Oettinger called for a 'Gigabit infrastructure for the Gigabit economy'.[54] He stated:

[48] See 'The World in 2030', available at http://europa.eu/espas/pdf/espas-outreach-leaflet.pdf (Accessed 13 September 2016).

[49] See European Commission, H2020-ICT-2015 Collective Awareness Platforms for Sustainability and Social Innovation, 2015

[50] See www2.warwick.ac.uk/fac/cross_fac/complexity/research/ (Accessed 13 September 2016).

[51] Mazzucato (2013).

[52] Score (2015).

[53] Grove (2010).

[54] CEBIT News (2016).

> Everyone should enjoy adequate connectivity to fully benefit from digital opportunities and from Digital Single Market. For me the adequate level of connectivity is a Gigabit society by 2025.

This requires universal fibre connection deployment in only eight years.[55]

Ofcom's November 2015 SamKnows traffic measurement showed that only the fastest UK consumer broadband product, VirginMedia's 'up to 200Mbps' fibre service, with the 'highest average actual download speed at 174.0Mbit/s' and the only fibre to the home option available, could offer UHD: 'Thirteen per cent of ADSL2+ packages streamed NetFlix videos reliably in Ultra High Definition (UHD), while this figure was over 90% for cable and FTTC services.'[56] Fibre is increasingly accepted as the route to that 'Gigabit economy', yet the UK continues to claim that industrial policy should play little part in private providers' deployment of higher speed connectivity, even though Ofcom-measured 'average download speeds in urban areas (50.5Mbit/s) were over three times those in rural areas (13.7Mbit/s). The main reasons for this difference were the lower availability of fibre and cable broadband in rural areas and slower average ADSL and fibre-to-the-cabinet (FTTC) connection speeds.'[57]

The need for fibre is evident, but the UK government is not investing in upgrading households beyond copper broadband service. The pursuit of sharing economy policy, inspired by Ayn Rand acolytes in venture capital, is sponsored by the British government but vociferously opposed and held in contempt by the Germans, and indeed much of the social democratic polity.[58] Instead, their platform protection against Silicon Valley venture capital-backed attempts to overturn European-regulated accommodation and taxi services, amongst others, is daily backed by those constituencies on which business travellers depend.

Net neutrality and the future of law

Lawyers have been challenged by net neutrality and digital communication regulation for three reasons:

1. technologies are fast-moving and require expert design choice to implement policy choices;
2. the technologies are typically international, if not of the mythical 'borderless' character ascribed to them by libertarians; and

[55] Oettinger (2016).

[56] Ofcom, UK home broadband performance, November 2015: the performance of fixed-line broadband delivered to UK residential customers, 2016.

[57] *Ibid.*

[58] Hill (2016).

3. enforcement of legislative will is difficult and uneven on the Internet, with the result that many more sophisticated types of legal instrument, including 'soft law' types, are required.

Each challenge and the stereotypical and failed legal responses are discussed in turn.

Regulation is a term of art used by lawyers to describe the broad set of attempts to control an environment using control systems, which has been extended to explain legal controls as one of a set of four modes acting on the environment to be regulated, together with architectural control (road planning, urban design or, in this case, technosocial construction of the software environment), social norms imposed by the community and economic forces acting on the exchange of goods and services (including reputation and other intangibles) online.[59] This gives law a much broader toolset with which to influence other environments than a narrow description of legislative and court-based prescriptions on particular behaviours based on 'law in books'. For instance, governments authorised by law may fund standards bodies to develop new technical standards that enable better protection of privacy, exchange of information or an architecture of control. The development of the public Internet was in part enabled by law using public funds in public universities under state law.

Law also facilitates technological innovation by authorising the creation of the science base, from the corporations and non-profit bodies for higher and secondary education, to compulsory primary and then secondary education, to the protection of competition and its primary private sector exemption, intellectual property. Authorising the creation of the science base is critical to the development of digital technologies.[60] Moglen explained that the corporate–political relationships:

> extend in many cases back to the period immediately after [World War 2]. They have merely grown with time. The technical facilities that were covered by the arrangements went from telegraph to telephone, through rebuilding of the communication network destroyed in Europe.[61]

That continues in the broadband space. I explained:

> The influence of the dominant super-power is greater than ever before, driven by ICTs. Where the British Empire was represented by telegraphs and railways, the US is represented by satellite television, Hollywood and the Internet .... ICT standards have driven the clustering of economic power within the most connected networks of private corporations globally.[62]

[59] Lessig (1999).
[60] Mazzucato (2013).
[61] Moglen (2013).
[62] Marsden (2001), p. 4.

Law also plays a role in ensuring the standardisation of technologies, and much of the early modern 'weights and measures' legislation was updated versions of Roman legal standards for measurement. Two examples of standard setting are P2P distribution networks for Internet content, and encryption. Standards have always been important. Legal standardisation plays a role in prohibiting the use of many non-standard technologies, which has created great legal controversy over time, notably in the adoption of Internet technology standards that, it has been argued, are markedly inferior to their defeated rivals: the QWERTY keyboard for the English language and standard gauge rail are good examples,[63] as is net neutrality according to some network engineers, as we saw in Chapters 3 and 6.[64]

Law can have its longest-run effect on technological innovation in institutionalising a dominant standard. The use of the IP suite for much of global communications was effected, not by commercial standardisation nor law initially, but by government-funded (mainly) university researchers, as was the E2E principle. Their adoption as standards was an unwelcome, and still unwelcomed, surprise to the data communications industry. Debates in the third millennium about 'network neutrality' are in fact debates about the attempts by the telecommunications carriers to return to a more clockwork-Cartesian pace of technological development in data traffic, away from the innovative chaos unleashed by the unheralded arrival of IP and E2E in the 1980s. Note that even in more traditional telecommunications standards bodies, such as the intergovernmental International Telecommunication Union, the overwhelming majority of actual technical standard setting and development is carried out by companies, universities and affiliated researchers, before eventually receiving official approval.

Net neutrality can be seen as another iteration of the universal access problem, involving great complexity and difficulty.

Economic-technical efficiency and human rights reasons predominate in the emergent areas studied, with net neutrality as a case in point. Solutions could involve non-traditional methods such as complexity, behavioural solutions, co-regulation, filtering, private censorship, licensing and contracts, security audit and liability rules. A new institutional analytical model is emerging that is based in policy networks literature, with epistemic communities built around issue areas such as net neutrality governance. Issues themselves have actors, accumulate people and define powers and instruments: applying an institutional context defines these emerging regulatory communities. Informational challenges for global public goods such as net neutrality are fundamental, including governance (with new venues for international state–firm diplomacy) and

[63] David (2001).

[64] Crowcroft (2015); Geddes (2015).

security (privacy and openness). The alarming lack of expertise revealed by the Snowden leaks make more risible the escalating political calls for 'cyberwar' and sanctions against nation states, and retaliatory strikes between states hide a much more interesting need for informed debate about censorship, encryption-by-default and liabilities for net neutrality violation. The rise of the policy agenda surrounding privacy in the wake of Snowden's revelations is often obscured by the surveillance industry's calls for greater DPI-based intrusion against real or inflated risks rather than sensible evidence-led policy. Global public goods are too important in this sphere to be left to corporate lobbyists without a robust independent scientific evidence base. There is a need to nurture the independence of researchers who can robustly analyse real rather than invented risks.

In conclusion I offer some thoughts on methods. In searching for hard regulatory cases, the net neutrality case studies attempt to understand actors with empirical analysis of method; this is work that needs to continue. For example: Is multi-stakeholderism a reality? Multi-stakeholderism requires legitimacy, and there is therefore a need to expand our methodological toolset so that we understand how this legitimacy is constructed, in order to avoid creating the conditions for Potemkin (sham or empty) multi-stakeholderism. The case studies highlight sites and methods for undertaking this further research, but other methodological and empirical sites and approaches can be identified. To construct the possibility for real multi-stakeholderism, we need to understand the technical and social aspects of governance and to work constructively on developing methods in this area. Scientists and governments must urgently address the need to strengthen the research base in this vital agenda.

References

Access Now (2015) 'Open letter to Mark Zuckerberg regarding Internet.org, net neutrality, privacy, and security', 18 May, available at www.facebook.com/notes/accessnoworg/open-letter-to-mark-zuckerberg-regarding-internetorg-net-neutrality-privacy-and-/935857379791271 (Accessed 25 May 2016).

Adalet McGowan, Müge, Dan Andrews, Chiara Criscuolo and Giuseppe Nicoletti (2015) *The Future of Productivity*, OECD, available at www.oecd.org/eco/growth/OECD-2015-The-future-of-productivity-book.pdf (Accessed 25 May 2016).

Advertising Standards Authority (2013) 'ASA ruling on British Telecommunications plc: Ruling against BT Infinity', available at www.asa.org.uk/Rulings/Adjudications/2013/9/British-Telecommunications-plc/SHP_ADJ_228190.aspx (Accessed 25 May 2016).

AGCOM (Autorità per le garanzie nelle comunicazioni) (2011), 'Delibera 40/11/CONS, Public consultation on Net Neutrality', 3 February.

Agence France-Presse (2013) 'Brazil to host Internet governance summit next year', 9 October.

Aisch, Gregor, Wilson Andrews and Josh Keller (2015) 'The cost of mobile ads on 50 news websites', *New York Times*, 1 October, available at www.nytimes.com/interactive/2015/10/01/business/cost-of-mobile-ads.html?_r=1 (Accessed 25 May 2016).

Akamai (2015a) *Improving Online Video Quality and Accelerating Downloads: the FastTCP Network Enhancement*, White Paper: Akamai, Cambridge, MA.

Akamai (2015b) *Streaming toward Television's Future: A Detailed Look at 4K Video and How Akamai Is Making it a Reality*, White Paper: Akamai, Cambridge, MA.

Allen, James (2012) 'Article 7a of the EU telecoms Framework Directive: first thoughts on second guessing a second guess', Analysys Masons, 4 July, available at www.analysysmason.com/About-Us/News/Newsletter/Article7a-EU-telecoms-AMQ-Jul2012/#sthash.JMZ2nYZo.dpuf (Accessed 25 May 2016).

Ammori, M. (2015) 'T-Mobile is likely violating net neutrality: by throttling all video for its subscribers', *Slate*, December, available at www.slate.com/articles/technology/future_tense/2015/12/t_mobile_s_binge_on_program_likely_violates_net_neutrality.2.html (Accessed 25 May 2016).

Andersen *et al.* (2010) Joint Reply: Comments of Various Advocates for the Open Internet, 4 November, 'Comments on Advancing Open Internet Policy through

Analysis Distinguishing Open Internet from Specialized Network Services', available at: www.scribd.com/document/41002510/On-Advancing-the-Open-Internet-by-Distinguishing-it-from-Specialized-Services or www.fcc.gov/ecfs/filing/6016060935 (Acessed 15 September 2016).

Andrews & Arnold Ltd (2016) 'Real internet connection', available at http://aaisp.net/kb-broadband-realinternet.html (Accessed 25 May 2016).

Ansip, Andrus (2015) 'Making the EU work for people: roaming and the open internet', blog post, 8 July, available at https://ec.europa.eu/commission/2014-2019/ansip/blog/making-eu-work-people-roaming-and-open-internet_en (Accessed 25 May 2016).

Antunes, Anderson (2015) 'Mark Zuckerberg meets with Brazil's president at the 7th Summit of the Americas, in Panama', *Forbes*, 11 April, available at www.forbes.com/sites/andersonantunes/2015/04/11/mark-zuckerberg-meets-with-brazils-president-at-the-7th-summit-of-the-americas-in-panama/#4289c2941878 (Accessed 25 May 2016).

APCOMMS (2014) 'Can we keep our hands off the net?', available at http://uk.practicallaw.com/4-500-4840?q=&qp=&qo=&qe= (Accessed 15 September 2016).

Apple (2012) 'About download times for the iTunes Store purchases and rentals', last modified 12 April, available at http://support.apple.com/kb/ht1577 (Accessed 25 May 2016).

ARCEP (2010a) *Discussion Points and Initial Policy Directions on Internet and Network Neutrality*, May, available at www.arcep.fr/uploads/tx_gspublication/consult-net-neutralite-200510-ENG.pdf (Accessed 25 May 2016).

ARCEP (2010b) *Internet and Network Neutrality: Proposals and Recommendations*, September, available at www.arcep.fr/uploads/tx_gspublication/net-neutralite-orientations-sept2010-eng.pdf (Accessed 25 May 2016).

ARCEP (2012a), *Report to Parliament and the Government on Net Neutrality*, available at www.arcep.fr/uploads/tx_gspublication/rapport-parlement-net-neutrality-sept2012-ENG.pdf (Accessed 25 May 2016).

ARCEP (2012b) Decision No. 2012-0366 of 29 March 2012, available at www.arcep.fr/fileadmin/reprise/dossiers/net-neutralite/12-0366-eng.pdf (Accessed 25 May 2016).

ARCEP (2013) Press Release, 10 July, available at http://arcep.fr/index.php?id=8571&tx_gsactualite_pi1%5Buid%5D=1616&tx_gsactualite_pi1%5Bannee%5D&tx_gsactualite_pi1%5Btheme%5D&tx_gsactualite_pi1%5Bmotscle%5D&tx_gsactualite_pi1%5BbackID%5D=26&cHash=af231efe682036dbe00ed2317f1a9dcc&L=1 (Accessed 25 May 2016).

Areeda, Philip and Donald F. Turner (1975) 'Predatory pricing and related practices under Section 2 of the Sherman Act', *Harvard Law Review*, 88:4, pp. 697–733.

Arnold, René *et al.* (2015) *The Value of Network Neutrality to European Consumers*, 43rd Research Conference on Communication, Information and Internet Policy, available at www.wik.org/fileadmin/Studien/2015/2015_BEREC_Summary_Report.pdf (Accessed 25 May 2016).

Arthur, Charles (2010) 'Vaizey insists he favours net neutrality – and agrees with Berners-Lee', *Guardian*, 22 November, available at www.theguardian.com/technology/blog/2010/nov/22/vaizey-net-neutrality-berners-lee (Accessed 25 May 2015).

Arthur, Charles (2011) 'Berners-Lee warns ISPs on net neutrality', *Guardian*, 16 March, available at www.theguardian.com/technology/2011/mar/16/tim-berners-lee-net-neutrality (Accessed 25 May 2015).

Atiyah, P. S. (1980) 'Liability for railway nuisance in the English common law: a historical footnote', *Journal of Law and Economics*, 23:1, pp. 191–196.

AUSCANNZUKUS (2013) History, available at www.auscannzukus.net/history.html (Accessed 16 September 2016).

Autorité de la concurrence (2012) '12-D-18: Décision du 20 septembre 2012 relative à des pratiques mises en œuvre dans le secteur des prestations d'interconnexion réciproques en matière de connectivité Internet', 20 September, available at www.autoritedelaconcurrence.fr/user/avisdec.php?numero=12D18 (Accessed 25 May 2015).

Balcells, J. *et al.* (coords.) (2013) *Big Data: Challenges and Opportunities*, Proceedings of the 9th International Conference on Internet, Law & Politics. Universitat Oberta de Catalunya, Barcelona. Barcelona: UOC-Huygens Editorial.

Baldwin, R. (2015) 'Nudge: three degrees of concern', SSRN Scholarly Paper No. 2573334. Social Science Research Network, Rochester, NY.

Baraniuk, Chris (2015) 'EU parliament set to vote on net neutrality rules', BBC Technology, 27 October, available at www.bbc.co.uk/news/technology-34641515 (Accessed 25 May 2015).

BBC (2009) 'Home Office "colluded with Phorm"', 28 April, available at http://news.bbc.co.uk/2/hi/technology/8021661.stm (Accessed 25 May 2015).

BBC (2013) 'Phone hacking: arrests by investigation', 13 June, available at www.bbc.co.uk/news/uk-politics-17014930 (Accessed 25 May 2015).

BBC (2014) 'Freedom of Information Act 2000 – RF1201-40419', Information Compliance, 4 April, explaining the use of ISP CDNs such as Sky and British Telecom, together with four commercial CDNs: Akamai, Atos, Level3 and Limelight, available at http://downloads.bbc.co.uk/foi/classes/disclosure_logs/digital_and_technology/RF1201-40419-iplayer-content.pdf (Accessed 25 May 2015).

Beaumont, Claudine (2010) 'Information Commissioner reprimands Talk Talk', *Daily Telegraph*, 8 September, available at www.telegraph.co.uk/technology/internet/7989262/Information-Commissioner-reprimands-Talk-Talk.html (Accessed 25 May 2015).

Bela, Germa and Francesc Trillas (2005) 'Privatization, corporate control and regulatory reform: the case of Telefonica', *Telecommunications Policy*, 29:1, pp. 25–51.

Belli, L. (2013) 'From "End-to-End" to the "Rule of Law": should network neutrality be enshrined into legislation?', MediaLaws, 3 June, available at www.medialaws.eu/from-%E2%80%98end-to-end%E2%80%99-to-the-%E2%80%98rule-of-law%E2%80%99-should-network-neutrality-be-enshrined-into-legislation (Accessed 25 May 2016).

Belli, Luca and Primavera De Filippi (eds) (2015) *Net Neutrality Compendium: Human Rights, Free Competition and the Future of the Internet*. Cham: Springer Verlag.

Belli, Luca and Nathalia Foditsch (2015) 'Network Neutrality: An Empirical Approach to Legal Interoperability', pp. 281–298 in Luca Belli and Primavera De Filippi (eds) *Net Neutrality Compendium: Human Rights, Free Competition and the Future of the Internet*, Cham: Springer Verlag.

Belli, L. and C. Marsden (2015) 'Not neutrality but "open internet" à l'Européenne', LSE Media Policy Blog, 29 October, available at http://blogs.lse.ac.uk/mediapolicyproject/2015/10/29/not-neutrality-but-open-internet-a-la-europeenne (Accessed 25 May 2016).

Bendrath, Ralf (2009) 'Deep packet inspection reading list', available at http://bendrath.blogspot.co.uk/2009/03/deep-packet-inspection-reading-list-and.html (Accessed 25 May 2016).

Berners-Lee, Tim (2006) 'Net neutrality: this is serious', blog post, 21 June, available at http://dig.csail.mit.edu/breadcrumbs/node/144 (Accessed 25 May 2016).

Berners-Lee, Tim (2010) 'Long live the Web: a call for continued open standards and neutrality', *Scientific American*, 22 November, available at www.scientificamerican.com/article.cfm?id=long-live-the-web (Accessed 25 May 2016).

Berners-Lee, Tim (2012) 'This is for everyone', Twitter, 27 July, available at https://twitter.com/timberners_lee/status/228960085672599552 (Accessed 25 May 2016).

Birnhack, Michael (2012) 'Reverse engineering informational privacy law', *Yale Journal of Law and Technology*, 15:1, pp. 24–91.

Bits of Freedom (2015) 'Net neutrality in the Netherlands: the state of play', unofficial translation of the provision in Article 7.4a of the Telecommunications Act and its underlying considerations, available at www.bof.nl/2011/06/15/net-neutrality-in-the-netherlands-state-of-play (Accessed 19 May 2016).

Blackman, Colin *et al.* (2010) *Towards a Future Internet: Interrelation between Technological, Social and Economic Trends*, European Commission, available at http://cordis.europa.eu/fp7/ict/fire/docs/tafi-final-report_en.pdf (Accessed 19 May 2016).

Blauberger, Michael and Berthold Rittberger (2015) 'Conceptualizing and theorizing EU regulatory networks', *Regulation & Governance*, 9:4, pp. 367–376.

Boeger , Nina and Joseph Corkin (2012) 'How Regulatory Networks Shaped Institutional Reform under the EU Telecoms Framework', Chapter 2, pp. 48–89 in Catherine Barnard and Markus Gehring with Iyiola Solanke (eds) *Cambridge Yearbook of European Legal Studies*, Vol. 14, 2011–2012, Oxford: Hart.

Booz Allen Hamilton and Raul Katz (2012) 'Maximising the Impact of Digitization', in *The Global Information Technology Report 2012: Living in a Hyperconnected World*, World Economic Forum, Davos, Switzerland.

Borger, Julian (2013) 'Inquiry into snooping laws as committee clears GCHQ', *Guardian*, 18 July, available at www.theguardian.com/world/2013/jul/17/prism-nsa-gchq-review-framework-surveillance (Accessed 19 May 2016).

Bowden, Caspar (2013a) 'The US National Security Agency (NSA) surveillance programmes (PRISM) and Foreign Intelligence Surveillance Act (FISA) activities and their impact on EU citizens' fundamental rights', European Parliament Civil Liberties Committee, 24 September.

Bowden, Caspar (2013b) 'PRISM: The EU must take steps to protect cloud data from US snoopers', *Independent*, 10 July, available at www.independent.co.uk/voices/comment/prism-the-eu-must-take-steps-to-protect-cloud-data-from-us-snoopers-8701175.html (Accessed 19 May 2016).

Brazilian Government (2015) 'Meeting with President of Facebook', available at www2.planalto.gov.br/centrais-de-conteudos/imagens/encontro-com-presidente-do-facebook (Accessed 19 May 2016).

Brazilian Ministry of Justice (2016) 'Civil Rights Framework for the Internet in Brazil: What is it?', information in English about the consultation, available at http://pensando.mj.gov.br/marcocivil/civil-rights-framework-for-the-internet-in-brazil/ (Accessed 19 May 2016).

Broadband Commission For Digital Development (2015) *The State Of Broadband 2015: Broadband As A Foundation For Sustainable Development*, available at www.broadbandcommission.org/documents/reports/bb-annualreport2015.pdf (Accessed 19 May 2016).

Broadband Delivery UK (2012) 'National Broadband Scheme for the UK. Guidance: Wholesale Access and Pricing', available at: http://webarchive.nationalarchives.gov.uk/20121204113822/http://www.culture.gov.uk/images/publications/State_aid_Guidance_Benchmarking.pdf (Accessed 19 May 2016).

Broadband Internet Technical Advisory Group (2011) By-laws of Broadband Internet Technical Advisory Group S. 7.1.

Broadband Initiatives Program (2009) Broadband Technology Opportunities Program Notice, 74 Fed. Reg. 33104, 33110–11, 9 July (Broadband NOFA).

Broadband Stakeholder Group (2009) 'BSG response to interim Digital Britain Report', available at www.broadbanduk.org/2009/03/12/bsg-submits-response-to-interim-digital-britain-report/ (Accessed 16 September 2016).

Broadband Stakeholder Group (2013b) 'Voluntary industry code of practice on traffic management transparency for broadband services', available at www.broadbanduk.org/wp-content/uploads/2013/08/Voluntary-industry-code-of-practice-on-traffic-management-transparency-on-broadband-services-updated-version-May-2013.pdf (Accessed 19 May 2016).

Broadband Stakeholder Group (2013c) 'BSG publishes details of voluntary process underpinning Open Internet code', 14 June, available at www.broadbanduk.org/2013/06/14/bsg-publishes-details-of-voluntary-process-underpinning-open-internet-code/ (Accessed 19 May 2016).

Broadband Stakeholder Group (2014) 'BSG Chairman's comment on net neutrality/open internet news articles', 17 November, available at www.broadbanduk.org/2014/11/17/bsg-chairmans-comment-on-net-neutralityopen-internet-news-articles/ (Accessed 19 May 2016).

Broadband Stakeholder Group (2015) 'Remaining ISPs commit to the UK's Open Internet Code', 19 January, available at www.broadbanduk.org/2015/01/19/remaining-isps-commit-to-the-uks-open-internet-code/ (Accessed 19 May 2016).

Brodkin, Jon (2014) 'Comcast to stop blocking HBO Go and Showtime on Roku streaming devices', Ars Technica, 17 December, available at http://arstechnica.com/business/2014/12/comcast-to-stop-blocking-hbo-go-and-showtime-on-roku-streaming-devices/ (Accessed 19 May 2016).

Brown, Ian (2013a) 'Expert Witness Statement for Big Brother Watch and Others Re: Large-Scale Internet Surveillance by the UK', Application No: 58170/13 to the European Court of Human Rights available at SSRN: http://ssrn.com/abstract=2336609 (Accessed 19 May 2016).

Brown, Ian (2013b) 'Lawful Interception Capability Requirements', available at www.scl.org/site.aspx?i=ed32980 (Accessed 19 May 2016).

Brown, Ian (2015) 'Regulation and the Internet of Things', GSR discussion paper, International Telecommunications Union, available at www.itu.int/en/ITU-D/Conferences/GSR/Pages/GSR2015/GSR15-discussion-paper.aspx (Accessed 19 May 2016).

Brown, I. and C. Marsden (2013a) *Regulating Code*. Cambridge, MA: MIT Press.

Brown, Ian and C. Marsden (2013b) 'Regulating Code: Towards Prosumer Law', pp. 101–126 in J. Balcells *et al.* (coords.) *Big Data: Challenges and Opportunities*, Proceedings of the 9th International Conference on Internet, Law & Politics, Universitat Oberta de Catalunya, Barcelona. Barcelona: UOC-Huygens Editorial.

Brown, I., L. Edwards and C. Marsden (2006) 'Legal and institutional responses to Denial of Service Attacks', Communications Research Network/Department for Trade and Industry joint seminar on Spam/DDoS, 13 November, available at www.communicationsresearch.net/object/download/1846/doc/marsden-edwards.ppt and on file with the author.

BT (2013) First Quarter results, available at www.btplc.com/Sharesandperformance/Quarterlyresults/ (Accessed 15 September 2016).

BT Pensions (2014) Crown Guarantee Update, 16 July, available at www.btpensions.net/192/364/crown-guarantee (Accessed 15 September 2016).

BT Press Releases (2015) BT announces results of pension fundingValuation DC15-36, 30 January 2015, available at www.btpensions.net/180/335/bt-announces-results-of-pension-funding-valuation (Accessed 15 September 2016).

Bubley, Dean (2015) 'Is there a regulatory elephant lurking in the SDN/NFV room?', LinkedIn, 16 October, available at www.linkedin.com/pulse/regulatory-elephant-lurking-sdn-nfv-room-dean-bubley?trk=hb_ntf_MEGAPHONE_ARTICLE_POST (Accessed 19 May 2016).

Burstein, D. (2011) 'Wireline costs and caps: a few facts' (copy on file with author).

Business Week International Online Extra (2005) 'At SBC, It's all about "scale and scope"', 7 November, available at: www.bloomberg.com/news/articles/2005-11-06/online-extra-at-sbc-its-all-about-scale-and-scope (Accessed 19 May 2016).

Cabinet Office (2010) 'Cabinet Office Spending Review settlement', 20 October, available at: www.gov.uk/government/news/cabinet-office-spending-review-settlement (Accessed 15 September 2016).

Caf, Dusan (2014) 'Zero-rating violates slovenian net neutrality law', *Competitive Analysis & Foresight*, 5 December, available at http://blog.caf.si/2014/12/zero-rating-violates-slovenian-net-neutrality-law.html (Accessed 19 May 2016).

Caf, Dusan (2015) 'Another win for net neutrality advocates in Slovenia: AKOS issues new decisions limiting zero-rating', *Competitive Analysis & Foresight*, 22 February, available at http://blog.caf.si/2015/02/another-win-for-net-neutrality-advocates-in-slovenia-akos-issues-new-decisions-limiting-zero-rating.html (Accessed 19 May 2016).

Cameron, David (2015) Hansard Column 344, 28 October 2015, available at www.publications.parliament.uk/pa/cm201516/cmhansrd/cm151028/debtext/151028-0001.htm (Accessed 19 May 2016).

Campbell, Duncan (1999) 'The state of the art in communications Intelligence (COMINT) of automated processing for intelligence purposes of intercepted broadband multi-language leased or common carrier systems, and its applicability to COMINT targetting and selection, including speech recognition', European Parliament, Ref.: EP/IV/B/STOA/98/1401.

Candeub, Adam (2015) 'Is there anything new to say about network neutrality?' *Michigan State Law Review*, Issue 2, pp. 455–463, available at http://digitalcommons.law.msu.edu/cgi/viewcontent.cgi?article=1114&context=lr (Accessed 19 May 2016).

Candeub, Adam and Daniel John McCartney (2012) 'Law and the Open Internet', *Federal Communications Law Journal* 64:3, pp. 493–548.

Cannon, Robert (2003) 'The legacy of the FCC's computer inquiries', *Federal Communications Law Journal* 55:2, pp. 167–205.

Cave, Jonathan and Ben Cave (2012) 'Nudging eConsumers: Online Ecolabelling as Part of the Green Internet', available at http://ssrn.com/abstract=2141967 and http://dx.doi.org/10.2139/ssrn.2141967 (Accessed 19 May 2016).

Cave, M. (2011) DAF/COMP/WP2 4 Directorate For Financial And Enterprise Affairs: Competition Committee Working Party No. 2 On Competition And Regulation: Hearing On Network Neutrality Paper.

CBC News (2015) 'CRTC backs net neutrality in ruling against apps that favour certain content', CBC News, 29 January, available at www.cbc.ca/news/business/crtc-backs-net-neutrality-in-ruling-against-apps-that-favour-certain-content-1.2936358 (Accessed 19 May 2016).

CEBIT News (2016) 'We need the Gigabit infrastructure for the Gigabit economy', 14 March, available at www.cebit.de/en/news/news-details_27597.xhtml?utm_source=twitter&utm_medium=social&utm_content=DGConnect&utm_campaign=DigitalSingleMarket (Accessed 19 May 2016).

Cerda, Alberto (2013) 'An evaluation of the net neutrality law in Chile', Digital Rights LAC, 17 July, available at www.digitalrightslac.net/en/una-evaluacion-de-la-ley-de-neutralidad-de-la-red-en-chile (Accessed 19 May 2016).

CGI (2009) 'Resolução CGI.br/2009/03', available at www.cgi.br/resolucoes/documento/2009/003 (Accessed 19 May 2016).

Cherry, Barbara A. (2006) 'Misusing network neutrality to eliminate common carriage threatens free speech and the postal system', *Northern Kentucky Law Review*, 33, pp. 483–511.

Cherry, Barbara A. (2008) 'Back to the future: how transportation deregulatory policies foreshadow evolution of communications policies', *The Information Society*, 24, pp. 273–291.

Cherry, Barbara A. and Jon M. Peha (2014) 'The Telecom Act of 1996 Requires the FCC to Classify Commercial Internet Access as a Telecommunications Service', available at http://ssrn.com/abstract=2602091 and http://dx.doi.org/10.2139/ssrn.2602091 (Accessed 19 May 2016).

Chilvarquer, Marcelo (2015) 'Debate Público Regulamentação do Marco Civil da Internet, Secretaria de Legislativos Assuntos, Ministeria da Justicia', paper presented at Conferência Internacional sobre a Elaboração de Regras de Neutralidade de Rede', FGV Rio de Janeiro, 8 June 2015, available at http://direitorio.fgv.br/eventos/Conferencia-Internacional-sobre-a-Elaboracao-de-Regras-de-Neutralidade-de-Rede (Accessed 19 May 2016).

Chirico, Filomena, Ilse Van der Haar and Pierre Larouche (2007) 'Network neutrality in the EU', TILEC Discussion Paper, available at http://ssrn.com/abstract=1018326 (Accessed 19 May 2016).

Churchill, Winston S. (1942) Mansion House speech, 10 November, available at www.churchill-society-london.org.uk/EndoBegn.html (Accessed 19 May 2016).

Cisco (2012) Visual Networking Index, available at www.cisco.com/c/en/us/solutions/service-provider/visual-networking-index-vni/index.html (Accessed 19 May 2016).
Cisco (2015) VNI Forecast Highlights, May, available at www.cisco.com/c/en/us/solutions/service-provider/visual-networking-index-vni/vni-forecast.html (Accessed 19 May 2016).
Clark, David D. and K. C. Claffy (2015) 'Anchoring policy development around stable points: an approach to regulating the co-evolving ICT ecosystem', *Telecommunications Policy*, 39:10, pp. 848–860.
Clayton, R. (2008) The Phorm 'Webwise' System, available at www.cl.cam.ac.uk/~rnc1/080404-phorm.pdf and (later version) www.cl.cam.ac.uk/~rnc1/080518-phorm.pdf (Accessed 19 May 2016).
Coates, K. (2011) *Competition Law and Regulation of Technology Markets*. New York: Oxford University Press.
Collins, Barry (2008) 'Sam shines a light on BT's traffic shaping', Alphr, 4 August, available at www.alphr.com/news/internet/216252/sam-shines-a-light-on-bts-traffic-shaping (Accessed 19 May 2016).
Conradi, Mike and Eamon Holley (2014) 'Appealing a Telecoms Regulatory Decision', 6 January, DLA Piper, available at www.technologyslegaledge.com/2014/01/appealing-a-telecoms-regulatory-decision/ (Accessed 19 May 2016).
Consumer Focus (2012) 'Lost on the broadband super highway', available at https://ec.europa.eu/digital-single-market/en/news/lost-broadband-super-highway-consumer-understanding-information-traffic-management (Accessed 15 September 2016).
Cooper, Alissa (2013) 'How Regulation and Competition Influence Discrimination in Broadband Traffic Management: A Comparative Study of Net Neutrality in the United States and the United Kingdom', Thesis submitted for the degree of DPhil, University of Oxford, September 2013.
Cooper, Alissa and Ian Brown (2015) 'Net neutrality: discrimination, competition, and innovation in the UK and US', *ACM Transactions on Internet Technology*, 15:1, pp. 1–21.
Cooper, Alissa and Alison Powell (2011) 'Net neutrality discourses: comparing advocacy and regulatory arguments in the US and the UK,' *The Information Society*, 27:5, pp. 311–325, available at www.alissacooper.com/wp-content/uploads/2011/10/NN-Discourses-pre-print.pdf (Accessed 19 May 2016).
Costa-Cabral, Francisco and Orla Lynskey (2015) 'The Internal and External Constraints of Data Protection on Competition Law in the EU', LSE Legal Studies Working Paper 25/2015, available at http://eprints.lse.ac.uk/64887/ (Accessed 15 September 2016).
Council of Europe (2001) ETS No. 185 Convention on Cybercrime, available at www.coe.int/en/web/conventions/full-list/-/conventions/treaty/185 (Accessed 16 September 2016).
Council of Europe (2003) Declaration on freedom of communication on the Internet, adopted on 28 May 2003 at the 840th meeting of the Ministers' Deputies.
Council of Europe (2010) Declaration of the Committee of Ministers on network neutrality adopted 29/9/2010: 1094th meeting of the Ministers' Deputies.

Council of Europe (2013) 29–30 May, Multi-stakeholder dialogue on network neutrality communicated to the Council of Europe Steering Committee on Media and Information Society (CDMSI).

Council of Europe (2016) Recommendation CM/Rec1 of the Committee of Ministers to member States on protecting and promoting the right to freedom of expression and the right to private life with regard to network neutrality (Adopted by the Committee of Ministers on 13 January 2016, at the 1244th meeting of the Ministers' Deputies).

Council of the European Union (2015), Interinstitutional File: 2013/0309 (COD) 12279/15 CODEC 1226 TELECOM 177 COMPET 418 MI 576 CONSOM 152 From: General Secretariat of the Council To: Permanent Representatives Committee/Council Subject: Draft Regulation laying down measures concerning the European single market for electronic communications and to achieve a Connected Continent, and amending Directives 2002/20/EC, 2002/21/EC and 2002/22/EC and Regulations (EC) No 1211/2009 and (EU) No 531/2012 (first reading) Adoption a) of the Council's position b) of the statement of the Council's reasons Brussels, 29 September 2015, available at http://data.consilium.europa.eu/doc/document/ST-12279-2015-ADD-1-REV-1/en/pdf.

Crawford, Susan (2011) 'The communications crisis in America', *Harvard Law & Policy Review*, 5, pp. 244–263.

Crowcroft, Jon (2015) 'The UK doesn't yet need net neutrality regulations', 4 March, The Conversation, available at http://phys.org/news/2015-03-uk-doesnt-net-neutrality.html (Accessed 19 May 2016).

Crown Prosecution Service (2011) 'CPS decides no prosecution of BT and Phorm for alleged interception of browsing data', 8 April, available at http://blog.cps.gov.uk/2011/04/no-prosecution-of-bt-and-phorm-for-alleged-interception-of-browsing-data.html (Accessed 19 May 2016).

Cruz, Francisco Carvalho de Brito, Jonas Coelho Marchezan and Maike Wile dos Santos (2015) 'What is at stake in the regulation of the Marco Civil da Internet?', Final Report On The Public Debate Sponsored By Ministry Of Justice On Regulation Of Law 12.965/2014, Internet Lab, Rua Augusta, 2690, Galeria Ouro Fino, Loja 326, available at www.internetlab.org.br/en/news/what-is-at-stake-in-the-regulation-of-the-marco-civil/ (Accessed 19 May 2016).

Cukier, Kenn (1997) 'Peering and Fearing: ISP Interconnection and Regulatory Issues, Internet and Telecommunications Policy'. Harvard University, Kennedy School of Government, Information Infrastructure Project, December.

Curien, N. and W. Maxwell (2010) 'Net Neutrality in Europe: An Economic and Legal Analysis', *Concurrences, Review of competition laws*, No. 4.

Currie, David (2015) 'Homo economicus and Homo sapiens: the CMA experience of behavioural economics', speech given by CMA Chairman, David Currie, at a New Zealand Commerce Commission public lecture, 21 April, available at www.gov.uk/government/speeches/david-currie-speaks-about-the-cma-experience-of-behavioural-economics (Accessed 19 May 2016).

D'Ignazioa, Alessio and Emanuele Giovannetti (2015) 'Predicting internet commercial connectivity wars: the impact of trust and operators' asymmetry', *International Journal of Forecasting*, 31:4, pp. 1127–1137.

David, Paul (2001) 'The evolving accidental information super-highway', *Oxford Review of Economic Policy*, 17:2, pp. 159–187.

Davies, Simon (2013) 'European Parliament votes to hold inquiry into US spying', 4 July, available at www.privacysurgeon.org/blog/incision/european-parliament-votes-to-hold-full-inquiry-into-us-spying/ (Accessed 19 May 2016).

De Guzman, Noelle Francesca (2014) 'Zero rating: enabling or restricting Internet access?', Asia Pacific Bureau: Internet Society, 24 September, available at www.internetsociety.org/blog/asia-pacific-bureau/2014/09/zero-rating-enabling-or-restricting-internet-access (Accessed 19 May 2016).

de Sola Pool, I. (1983) *Technologies of Freedom*. Cambridge, MA: Belknap.

Deering, S. (1989) 'Host extensions for IP multicasting', STD 5, RFC 1112, August 1989.

Department for Culture, Media and Sport (2009) 'Digital Britain', Chapter 4: Creative Industries in the Digital World, available at www.gov.uk/government/uploads/system/uploads/attachment_data/file/228844/7650.pdf (Accessed 15 September 2016).

Department for Culture, Media and Sport (2013a) 'UK Broadband Impact Study: Literature Review', produced by SQW Consultants, available at www.gov.uk/government/publications/uk-broadband-impact-study (Accessed 19 May 2016).

Department for Culture, Media and Sport (2013b) 'Connectivity, content and consumers: Britain's digital platform for growth', 30 July, available at www.gov.uk/government/publications/connectivity-content-and-consumers-britains-digital-platform-for-growth (Accessed 19 May 2016).

Department for Culture, Media and Sport (2015a) 'Government plans to make sure no-one is left behind on broadband access', Prime Minister's Office, 10 Downing Street, The Rt Hon. David Cameron MP and The Rt Hon. John Whittingdale MP, 7 November, available at www.gov.uk/government/news/government-plans-to-make-sure-no-one-is-left-behind-on-broadband-access (Accessed 19 May 2016).

Department for Culture, Media and Sport (2015b) 'UK NonPaper: Review of the Electronic Communications Regulatory Framework', September.

Department of Justice (2012) 'Justice Department Charges Leaders of Megaupload with Widespread Online Copyright Infringement', Office of Public Affairs.

Dewenter, Ralf and Jörn Kruse (2011) 'Calling party pays or receiving party pays? The diffusion of mobile telephony with endogenous regulation', *Information Economics and Policy*, 23:1, pp. 107–117.

D'Ignazio, A. and E. Giovannetti (2015) 'Predicting Internet commercial connectivity wars: the impact of trust and operators asymmetry', *International Journal of Forecasting*, 31:4, pp. 1127–1137.

Doval, Pankaj (2015) 'Zero-rating plans must be open to all users: DoT panel member', *Times of India*, 20 July, available at http://timesofindia.indiatimes.com/tech/tech-news/Zero-rating-plans-must-be-open-to-all-users-DoT-panel-member/articleshow/48138850.cms (Accessed 19 May 2016).

DSL Prime (2012) 'France Telecom, free to google YouTube: you're blocked unless you pay', 27 December (copy on file with author).

Dunstone, C. (2006) 'Presentation by Carphone Warehouse/TalkTalk CEO at the 2006 Ofcom conference' (copy on file with author).

EaPeReg (2015) Summit Press Release, 4 July, available at http://berec.europa.eu/eng/document_register/subject_matter/berec/press_releases/5081-press-release-from-the-berec-8211-emerg-8211-eapereg-regulatel-summit-2-3-july-2015-barcelona (Accessed 15 September 2016) (copy on file with author).

Easterbrook, Frank H. (1984) 'Limits of antitrust', *Texas Law Review*, 63:1, pp.1–40.

Economic Times (2015) 'Trai asks Reliance Communications to put Facebook's Free Basics service on hold till it approves', 24 December, available at http://articles.economictimes.indiatimes.com/2015-12-24/news/69282660_1_consultation-paper-telecom-service-providers-telecom-sector-regulator (Accessed 19 May 2016).

Economides, Nicholas (2015) 'Testimony on network neutrality to US Congress', available at http://works.bepress.com/economides/53/ (Accessed 19 May 2016).

Economides, N. and J. Tåg (2007) 'Net Neutrality on the Internet: A Two-Sided Market Analysis', working paper, NYU Center for Law and Economics, New York.

EDRi (2015) Net neutrality: document pool II, available at https://edri.org/net-neutrality-document-pool-2/.

EDRi (2016) 'Final consultation to save the open Internet in Europe', 6 April, available at https://edri.org/final-consultation-to-save-the-open-internet-in-europe/ (Accessed 19 May 2016).

EINS (2015) 'Governance, regulation and standards (JRA4): about this Working Group', available at www.internet-science.eu/groups/governance-regulation-and-standards (Accessed 19 May 2016).

Eisenach, Jeffrey (2015) *Economics of Zero Rating*, National Economic Research Associates, March, available at www.nera.com/content/dam/nera/publications/2015/EconomicsofZeroRating.pdf (Accessed 19 May 2016).

Eisenberg, R. E. (1988) 'Academic freedom and academic values in sponsored research', *Texas Law Review*, 66, pp.1363–1404, available at http://heinonline.org/HOL/LandingPage?handle=hein.journals/tlr66&div=65&id=&page= (Accessed 19 May 2016).

Eldar Shafir (ed.) (2012) *The Behavorial Foundations of Public Policy*, Princeton: Princeton University Press.

Emert, Monika (2013) 'Germany's new government to move forward with a mixed digital agenda', *Internet Policy Review*, 11 December, available at http://policyreview.info/articles/news/germanys-new-government-move-forward-mixed-digital-agenda/224 (Accessed 19 May 2016).

Emmott, Christopher and Lydia Harris (2015) 'UK Broadband Infrastructure', POSTnotes POST-PN-0494, UK Parliametnary Office of Science and Technology, 29 May, available at http://researchbriefings.parliament.uk/ResearchBriefing/Summary/POST-PN-0494 (Accessed 19 May 2016).

ESPAS (2015) 'The World in 2030', available at http://europa.eu/espas/pdf/espas-outreach-leaflet.pdf (Accessed 19 May 2016).

ETICS (2011) Deliverable 3.4 at Chapters 4–5: www.ict-etics.eu/fileadmin/documents/publications/deliverables/D3.4_Master_Document_v1.0_final_20120517.pdf p211.

ETICS (2012) Deliverable 8.4 (copy on file with author).

Europa (2014) 'Rome declaration on fostering RRI', 21 November, available at http://ec.europa.eu/research/swafs/index.cfm?pg=newspage&item=141217 (Accessed 19 May 2016).

European Scrutiny Committee (2015) 'Documents considered by the Committee on 21 July 2015 – No. 11 The Telecommunications Single Market', available at www.publications.parliament.uk/pa/cm201516/cmselect/cmeuleg/342-i/34214.htm (Accessed 19 May 2016).

European Telecommunications Network Operators' Association (2015) *Annual Economic Report 2015*, available at https://etno.eu/datas/publications/economic-reports/AER2015_Final.pdf (Accessed 19 May 2016).

Facebook (2015a) 'Response to Free Basics Opponents, Item 6', available at https://www.facebook.com/internetdotorg.india/posts/1676050986006469?comment_id=1677305315881036&comment_tracking=%7B%22tn%22%3A%22R0%22%7D (Accessed 15 September 2016).

Facebook (2015b) Q3-2015 Earnings Call 4 November 2015 2:00 p.m. PT, p. 13, available at http://investor.fb.com/results.cfm (Accessed 19 May 2016).

Faratin, P. *et al.* (2008) 'The growing complexity of internet interconnection', *Communications & Strategies*, 72, pp.51–72, 4th Quarter.

Faris, Robert *et al.* (2015) 'Score another one for the Internet? The role of the networked public sphere in the U.S. net neutrality policy debate', 10 February, Berkman Center Research Publication No. 2015-4, available at: http://ssrn.com/abstract=2563761 or http://dx.doi.org/10.2139/ssrn.2563761.

Farman, J. (2015a) 'Infrastructures of mobile social media', *Social Media + Society*, April–June 2015 1: 2056305115580343, first published on 11 May 2015 doi:10.1177/2056305115580343.

Farman, J. (2015b) 'The materiality of locative media: on the invisible infrastructure of mobile networks', pp. 45–59 in A. Herman, J. Hadlaw and T. Swiss (eds), *Theories of the Mobile Internet: Materialities and Imaginaries*. New York: Routledge Press.

Fatas, Harker *et al.* (2013) *Behavioural Economics in Competition and Consumer Policy*, ESRC Centre for Competition Policy, University of East Anglia, Norwich.

Federal Trade Commmission (2011) Google, Inc., FTC File No. 102 3136G. 30 March consent order accepted for public comment, available at: www.ftc.gov/opa/2011/03/google.shtm (Accessed 16 September 2016).

Feld, H. (2015) 'Net neutrality in court this week: the story of how we got here', Public Knowledge, 2 December, available at www.publicknowledge.org/news-blog/blogs/net-neutrality-in-court-this-week-the-story-of-how-we-got-here/ (Accessed 19 May 2016).

Felten, E. (2008) 'Comcast and BitTorrent: Why You Can't Negotiate with a Protocol', Freedom to Tinker 28 March, available at www.freedom-to-tinker.com/blog/felten/comcast-and-bittorrent-why-you-cant-negotiate-protocol (Accessed 5 September 2016).

Ferguson, Andrew (2013) 'More faster fibre broadband from Openreach', ThinkBroadband, 24 September, available at www.thinkbroadband.com/news/6052-more-faster-fibre-broadband-from-openreach.html (Accessed 19 May 2016).

Ferguson, Andrew (2015) 'Openreach line rental charges continue to diverge from retail price', Think Broadband, 2 January, available at www.thinkbroadband.com/news/6772-openreach-line-rental-charges-continue-to-diverge-from-retail-price.html (Accessed 19 May 2016).

Fierce Wireless (2011) 'Predictability is the key to successful wireless data pricing', available at www.fiercewireless.com/wireless/predictability-key-to-successful-wireless-data-pricing (Accessed 15 September 2016).

FIPR (2008) 'Continuing concerns about Phorm', 6 April, available at www.fipr.org/press/080406phorm.html (Accessed 19 May 2016).

Flyvbjerg, Bent (1998) 'Habermas and Foucault: thinkers for civil society?', *British Journal of Sociology*, 49:2, pp. 210–233.

Freedom House (2015) 'Freedom on the Net Report 2015: Germany', available at https://freedomhouse.org/report/freedom-net/2015/germany (Accessed 19 May 2016).

Freischlad, Nadine (2015) 'Soon everyone will be able to afford a smartphone. But what about data?', Tech In Asia, 24 July, available at www.techinasia.com/smartphones-are-getting-cheaper-but-what-about-data/ (Accessed 19 May 2016).

Frieden, R. (2006) 'What do pizza delivery and information services have in common? Lessons from recent judicial and regulatory struggles with convergence', *Rutgers Computer & Technology Law Journal*, 32, pp. 247–296.

Frieden, R. (2008) 'Internet packet sniffing and its impact on the network neutrality debate and the balance of power between intellectual property creators and consumers', *Fordham Intellectual Property, Media & Entertainment Law Journal*, 18:3, pp. 633–675.

Frieden, R. (2010) 'Invoking and avoiding the First Amendment: how internet service providers leverage their status as both content creators and neutral conduits', *University of Pennsylvania Journal of Constitutional Law*, 12:5, pp. 1279–1324.

Frieden, R. (2012) 'Rationales for and against regulatory involvement in resolving internet interconnection disputes', *Yale Journal of Law and Technology*, 14, pp. 266–313.

Frieden, R. (2015a) 'What's new in the network neutrality debate', *Michigan State Law Review*, Issue 2, pp. 739–786.

Frieden, R. (2015b) 'Déjà vu all over again: questions and a few suggestions on how the FCC can lawfully regulate internet access', *Federal Communications Law Journal*, 3, pp. 325–376.

Frieden, R. (2015c) 'Internet protocol television and the challenge of "mission critical" bits', *Cardozo Arts & Entertainment Law Journal*, 33:1, pp. 47–87.

Frischmann, Brett and Barbara van Schewick (2007) 'Yoo's Frame and What it ignores: network neutrality and the economics of an information superhighway', JURIMETRICS J., 47, pp. 383–428.

Frischmann, Brett and Spencer Weber Waller (2008) 'Revitalizing Essential facilities', *Antitrust Law Journal*, 75:1, pp. 1–65.

Galpaya , Helani (2015) 'Zero rating: are we in danger of killing the goose before knowing if its eggs are golden?', Council on Foreign Relations, 5 October, available at http://blogs.cfr.org/cyber/2015/10/05/zero-rating-are-we-in-danger-of-killing-the-goose-before-knowing-if-its-eggs-are-golden/ (Accessed 19 May 2016).

Garcia-Algarra, J. (2010) 'The American influence in Telefónica's public relations strategy during the 20's and 30's', (HISTELCON), 2010 Second IEEE Region 8 Conference on the History of Telecommunications Conference, 3–5 November, pp. 1–6.

Geddes, Martin (2015) 'Ofcom publishes scientific report on net neutrality', LinkedIn, 12 August, available at www.linkedin.com/pulse/ofcom-publishes-scientific-report-net-neutrality-martin-geddes?trk=hp-feed-article-title-share (Accessed 19 May 2016).

Geist, Michael (2011) 'Canada's net neutrality enforcement failure', 8 July, available at www.michaelgeist.ca/2011/07/net-neutrality-enforcement-fail/ (Accessed 19 May 2016).

Geist, Michael (2015) 'Why Canada's net neutrality enforcement is going at half-throttle', 10 August, available at www.michaelgeist.ca/2015/08/why-canadas-net-neutrality-enforcement-is-going-at-half-throttle/ (Accessed 19 May 2016).

Genna, Innocenzo (2015) 'Zero-rating: the European Parliament washing hands like Pontius Pilate', Radiobruxelleslibera, 26 October, available at https://radiobruxelles-libera.wordpress.com/2015/10/26/zero-rating-the-european-parliament-washing-hands-like-pontius-pilate/ (Accessed 19 May 2016).

Goldstein, Phil (2015) 'Net neutrality rules won't force carriers to get FCC permission for new plans, officials say', Fierce Wireless, 26 February, available at www.fiercewire-less.com/story/net-neutrality-rules-wont-force-carriers-get-fcc-permission-new-plans-offic/2015-02-26 (Accessed 19 May 2016).

Goldstone, David and Michael O'Leary (2001) 'Novel criminal copyright infringement issues related to the Internet', *United States Attorneys' Bull.*, 49:5.

Graef, Inge (2014) 'Why not "go Dutch" and protect net neutrality without defining specialised services?', LSE Media Policy Project, available at http://blogs.lse.ac.uk/mediapolicyproject/2014/04/04/why-not-go-dutch-and-protect-net-neutrality-without-defining-specialised-services/ (Accessed 19 May 2016).

Greenbaum, Eli (2014) 'Net Neutrality II', Israel Technology Law Blog (copy on file with author).

Grove, Andy (2010) 'Andy Grove: how America can create jobs', Bloomberg, 1 July, available at www.bloomberg.com/news/articles/2010-07-01/andy-grove-how-america-can-create-jobs#p1 (Accessed 19 May 2016).

Guadamuz, Andrés (2011) 'Networks, complexity and internet regulation: scale-free law'. PhD Thesis. Edward Elgar, Cheltenham, UK; Northampton, MA. ISBN 9781848443105.

Guha, R. and G. Aulakh (2016) 'Trai bars Facebook Free Basics, Airtel Zero; releases notification on differential data pricing', Telecom, 8 February, available at http://tel-ecom.economictimes.indiatimes.com/news/trai-bars-differential-pricing-of-data-services/50899934 (Accessed 19 May 2016).

Haddadi, Hamed *et al.* (2009) 'Analysis of the Internet's structural evolution', Technical Report Number 756 Computer Laboratory UCAM-CL-TR-756 ISSN 1476–2986.

Hahn, Robert and Scott Wallsten (2006) 'The economics of net neutrality', AEI Brookings Joint Center for Regulatory Studies: Washington DC.

Hall, Kat (2016) 'BT and EE, O2 and Three: are we in for a year of Euro telco mega-mergers?', The Register, 6 January, available at www.theregister.co.uk/2016/01/06/will_2016_be_the_year_of_european_telco_consolidation/ (Accessed 19 May 2016).

Harbour, Malcolm (2013) Opinion on the Implementation Report on the regulatory framework for electronic communications 09-09-2013 IMCO_AD(2013)510798PE 510.798v02-00 IMCO.

Hart, Jeffrey A. (2011) 'The net neutrality debate in the United States', *Journal of Information Technology & Politics*, 8:4, pp. 418–443.

Havergal, Naomi (2013) 'UK at-home workers unhappy with their broadband', Broadband Finder, 23 December, available at www.broadband-finder.co.uk/news/uk-at-home-workers-unhappy-with-their-broadband (Accessed 19 May 2016).

Heath, Ryan (2013) '10 mistakes & myths in a single La Quadrature du Net article', available at http://ecspokesryan-blog.tumblr.com/post/60176599392/10-mistakes-myths-in-a-single-la-quadrature-du (Accessed 19 May 2016).

Herrera Anchustegui, Ignacio (2015) 'Competition and buyer power through an ordo-liberal lens', available at http://dx.doi.org/10.2139/ssrn.2579308 (Accessed 19 May 2016).

HEVC Advance (2015) 'HEVC Advance announces key milestones to support next era of digital video creation and distribution', Press Release, 22 July, available at www.hevcadvance.com/pdf/HEVCPressRelease_22_July_2015.pdf (Accessed 19 May 2016).

Hill, Steven (2016) 'Good riddance, gig economy: Uber, Ayn Rand and the awesome collapse of Silicon Valley's dream of destroying your job', Salon.com, 27 March, available at www.salon.com/2016/03/27/good_riddance_gig_economy_uber_ayn_rand_and_the_awesome_collapse_of_silicon_valleys_dream_of_destroying_your_job/ (Accessed 19 May 2016).

Hills, Jill (2002) *The Struggle for Control of Global Communication: The Formative Century*. Champaign, IL: University of Illinois Press.

Hirst, Nicholas (2015) 'Push for telecoms deal', European Voice, Politico, 19 January, available at www.politico.eu/article/push-for-telecoms-deal/ (Accessed 19 May 2016).

HM Treasury (2009) 'The Next Phase of Broadband UK: Action Now for Long Term Competitiveness', available at www.umic.pt/images/stories/publicacoes2/file47788.pdf (Accessed 15 September 2016).

HM Treasury (2010) Spending Review 2010 20 October – The Department for Business Innovation and Skills.

Hodges, Sir William and Charles Manley Smith (1876) *A Treatise on the Law of Railways*, 6th Edition by C. M. Smith and H. Sweet, available at National Archives, available at http://discovery.nationalarchives.gov.uk/details/record?catid=3205002&catln=6 (Accessed 19 May 2016).

Hooper Richard (2015) 'Preserving the open internet – the UK approach', Broadband Stakeholder Group (BSG), 20 October, Broadband World Forum 2015, Excel, London, available at www.broadbanduk.org/wp-content/uploads/2015/10/BBWF-RH-Speech.pdf (Accessed 19 May 2016).

Horten, Monica (2009) 'Telecoms Package – sealed but not with a kiss', 5 November, available at: www.iptegrity.com (Accessed 16 September 2016).

Horten, Monica (2011) *The Copyright Enforcement Enigma: Internet Politics and the 'Telecoms Package'*. Basingstoke: Palgrave Macmillan.

Horten, Monica (2015) 'EU drops net neutrality principle – will it mean content restrictions?', Iptegrity.com, 29 October, available at www.iptegrity.com/index.php/telecoms-package/net-neutrality/1013-eu-drops-net-neutrality-principle-will-it-mean-content-restrictions (Accessed 15 September 2016).

Höttges, Timotheus (2015) 'Net neutrality: finding consensus in the minefield', Deutsche Telekom AG, 28 October, available at www.telekom.com/media/management-to-the-point/291728 (Accessed 19 May 2016).

House of Commons 571 (2012) Annual Report of the Interception of Communications Commissioner, Ordered by the House of Commons to be printed on 18th July 2013, SG/2013/131.

House of Commons 571, Business, Innovation and Skills Committee (2016) 'Oral evidence: The Digital Economy', 22 March, Q493, available at http://data.parliament.uk/writtenevidence/committeeevidence.svc/evidencedocument/business-innovation-and-skills-committee/the-digital-economy/oral/30984.html (Accessed 19 May 2016).

Hugenholtz, P. Bernt (2008) 'Re: Open Letter concerning European Commission's "Intellectual Property Package"', Amsterdam, 18 August, available at www.scribd.com/document/14561635/Open-Letter-EC (Accessed 15 September 2016).

Independent Regulators Group (2015) 'About Us/Members', available at www.irg.eu.

Indian Department of Telecommunications (2015) Report of Committee on Net Neutrality, May, available at www.dot.gov.in/reports-statistics/report-committee-net-neutrality-0 (Accessed 15 September 2016).

Intelligence and Security Committee of Parliament (2013) 'Statement on GCHQ's Alleged Interception of Communications under the US PRISM Programme', available at www.gov.uk/government/uploads/system/uploads/attachment_data/file/225459/ISC-Statement-on-GCHQ.pdf (Accessed 19 May 2016).

Interception of Communications Commissioner (2011) Investigation of Unintentional Electronic Interception: Monetary Penalty Notice, Exercise of Powers under Section 1a and Schedule A1 of the Regulation of Investigatory Powers Act 2000, available at www.intelligencecommissioners.com/docs/Interception_Commissioner_Guidance_RIPA.pdf.

Interception of Communications Commissioner (2013) Sir Anthony May's response to the article published in the Independent, 16 July 13, available at www.iocco-uk.info/sections.asp?sectionID=8&chapter=4&type=top.

Jackson, Mark (2013a) 'National Audit Office scalds 2 years late Broadband Delivery UK project', ISP Review, 5 July, available at www.ispreview.co.uk/index.php/2013/07/national-audit-office-report-scathes-broadband-delivery-uk-project.html (Accessed 19 May 2016).

Jackson, Mark (2013b) 'Broadband Delivery UK invite BT and smaller ISPs to crucial in-fill consultation', ISP Review, 3 October, available at www.ispreview.co.uk/index.php/2013/10/broadband-delivery-uk-invites-bt-smaller-isps-crucial-fill-consultation.html (Accessed 19 May 2016).

Jackson, Mark (2014) 'Diagram of how BT's new FTTrN superfast broadband technology works', 1 September, available at www.ispreview.co.uk/index.php/2014/09/diagram-bts-new-fttrn-broadband-technology-works.html (Accessed 19 May 2016).

Jackson, Mark (2015) 'BT top 7.88m internet subs as fibre broadband covers 24m UK premises', ISP Preview, 29 October, available at www.ispreview.co.uk/index.php/2015/10/bt-top-7-88m-internet-subs-as-fibre-broadband-covers-24-million-premises.html (Accessed 19 May 2016).

Jain, Rekha with Radha Ravattu, Rishabh Dara, Pranesh Prakash (2015) 'Response to TRAI Consultation Paper on Regulatory Framework for Over-the-top (OTT) Services', 27 March, Centre for Internet Studies.

Jasserand, Catherine (2013) 'Critical views on the French approach to "net neutrality"', *Journal of Internet Law*, 16:9, pp. 18–28.

Jitsuzumi, T. (2012) 'An analysis of prerequisites for Japan's approach to network neutrality', Proceedings of the 38th Research Conference on Communication, Information and Internet policy (TPRC).

Jitsuzumi, Toshiya (2015) 'Recent development of net neutrality conditions in Japan', presentation slides at the ITS Europe 2015 @San Lorenzo de El Escorial, Spain, 25 June, available at www.slideshare.net/toshiyajitsuzumi/recent-development-of-net-neutrality-conditions-in-japan?qid=aa9e9595-f430-434a-b6f2-1b2444237266&v=-default&b=&from_search=1 (Accessed 19 May 2016).

Johnson, Eric J. and Daniel G. Goldstein (2012) 'Decisions by Default,' in Eldar Shafir (ed.) *The Behavorial Foundations of Public Policy.* Princeton, NJ: Princeton University Press.

Johnson, D. and Post, D. (1996) 'Law and Borders: The Rise of Law in Cyberspace', *First Monday*, 1:4, available at http://firstmonday.org/article/view/468/389 (Accessed 19 May 2016).

Jolls, Christine, Cass R. Sunstein and Richard Thaler (1998) 'A behavioral approach to law and economics', *Stanford Law Review*, 50, pp. 1471–1550.

Jones, Sir William and William Theobald (1833) *An Essay on the Law of Bailments*. London: S. Sweet.

Kahn-Freund, Otto (1963) 'Transport Act, 1962', *Modern Law Review*, 26:2, pp. 174–184.

Kang, Cecilia and Hayley Tsukayama (2011), 'AT&T to throttle data speeds for heaviest wireless users', *Washington Post*, 1 August, available at www.washingtonpost.com/business/technology/atandt-to-throttle-data-speeds-for-heaviest-wireless-users/2011/08/01/gIQAh0HBoI_story.html (Accessed 19 May 2016).

Karpinski, R. (2009) 'Comcast's congestion catch-22', 23 January, available at www.benton.org/node/21282 (Accessed 15 September 2016).

Kaye, David and Brett Solomon (2015) 'Merely connecting the developing world to the Internet isn't enough', Slate: Future Tense, 13 October, available at www.slate.com/blogs/future_tense/2015/10/13/the_u_n_wants_to_connect_the_world_to_the_internet_that_s_not_enough.html (Accessed 19 May 2016).

Keizer, Gregg (2015) 'An incredibly shrinking Firefox faces endangered species status', Computerworld, 5 March, available at www.computerworld.com/article/2893514/an-incredibly-shrinking-firefox-faces-endangered-species-status.html (Accessed 19 May 2016).

Kelemen, R. D. and A. D Tarrant (2011) 'The political foundations of the Eurocracy', *West European Politics*, 34, pp. 922–947.

Kessell, Clive (2015) 'UK Railway Telecommunications 2015 Update', Rail Engineer 28 August, available at www.railengineer.uk/2015/08/12/uk-railway-telecommunications-2015-update/ (Accessed 19 May 2016).

Kiedrowski, T. (2007) 'Net neutrality: Ofcom's view', available at http://media.ofcom.org.uk/speeches/2007/joint-ceps-and-progress-for-freedom-conference/ (Accessed 15 September 2016).

Kleinsteuber Hans J. (2004) *The Media Freedom Internet Cookbook*, OSCE, available at www.osce.org/fom/13836 (Accessed 19 May 2016).

Koops, B.-J. and J. P. Sluijs (2012), 'Network neutrality and privacy according to Art. 8 ECHR', *European Journal of Law and Technology*, 3:2, pp. 1–23.

Kosmopolit (2009) 'Oettinger German EU Commissioner – Is the Spree freezing over?', 24 October, available at http://grahnlaw.blogspot.co.uk/2009/10/oettinger-german-eu-commissioner-is.html (Accessed 15 September 2016).

Kron, J. (2012) 'Open source politics: the radical promise of Germany's Pirate Party', *The Atlantic*, 21 September, available at www.theatlantic.com/international/archive/2012/09/opensource-politics-the-radical-promise-of-germanys-pirate-party/262646/ (Accessed 19 May 2016).

La Rue, Frank (2011) 'Report of the Special Rapporteur on the promotion and protection of the right to freedom of opinion and expression', Human Rights Council Seventeenth session Agenda item 3, A/HRC/17/27.

Ladurantaye, Steve (2013) 'The CRTC's Jean-Pierre Blais: the regulator who speaks truth to power', *Globe and Mail*, 1 November, available at www.theglobeandmail.com/report-on-business/careers/careers-leadership/the-crtcs-jean-pierre-blais-speaking-truth-to-power/article15224721/?page=2 (Accessed 19 May 2016).

Lamadrid, Alfonso de Pablo (2015) 'Regulating platforms? A competition law perspective', Chillin'Competition, 24 November, available at http://chillingcompetition.com/2015/11/24/regulating-platforms-a-competition-law-perspective/#more-9329 (Accessed 19 May 2016).

Landes, William M. and Richard A. Posner (1981) 'Market power in antitrust cases', *Harvard Law Review*, 94, pp.937–996.

Lardinois, Frederic (2012) 'Facebook and FTC settle privacy charges – no fine, but 20 years of privacy audits', Tech Crunch, 10 August, available at http://techcrunch.com/2012/08/10/facebook-ftc-settlement-12/ (Accessed 19 May 2016).

Law Professors (2011) Letter in opposition to 'Preventing Real Online Threats to Economic Creativity and Theft of Intellectual Property Act of 2011', draft 27 June, available at www.scribd.com/doc/59241037/PROTECT-IP-Letter-Final (Accessed 19 May 2016).

Lee, Timothy B. (2012) 'Sony: internet video service on hold due to Comcast data cap', Ars Technica, 2 May, available at http://arstechnica.com/tech-policy/2012/05/sony-warns-comcast-cap-will-hamper-video-competition/ (Accessed 19 May 2016).

Lemley, M. A. and L. Lessig (1999) Ex Parte Declaration of Professor Mark A. Lemley and Professor Lawrence Lessig in the Matter of: Application for Consent to the Transfer of Control of Licenses of MediaOne Group, Inc. to AT&T Corp CS, Docket No. 99–251 before the Federal Communications Commission, Washington DC 20554.

Lemley, M. A. and L. Lessig (2000) 'The End of the E2E: preserving the architecture of the Internet in the broadband era', UC Berkeley Law & Econ Research Paper No. 2000–19; Stanford Law & Economics Olin Working Paper No. 207; UC Berkeley Public Law Research Paper No. 37.

Lessig, Lawrence (1998) 'The New Chicago School', *Journal of Legal Studies*, 27:S2, pp. 661–691.

Lessig, L. (1999) *Code and Other Laws of Cyberspace*. New York: Basic Books.

Leveson, Brian L. J. (2012) *An Inquiry into the Culture and Ethics of the Press, Politicians and Police*, Part 1, Final report.

Lin, A., B. Davie and F. Baker (1996) 'Tag switching support for classes of service', Internet Engineering Task Force, available at http://tools.ietf.org/html/draft-lin-tags-cos-00 (Accessed 19 May 2016).

LINX Public Affairs (2009) 'Home Office "colluded with Phorm"', 29 April, available at https://publicaffairs.linx.net/news/?p=993 (Accessed 19 May 2016).

Lokhandwala, Taha (2014) 'UK government no longer on hook for BT Pension Scheme buyout', Investment & Pensions Europe, 17 July, available at www.ipe.com/news/uk-government-no-longer-on-hook-for-bt-pension-scheme-buyout/10002523.fullarticle (Accessed 19 May 2016).

Longley, H. (1967) *Common Carriage of Cargo*. New York: Matthew Bender & Co.

Lynn, Barry (2010) *Cornered: The New Monopoly Capitalism and the Economics of Destruction*. Hoboken, NJ: John Wiley & Sons, Inc.

McCarthy, Kieren (2001) 'BT admits to bandwidth restrictions for file-sharing sites: forced to come clean after evidence builds up', The Register, 8 October, available at www.theregister.co.uk/2001/10/08/bt_admits_to_bandwidth_restrictions (Accessed 20 May 2016).

McCarthy, Kieren (2015) 'Net neutrality debate: if startups want to rival Google, they must show some green to telcos: so says the CEO of Deutsche Telekom, a, er, telco giant', The Register, 30 October, available at www.theregister.co.uk/2015/10/30/deutsche_telekom_starts_tiered_pricing (Accessed 20 May 2016).

McGregor, Richard (2011) 'Zhou's cryptic caution lost in translation', *Financial Times*, 10 June, available at www.ft.com/cms/s/0/74916db6-938d-11e0-922e-00144feab49a.html (Accessed 20 May 2016).

McNamee, Joe (2015) 'ENDitorial: a system you never heard of undermined net neutrality', EDRi Newsletter, 4 November, available at https://edri.org/system-you-never-heard-of-undermined-net-neutrality (Accessed 20 May 2016).

McTaggart, Craig (2008) 'Net neutrality and Canada's Telecommunications Act', available at http://dx.doi.org/10.2139/ssrn.1127203 (Accessed 20 May 2016).

Madiega, Tambiama (2015) 'The EU rules on network neutrality: key provisions, remaining concerns', European Parliamentary Research Service Briefing November 2015 PE 571.318, available at www.europarl.europa.eu/thinktank/en/document.html?reference=EPRS_BRI%282015%29571318 (Accessed 19 May 2016).

Maillé, Patrick and Bruno Tuffin (2014) *Telecommunication Network Economics: From Theory to Applications.* Cambridge: Cambridge University Press.

Make the Net Work (2014) 'Make the Net Work for Europe: A healthy, thriving digital economy is critical to social and economic progress in Europe', available at www.gsma.com/gsmaeurope/tag/make-the-net-work/ (Accessed 15 September 2016).

Maniadaki, Katerina (2015) *EU Competition Law, Regulation and the Internet: The Case of Net Neutrality*, International Competition Law Series Vol. 59. Aalphen an den Rijn: Wolters Kluwer Law.

Manjoo, Farhad (2016) 'Tech's "Frightful 5" will dominate digital life for foreseeable future', *New York Times*, 20 January, available at www.nytimes.com/2016/01/21/technology/techs-frightful-5-will-dominate-digital-life-for-foreseeable-future.html (Accessed 19 May 2016).

Mankotia, A. S. (2016) 'PMO displeased with Facebook's reaction to Trai's consultation paper', *The Economic Times*, available at http://articles.economictimes.indiatimes.com/2016-02-04/news/70343830_1_net-neutrality-consultation-paper-digital-india (Accessed 19 May 2016).

Marcus, J. Scott and Martin Waldburger (2015) 'Identifying harm to the best efforts Internet', available at http://ssrn.com/abstract=2624604 or http://dx.doi.org/10.2139/ssrn.2624604 (Accessed 19 May 2016).

Marques, Camila, Laura Tresca, Luiz Alberto Perin Filho, Mariana Rielli and Pedro Iorio (2015) 'Marco Civil da Internet: seis meses depois, em que pé que estamos?', Article 19, 28 January, available at http://artigo19.org/blog/analise-marco-civil-da-internet-seis-meses-depois-em-que-pe-que-estamos/ (Accessed 19 May 2016).

Marsden, C. (1999) 'Pluralism in the multi-channel market: suggestions for regulatory scrutiny', Council of Europe Human Rights Commission, Mass Media Directorate,

MM-S-PL(1999)012, available at http://citeseerx.ist.psu.edu/viewdoc/download?-doi=10.1.1.471.9049&rep=rep1&type=pdf (Accessed 19 May 2016).

Marsden, C. (2001) 'Towards the hyperglobalisation of the individual: how the ubiquitous Internet will make the international political economy increasingly dynamically unstable', available at http://ssrn.com/abstract=1578203 or http://dx.doi.org/10.2139/ssrn.1578203 (Accessed 19 May 2016).

Marsden, C. (2002) Ref: Carrier Pre Selection Process Letter to Caroline Wallace, Oftel of 10 June (copy on file with author).

Marsden, C. (2004) 'Hyperglobalized individuals: the Internet, globalization, freedom and terrorism', *Foresight*, 6:3, pp. 128–140.

Marsden, C. (2005) 'Free, Open or closed – approaches to the information ecology', *info*, 7:5, pp. 6–19.

Marsden, C. (2009) 'The net neutrality zombie and net neutrality "lite"', *Computers & Law*, 19:6, pp.36–38.

Marsden, C. (2010) *Network Neutrality: Towards a Co-regulatory Solution*. London: Bloomsbury Academic.

Marsden, C. (2011) *Internet Co-Regulation*. Cambridge: Cambridge University Press.

Marsden, C. (2012a) 'Regulating Intermediary Liability and Network Neutrality', pp. 701–750 in I. Walden (ed.) *Telecommunications Law and Regulation*, 4th Edition. Oxford: Oxford University Press.

Marsden, C. (2012b) *Oxford Bibliography of Internet Law.* New York: Oxford University Press, section 'Origins of Internet Law'.

Marsden, C. (2013a) 'Network Neutrality: A Research Guide', chapter 16, pp. 419–444 in I. Brown (ed.) *Handbook of Internet Research*. Cheltenham, UK and Northampton, USA: Edward Elgar.

Marsden, C. (2013b) 'Net Neutrality Law: Past Policy, Present Proposals, Future Regulation?', Proceedings of the United Nations Internet Governance Forum: Dynamic Coalition on Network Neutrality, Nusa Dua Bali, Indonesia, 25 October, available at http://ssrn.com/abstract=2335359 (Accessed 19 May 2016).

Marsden, C. (2013c) 'The road to monopoly is littered with good intentions: how the EC let Google win the search war', Regulating Code Blog, 29 April, available at http://regulatingcode.blogspot.co.uk/2013/04/the-road-to-monopoly-is-littered-with.html (Accessed 19 May 2016).

Marsden, C. (2014a) 'Net neutrality regulation in the UK: more transparency and switching', *Journal of Law and Economic Regulation*, 7:1, pp. 44–68 (Seoul, Korea).

Marsden, C. (2014b) 'Hyper-power and private monopoly: the unholy marriage of (neo) corporatism and the imperial surveillance state', *Critical Studies in Media Communication*, 31:2, pp. 100–108.

Marsden, C. (2015) 'Technology and the Law' in *The International Encyclopedia of Digital Communication & Society*. Chichester: Wiley-Blackwell.

Marsden, C. (2016a) 'Comparative case studies in implementing net neutrality: a critical analysis of zero rating', *SCRIPT-Ed*, 13:1, available at https://script-ed.org/article/comparative-case-studies-in-implementing-net-neutrality-a-critical-analysis-of-zero-rating/ (Accessed 19 May 2016).

Marsden, C. (2016b) 'Zero rating and mobile net neutrality', pp. 241–260 in Luca Belli and Primavera de Filippi (eds) *Net Neutrality Compendium: Human Rights, Free Competition and the Future of the Internet*. Cham: Springer Verlag.

Marsden, C. and J. Cave (2007) 'Beyond the "net neutrality" debate: price and quality discrimination in next generation internet access', 35th Telecoms Policy Research Conference, Alexandria, Virginia, 29 September.

Marsden, C. *et al.* (2006) *Assessing Indirect Impacts of the EC Proposals for Video Regulation*, TR-414, Ofcom, Santa Monica, CA: RAND.

Marsden, Chris (2009) 'Summary of October events – regulators are getting proactive on net neutrality', blog post, 21 October, available at http://chrismarsden.blogspot.co.uk/2009/10/summary-of-october-events-regulators.html (Accessed 19 May 2016).

Marsden, Chris (2010) 'Canadian net neutrality: Geist critique', blog post, 14 July, available at http://chrismarsden.blogspot.co.uk/2010/07/canadian-net-neutrality-geist-critique.html (Accessed 19 May 2016).

Marsden, Chris (2011) 'Ofcom three wise monkeys: competition/switching mantra even for recalcitrant ISPs', blog post, 24 November, available at http://chrismarsden.blogspot.co.uk/2011/11/ofcom-three-wise-monkeys.html.

Marsden, Chris (2012) 'Analyzing the UK Voluntary Code of Conduct: shadowing co-regulation?', blog post, 27 July, available at http://chrismarsden.blogspot.co.uk/2012/07/analyzing-uk-voluntary-code-of-conduct.html (Accessed 19 May 2016).

Marsden, Chris (2013a) 'Ofcom workplan – net neutrality to be addressed in November 2013 infrastructure report', blog post, 21 January, available at http://chrismarsden.blogspot.co.uk/2013/01/ofcom-workplan-net-neutrality-to-be.html (Accessed 19 May 2016).

Marsden, Chris (2013b) 'Guaranteeing competition and the open internet in Europe', speech, 4 June, ALDE YouTube Channel, published 8 April 2015, www.youtube.com/watch?v=JDEY_dMoZa4 (Accessed 19 May 2016).

Marsden, Chris (2013c) 'Presentation on net neutrality at Internet Governance Forum', blog post, 27 October, available at http://chrismarsden.blogspot.co.uk/2013/10/presentation-on-net-neutrality-at.html (Accessed 19 May 2016).

Marsden, Chris (2013d) 'Internet Co-Regulation, EC Code Of Practice Agora', Brussels, 10 December, available at https://ec.europa.eu/digital-agenda/sites/digital-agenda/files/Presentation%20Chris%20Marsden.pdf (Accessed 19 May 2016).

Marsden, Chris, (2013e) 'Freedom of expression, the Council of Europe and net neutrality', blog post, 18 June, available at http://chrismarsden.blogspot.co.uk/2013/06/freedom-of-expression-council-of-europe.html (Accessed 19 May 2016).

Marsden, Chris (2014a) 'Will Commissioner Kroes be able to Skype her grandchildren's mobiles in retirement?', LSE Media Policy Blog, 4 April, available at http://blogs.lse.ac.uk/mediapolicyproject/2014/04/04/will-commissioner-kroes-be-able-to-skype-her-grandchildrens-mobiles-in-retirement/ (Accessed 19 May 2016).

Marsden, Chris (2014b) 'Reminder: UK isolated in EU Council – may lead to 2015 compromise', 28 November, available at http://chrismarsden.blogspot.co.uk/2014/11/reminder-uk-isolated-in-eu-council-may.html (Accessed 20 May 2016).

Marsden, Chris (2014c) 'Kroes' anti-neutrality all part of a cunning plan – priceless comments', 1 May, available at http://chrismarsden.blogspot.co.uk/2014/05/kroes-anti-neutrality-all-part-of.html (Accessed 20 May 2016).

Marsden, Chris (2015a) 'Not neutrality please, we're British!' 18 November, available at http://chrismarsden.blogspot.co.uk/2015/11/not-neutrality-please-were-british.html.

Marsden, Chris (2015b) '@CableEurope argues #coregulation #QoS #transparency as lack of @BERECeuropaeu resource to implement #netneutrality', Twitter, 8 December, available at https://twitter.com/ChrisTMarsden/status/674174024281415680 (Accessed 20 May 2016).

Marsden, Chris (2016) 'Zero-rating plans are a serious threat to the open internet: US advocacy letter', 31 March, available at http://chrismarsden.blogspot.co.uk/2016/03/zero-rating-plans-are-serious-threat-to.html (Accessed 20 May 2016).

Mazzucato, Mariana (2013) *The Entrepreneurial State: Debunking Private vs. Public Sector Myths*. London: Anthem Press.

Meisel, J. P. (2010) 'Trinko and mandated access to the Internet', *info*, 12:2, pp. 9–27.

Miller, Jeff (2012) 'Net-neutrality regulation in canada: assessing the CRTC's statutory competency to regulate the Internet', *Appeal*, 17, pp. 47–62.

Moglen, E. (2013) 'Snowden and the future. Part I: westward the course of empire', speech at Columbia Law School, 9 October, available at http://snowdenandthefuture.info (Accessed 20 May 2016).

Mueller, M. (1998) *Universal Service: Competition, Interconnection, and Monopoly in the Making of the American Telephone System*. Washington DC: AEI Press.

Mueller, M. (2007) 'Net neutrality as global principle for internet governance', Internet Governance Project Paper IGP07-003, available at http://internetgovernance.org/pdf/NetNeutralityGlobalPrinciple.pdf (Accessed 20 May 2016).

Murphy, David (2011) 'Google paying Mozilla almost $1b for Firefox search: why?', *PC Mag*, 24 December, available at www.pcmag.com/article2/0,2817,2398046,00.asp (Accessed 20 May 2016).

Musiani, F., Derrick L. Cogburn, Laura DeNardis and Nanette S. Levinson (2015) *The Turn to Infrastructure in Internet Governance*. New York: Palgrave Macmillan.

National Audit Office (2013) Rural Broadband Programme Report, available at www.nao.org.uk/wp-content/uploads/2013/07/10177-001-Rural-Broadband_HC-535.pdf (Accessed 20 May 2016).

Nature (2015) 'Time for the social sciences: governments that want the natural sciences to deliver more for society need to show greater commitment towards the social sciences and humanities', *Nature*, 517, p. 5.

NKOM (2015) 'BEREC and net neutrality', available at http://eng.nkom.no/technical/internet/net-neutrality/berec-and-net-neutrality (Accessed 20 May 2016).

Noam, E. (1994) 'Beyond liberalization II: the impending doom of common carriage', *Telecommunications Policy*, 18:6, pp. 435–452.

Northrup, Laura (2015) 'T-Mobile now exempts 33 streaming music services from data limits, adds Apple Music', Consumerist, 28 July, available at http://consumerist.com/2015/07/28/t-mobile-now-exempts-33-streaming-music-services-from-data-limits-adds-apple-music/ (Accessed 20 May 2016).

Nunziato, Dawn C. (2009) *Virtual Freedom: Net Neutrality and Free Speech in the Internet Age*. Stanford, CA: Stanford University Press.

Obama, Barack H. (2014) President Obama's Statement on Keeping the Internet Open and Free, 10 November, available at www.youtube.com/watch?v=uKcjQPVwfDk (Accessed 20 May 2016).

Odlyzko, Andrew (1998) 'The economics of the Internet: utility, utilization, pricing, and quality of service', available at www.dtc.umn.edu/~odlyzko/doc/internet.economics.pdf (Accessed 20 May 2016).

Odlyzko, Andrew (2004) 'The many paradoxes of broadband', *First Monday*, 8:9, available at http://firstmonday.org/ojs/index.php/fm/article/view/1072 (Accessed 20 May 2016).

Odlyzko, Andrew (2010) 'Bubbles, gullibility, and other challenges for economics, psychology, sociology, and information sciences', *First Monday*, 15:9, available at http://firstmonday.org/ojs/index.php/fm/article/view/3142/2603 (Accessed 20 May 2016).

Odlyzko, Andrew (2012) 'Will smart pricing finally take off?', Keynote at the IEEE workshop on Smart Data Pricing, Princeton University, 30 July, available at http://scenic.princeton.edu/SDP2012/program.html (Accessed 20 May 2016).

Odlyzko, Andrew (2014a) 'Will smart pricing finally take off?', Accession Number: ADA613589, Technical paper, available at www.dtc.umn.edu/~odlyzko/doc/smart.pricing.pdf (Accessed 20 May 2016).

Odlyzko, Andrew (2014b) 'This time is different: an example of a giant, wildly speculative, and successful investment mania', *The BE Journal of Economic Analysis & Policy*, 2010, available at http://econpapers.repec.org/article/bpjbejeap/v_3a10_3ay_3a2010_3ai_3a1_3an_3a60.htm (Accessed 20 May 2016).

Odlyzko, Andrew, Bill St. Arnaud, Erik Stallman and Michael Weinberg (2012) 'Know your limits: considering the role of data caps and usage based billing in internet access service', Public Knowledge, 23 April, available at www.publicknowledge.org/documents/know-your-limits-considering-the-role-of-data-caps-and-usage-based-billing (Accessed 20 May 2016).

O'Donoghue, Robert and Tom Pascoe (2016) 'Net neutrality in the EU: unresolved issues under the new regulation', 15 March, available at http://ssrn.com/abstract=2741173 (Accessed 20 May 2016).

OECD (2007) 'Recommendation on Cross-Border Co-operation in the Enforcement of Laws Protecting Privacy', available at www.oecd.org/sti/ieconomy/38770483.pdf (Accessed 20 May 2016).

OECD (2008) 'The Seoul Declaration for the Future of the Internet Economy', available at www.oecd.org/sti/40839436.pdf (Accessed 20 May 2016).

OECD (2011) 'Recommendation on Principles for Internet Policy Making', available at www.oecd.org/internet/ieconomy/49258588.pdf (Accessed 20 May 2016).

OECD (2013a) Guidelines on the Protection of Privacy and Transborder Flows of Personal Data, available at www.oecd.org/internet/ieconomy/oecdguidelinesontheprotectionofprivacyandtransborderflowsofpersonaldata.htm (Accessed 13 September 2016).

OECD (2013b) 'Recommendation of the Council concerning Guidelines governing the Protection of Privacy and Transborder Flows of Personal Data', [C(80)58/

FINAL, as amended on 11 July 2013 by C(2013)79], available at www.oecd.org/sti/ieconomy/2013-oecd-privacy-guidelines.pdf (Accessed 20 May 2016).

OECD (2015) OECD 'Digital Economy Outlook 2015: Main trends in communication policy and regulation', pp. 187–192, DOI:10.1787/9789264232440-6-en, available at www.keepeek.com/Digital-Asset-Management/oecd/science-and-technology/oecd-digital-economy-outlook-2015/main-trends-in-communication-policy-and-regulation_9789264232440-6-en#page22 (Accessed 20 May 2016).

OECD.stat (2015) Database Live, available at https://stats.oecd.org/Index.aspx?DataSetCode=PDB_LV (Accessed 20 May 2016).

Oettinger, G. (2016) 30 March 10.50am, Twitter post, https://twitter.com/GOettingerEU/status/715113943056957441 (Accessed 20 May 2016).

Olsen, Parmy (2015) 'This app is cashing in on giving the world free data', *Forbes*, 29 July, available at www.forbes.com/sites/parmyolson/2015/07/29/jana-mobile-data-facebook-internet-org/ (Accessed 20 May 2016).

Olsen, Torstein (2015) 'Net neutrality activities at BEREC and Nkom', Norwegian Communications Authority, 3 July, available at http://berec.europa.eu/files/doc/2015-07-13_09_56_36_3.%20Noruega%20Nkom%20net%20neutrality%20-%20Summit%20BEREC-EaPeReg-REGULATEL-EMERG.pdf (Accessed 20 May 2016).

Ozer, Jan (2015) 'New HEVC patent pool: what are the implications?', Frost & Sullivan Streaming Media Blog, 1 April, available at www.streamingmedia.com/Articles/Editorial/Featured-Articles/New-HEVC-Patent-Pool-What-Are-the-Implications-103042.aspx (Accessed 20 May 2016).

Pahwa, Nikhil (2015) 'Facebook's Internet.org platform is a privacy nightmare: tracks users on partner sites, allows telcos to track', Medianama, 4 May, available at www.medianama.com/2015/05/223-facebooks-internet-org-privacy/ (Accessed 20 May 2016).

Parnwell, Marcus (2014) 'New BBC iPlayer for connected TVs: update', BBC Internet Blog, 8 August, available at www.bbc.co.uk/blogs/internet/entries/88e41b19-17dd-33ed-9e00-5ee508c4d045 (Accessed 20 May 2016).

Pasquale Frank (2010) 'Dominant Search Engines: An Essential Cultural & Political Facility', pp. 401–418 in Berin Szoka and Adam Marcus (eds) *The Next Digital Decade: Essays on the Future of the Internet*. Washington DC: TechFreedom.

Pollock, Rufus (2010) 'Is Google the next Microsoft: competition, welfare and regulation in online search', *Review of Network Economics*, 9:4, ISSN (Online) 1446–9022, DOI: 10.2202/1446–9022.1240.

Posner, Richard A. (1974) 'Theories of economic regulation', *Bell Journal of Economics and Management Science*, 5:2, pp. 335–358.

Posner, Richard A. (1979) 'The Chicago School of Antitrust Analysis', *University of Pennsylvania Law Review*, 127:4, 925–958.

Powell, A. (2013) 'Assessing the influence of online activism on internet policy-making: SOPA and ACTA', at GIGANet conference, Bali Indonesia, 17 October.

Powell, Alison B. (2015) 'Network exceptionalism: online action, discourse and the opposition to SOPA and ACTA', *Information, Communication & Society*, 19:2, pp. 249–263.

Powell, M. (2004) Four Freedoms speech, available at http://hraunfoss.fcc.gov/edocs_public/attachmatch/DOC-243556A1.pdf (Accessed 20 May 2016).

Powles, Julia (2015) 'The case that won't be forgotten', *Loyola University Chicago Law Journal*, 47, pp. 583–615.

Predictable Network Solutions Limited (2015) 'A Study of Traffic Management Detection Methods and Tools', MC 316, June, available at http://t.co/rkVY62oRuf (Accessed 20 May 2016).

Prescott, Roberta (2015) 'LatAm: Claro Brazil resumes zero-rating plans', RC Wireless, 18 June, available at www.rcrwireless.com/20150618/americas/latam-claro-brazil-resumes-zero-rating-plans (Accessed 20 May 2016).

Privacy Commissioner of Canada (2009) Letter to Robert A. Morin, Secretary General, Canadian Radiotelevision and Telecommunications Commission, Re: Telecom Public Notice CRTC 2008–19 – Review of the Internet traffic management practices of Internet service providers, Canadian Radio-television and Telecommunications Commission, 18 February, available at www.crtc.gc.ca/public/partvii/2008/8646/c12_200815400/1027577.PDF (Accessed 20 May 2016).

Public Accounts Committee (2013) 'Twenty-Fourth Report: The rural broadband programme', ordered by the House of Commons to be printed 11 September 2013, available at www.publications.parliament.uk/pa/cm201314/cmselect/cmpubacc/474/47402.htm (Accessed 20 May 2016).

Public Knowledge (2013) Letter. RE: Public Knowledge Petition in MB Docket No. 10–56, Application of Comcast Corporation, General Electric Company and NBC Universal, Inc. For Consent to Assign Licenses and Transfer Control of Licenses, available at www.publicknowledge.org/files/PK%201%20Year%20Letter%20on%20Comcast%20Xbox%20Petition.pdf (Accessed 20 May 2016).

Quatrocchi, Matt (2015) 'Nobody wants to tak about transparency', Ripe for Discussion Blog, 28 August (copy on file with author).

Radaelli, Claudio M., Claire A. Dunlop and Oliver Fritsch (2013) 'Narrating impact assessment in the European Union', *European Political Science*, 12, pp. 500–521.

Radu, Roxana, Nicolo Zingales and Enrico Calandro (2015) 'Crowdsourcing ideas as an emerging form of multistakeholder participation in internet governance', *Policy & Internet*, 7:3, pp. 362–382.

Ramos, Pedro Henrique Soares (2014) 'Towards a developmental framework for net neutrality: the rise of sponsored data plans in developing countries', TPRC Conference Paper, available at http://ssrn.com/abstract=2418307 (Accessed 20 May 2016).

Rauhofer, Judith (2009) 'The Retention of Communications Data in Europe and the UK' in Lillian Edwards and Charlotte Waelde (eds), *Law and the Internet*, 3rd Edition. Oxford: Hart Publishing.

Rauhofer, Judith and Caspar Bowden (2013) 'Protecting Their own: fundamental rights implications for EU data sovereignty in the cloud', Edinburgh School of Law Research Paper No. 2013/28, available at http://ssrn.com/abstract=2283175 or http://dx.doi.org/10.2139/ssrn.2283175 (Accessed 20 May 2016).

Ray, Bill (2013) 'Net neutrality? We've heard of it, says Ofcom. You want mobile Skype, choose your network with care', The Register, 2 April, available at www.theregister.co.uk/2013/04/02/ofcom_annual_plan/ (Accessed 20 May 2016).

Rayburn, Dan (2015a) 'The adoption of 4K streaming will be stalled by bandwidth, not hardware & devices', Frost & Sullivan Streaming Media Blog, 14 January, available at http://blog.streamingmedia.com/2015/01/4k-streaming-bandwidth-problem.html (Accessed 20 May 2016).

Rayburn, Dan (2015b) 'New patent pool wants 0.5% of every content owner/distributor's gross revenue for higher quality video', Frost & Sullivan Streaming Media Blog, 22 July, available at http://blog.streamingmedia.com/2015/07/new-patent-pool-wants-share-of-revenue-from-content-owners.html (Accessed 20 May 2016).

Reda, Julia (2015) 'Net neutrality is a "Taliban-like issue", says Europe's top digital policymaker', 5 March, available at https://juliareda.eu/2015/03/oettinger-net-neutrality-taliban-like/ (Accessed 20 May 2016).

Reed, C. (2007) 'Taking sides on technology neutrality', SCRIPTed, 4:3, available at www.law.ed.ac.uk/ahrc/script-ed/vol4-3/reed.asp (Accessed 20 May 2016).

Rekhter, Y. *et al.* (1997) 'Tag switching architecture overview,' *Proceedings of the IEEE*, 85:12, pp. 1973–1983.

Richards, Ed (2008) Institution of Engineering and Technology Speech, 'Broadband Britain – towards the next generation', 16 April 2008, available at http://media.ofcom.org.uk/speeches/2008/institution-of-engineering-and-technology-speech-broadband-britain-towards-the-next-generation-wednesday-16-april-2008/ (Accessed 20 May 2016).

Richards, Ed (2010) Speech on the Internet and consumer protection in the digital age, Ofcom, 19 July, available at http://media.ofcom.org.uk/speeches/2012/speech-on-the-internet-and-consumer-protection-in-the-digital-age (Accessed 20 May 2016).

Richards, Ed (2012) 'Internet and consumer protection in the digital age', Speech to KCC and KAIT International Communications Conciliatory forum, 11 October, available at http://media.ofcom.org.uk/speeches/2012/speech-on-the-internet-and-consumer-protection-in-the-digital-age/ (Accessed 15 September 2016).

Richards, Ed (2013) Speech for consumers and citizens in the communications sector conference, 16 September, available at http://media.ofcom.org.uk/speeches/2013/speech-for-consumer-conference/ (Accessed 15 September 2016).

Richards, Neil M. (2015) *Intellectual Privacy: Rethinking Civil Liberties in the Digital Age*. Oxford: Oxford University Press.

Richardson, Tim (2002) 'Microsoft's home workers to get BT broadband: you lucky, lucky people', The Register 15 April, available at www.theregister.co.uk/2002/04/15/microsofts_home_workers_to_get/ (Accessed 20 May 2016).

Richardson, Tim (2005) 'Ofcom confirms Bulldog probe: Billing and customer service issues cited', The Register 2 September, available at www.theregister.co.uk/2005/09/02/bulldog_probe/ (Accessed 7 September 2016).

Richelson, Jeffrey T. and Desmond Ball (1985) *The Ties that Bind: Intelligence Cooperation Between the UKUSA Countries*. London: Allen & Unwin.

Roa, Huichalaf and Pedro Mariano (2015) 'La Neutralidad de la Red: El Caso Chileno', Barcelona, 3 July, Subsecretario de Telecomunicaciones, Chile, available at http://berec.europa.eu/files/doc/2015-07-13_10_00_01_4.%20Neutralidad%20de%20la%20red%20versi+%7Cn%20final.%20(3).pdf (Accessed 20 May 2016).

Robinson, James (2012) 'ARCEP wants flexible approach to net neutrality', Ovum Update, 28 September, available at www.fiercewireless.com/europe/arcep-wants-flexible-approach-to-net-neutrality (Accessed 15 September 2016).

Rossini, Carolina and Taylor Moore (2015) 'Exploring zero-rating challenges: views from five countries', Public Knowledge, 28 July, available at www.publicknowledge.org/documents/exploring-zero-rating-challenges-views-from-five-countries (Accessed 20 May 2016).

Rushton, Katherine (2016) 'Older mobile phone users face bills hike if telecoms firms are allowed to merge, watchdog boss warns', *Daily Mail*, 2 January, available at www.dailymail.co.uk/news/article-3381629/Older-mobile-phone-users-face-bills-hike-telecoms-firms-allowed-merge-watchdog-boss-warns.html (Accessed 20 May 2016).

Saltzer, J. H., D. P. Reed and D. D. Clark (1984) 'End-to-End arguments in system design', 2 ACM Transactions on Computer Systems.

SamKnows (2015) 'Our regulatory clients', available at www.samknows.com/regulators (Accessed 20 May 2016).

Sampson, Anthony (1973) *The Sovereign State: The Secret History of IT&T*. London: Hodder & Stoughton Ltd.

Sandvine (2015) 'Global internet phenomena report', available at www.sandvine.com/trends/global-internet-phenomena/ (Accessed 20 May 2016).

Schaake, M. *et al.* (2014) Letter to Mr. Antonello Giacomelli, Undersecretary of State of the Ministry of Economic Development, Largo Pietro di Brazzà, 86 00187 Rome, 25 November, available at https://marietjeschaake.eu/wp-content/uploads/2014/11/2014-11-25-Letter-to-Mr-Giacomelli.pdf (Accessed 15 September 2016).

Score, Tim (2015) ARM company overview, 'Our financial strategy', available at http://ir.arm.com/phoenix.zhtml?c=197211&p=irol-finstrat (Accessed 20 May 2016).

Sharp, Alastair (2015) 'BCE media executive apologizes for editorial intrusion', Reuters, 25 March, available at www.reuters.com/article/2015/03/25/us-canada-broadcasting-regulator-idUSKBN0ML2JN20150325 (Accessed 20 May 2016).

Shepardson, David (2015) 'FCC wants details on AT&T, Comcast, T-Mobile data plans', Reuters, 17 December, available at www.reuters.com/article/us-regulations-internet-data-idUSKBN0U02IH20151217 (Accessed 20 May 2016).

Shin, Dong-Hee and Eun-Kyung Han (2012) 'How will net neutrality be played out in Korea?', *Government Information Quarterly*, 29:2, pp. 243–251.

Shubber, Kadhim (2013) 'A simple guide to GCHQ's internet surveillance programme Tempora', Wired.com, 24 June, available at www.wired.co.uk/news/archive/2013-06/24/gchq-tempora-101 (Accessed 20 May 2016).

Sieradzki, D.L. and W. Maxwell (2008) 'The FCC's network neutrality ruling in the Comcast case: towards a consensus with Europe?', *Communications & Strategies*, 72, pp. 73–88.

Sikka, Prem (2013) 'Smoke and mirrors: corporate social responsibility and tax avoidance – a reply to Hasseldine and Morris', *Accounting Forum*, 37:1, pp. 15–28.

Sikka, Prem (2015) 'The corrosive effects of neoliberalism on the UK financial crises and auditing practices: a dead-end for reforms', *Accounting Forum*, 39:1, pp. 1–18.

Simpson, S. (2013) 'The interactive nature of "soft" and "hard" governance in the EU information society: lessons from the EU Electronic Communications Regulatory Framework', *Information, Communication & Society*, 16, pp. 899–917.

Sluijs, J. P. (2012) 'From competition to freedom of expression: introducing Art. 10 ECHR in the European network neutrality debate', *Human Rights Law Review*, 12:3, pp. 509–554.

Sluijs, Jasper P., Pierre Larouche and Wolf Sauter (2011) 'Cloud computing in the EU policy sphere', *Journal of Intellectual Property, Information Technology and e-Commerce Law*, 3:1, pp. 12–32.

Sluijs, Jasper P., Florian Schuett and Bastian Henze (2010) 'Transparency regulation as a remedy for network neutrality concerns: experimental results', TILEC Discussion Paper No. 2010–039, available at http://papers.ssrn.com/sol3/papers.cfm?abstract_id=1709268 (Accessed 20 May 2016).

SMART 2010/0030 'Statistical methodologies on the Internet as a source of data gathering', available at http://ec.europa.eu/digital-agenda/en/news/statistical-methodologies-internet-source-data-gathering-smart-20100030 (Accessed 20 May 2016).

Sørensen, Frode (2013) 'The Norwegian model for net neutrality', NKOM, 5 March, available at http://eng.nkom.no/topical-issues/news/the-norwegian-model-for-net-neutrality (Accessed 20 May 2016).

Sørensen, Frode (2014a) 'On the origin of specialised services', NKOM, 6 June, available at http://eng.nkom.no/technical/internet/net-neutrality/on-the-origin-of-specialised-services (Accessed 20 May 2016).

Sørensen, Frode (2014b) 'Net neutrality in Norway: background and results', 23 January, available at https://ipfrode.wordpress.com/2014/01/23/net-neutrality-in-norway-background-and-results/ (Accessed 20 May 2016).

Speta, James B. (2004) 'FCC Authority to regulate the Internet: creating it and limiting it', *Loyola University Chicago Law Journal*, 35:15, pp. 15–39.

Srivas, A. (2016) 'What Facebook's spat with TRAI tells us about the ethics of digital lobbying', available at http://thewire.in/2016/01/15/what-facebooks-spat-with-trai-tells-us-about-the-ethics-of-digital-lobbying-19316/ (Accessed 20 May 2016).

Stastna, Kazi (2013) 'Bell's discounting of mobile TV against the rules, complaint claims: exempting own content from data caps gives telcos advantage over Netflix, YouTube and other content providers', CBC News, 16 December, available at www.cbc.ca/news/technology/bell-s-discounting-of-mobile-tv-against-the-rules-complaint-claims-1.2445059 (Accessed 20 May 2016).

Steinberg, Steve G. (1996) 'Netheads vs BellHeads', *Wired*, 4, available at http://archive.wired.com/wired/archive/4.10/atm.html?pg=10&topic= (Accessed 20 May 2016).

Sterling, Toby (2011) 'Skype: Dutch House says mobile carriers can't limit its use', *Christian Science Monitor*, 23 June, available at www.csmonitor.com/Business/Latest-News-Wires/2011/0623/Skype-Dutch-House-says-mobile-carriers-can-t-limit-its-use (Accessed 20 May 2016).

Sunstein, Cass R. (2011) 'Empirically informed regulation', *University of Chicago Law Review*, 78, pp. 1349–1429.

Sutherland, Ewan (2014) 'Rethinking theory – rigor and relevance in telecommunications policy research', TPRC Conference Paper, available at http://ssrn.com/abstract=2418695 or http://dx.doi.org/10.2139/ssrn.2418695 (Accessed 20 May 2016).

Swedish Post and Telecom Agency (2009) *Open Networks and Services*, available at www.pts.se/upload/Rapporter/Internet/2009/2009-32-open-networks-services.pdf (Accessed 20 May 2016).

Sweney, Mark (2016) 'BBC says Sky, Virgin and consoles were most popular for festive access to iPlayer', *Guardian*, 8 January, available at www.theguardian.com/media/2016/jan/08/bbc-sky-virgin-consoles-iplayer-eastenders (Accessed 20 May 2016).

Taleb, Nassim Nicholas (2007) *The Black Swan: The Impact of the Highly Improbable*. London: Allen Lane.

Tambini, Damian (2010) 'Ofcom needs sharper teeth: the industry regulator has stuck to a very narrow remit since the general election', *Guardian*, 4 October, available at www.theguardian.com/media/2010/oct/04/ofcom-media-watchdog (Accessed 20 May 2016).

Tambini, Damian (2012) 'Consumer representation in UK communications policy and regulation', *info*, 14:2, pp. 3–16, DOI http://dx.doi.org/10.1108/14636691211204833 (Accessed 20 May 2016).

Telecom Paper (2015) 'FCC set to approve AT&T's DirecTV takeover with conditions', 22 July, available at www.telecompaper.com/news/fcc-set-to-approve-atandts-directv-takeover-with-conditions/ (Accessed 20 May 2016).

Thaler, Richard (2015) *Misbehaving: The Making of Behavioral Economics*. New York: W. W. Norton & Company.

Thaler, Richard H. and Cass R. Sunstein (2008) *Nudge: Improving Decisions about Health, Wealth, and Happiness*. New Haven, CT: Yale University Press.

Thomas, Daniel (2015) 'Brussels lays out concerns over O2 merger with Three', *Financial Times*, 8 December, available at www.ft.com/cms/s/0/f51a2088-9d0e-11e5-8ce1-f6219b685d74.html#axzz3urIFUj1g (Accessed 20 May 2016).

Thöny, Andreas (2014) 'Swisscom's path from VDSL2 to G.fast', presented to DSL Seminar, 16–18 June 2014, available at www.docdroid.net/file/download/o7k3/andreas-thoeny.pdf (Accessed 15 September 2016).

TRAI (2006) Consultation Paper on Review of Internet Service, 27 December, available at www.trai.gov.in/WriteReaddata/ConsultationPaper/Document/consultation27dec06.pdf (Accessed 20 May 2016).

TRAI (2015) Consultation Paper on Regulatory Framework for Over-the-top (OTT) services, 27 March, available at http://trai.gov.in/WriteReaddata/ConsultationPaper/Document/OTT-CP-27032015.pdf (Accessed 20 May 2016).

Trossen, D. (2010) 'The EIFFEL Think Tank', available at www.slideshare.net/FIA2010/1-eiffel-fia-171210-6503523 (Accessed 20 May 2016).

Turk, Z. (2010) 'Project Europe 2030', available at www.slideshare.net/ziga.turk/project-europe-2030 (Accessed 20 May 2016).

Turk, Z. (2015) 'Case Study 3: net neutrality legislation – the case of Slovenia', Annex: pp. 23–31, in C. Marsden *et al.*, *Deliverable 4.3: Final Report*, Internet Science EINS Project FP7-288021, available at www.internet-science.eu/publication/1149 (Accessed 20 May 2016).

Tushnet, Mark (1998) 'Everything old is new again: early reflections on the New Chicago School', *Wisconsin Law Review*, 2, pp. 579–590.

Tversky, Amos and Daniel Kahneman (1974) 'Judgment under uncertainty: heuristics and biases', *Science*, 185:4157, pp. 1124–1231.

Ungerer, Herbert (2000) 'Access issues under EU regulation & anti-trust law – telecoms & internet markets', Research Paper, Weatherhead Center for International Affairs, Harvard University.

Ungerer, Herbert (2013) 'Back to the roots: the 1987 telecom green paper 25 years after – has European telecom liberalization fulfilled its promise for Europe in the internet age?', *info*, 15:2, pp. 14–24.

Universities Superannuation Scheme (2014) Actuarial Valuation March 2014, available at www.uss.co.uk/how-uss-is-run/running-uss/funding-uss/actuarial-valuation (Accessed 15 September 2016).

US Attorney (2006) Prosecuting Intellectual Property Crimes Manual, Criminal Copyright Infringement Issues, §II.B.2, available at www.justice.gov/sites/default/files/criminal-ccips/legacy/2015/03/26/prosecuting_ip_crimes_manual_2013.pdf (Accessed 15 September 2016).

Vaizey, Ed (2010a) The Open Internet Speech, available at image.guardian.co.uk/sys-files/Media/documents/2010/11/17/EdVaizey.pdf (Accessed 20 May 2016).

Vaizey, Ed (2010b) Hansard Column c.409W, Written Answers to Questions, 25 November, available at www.publications.parliament.uk/pa/cm201011/cmhansrd/cm101125/text/101125w0001.htm#10112536000913 (Accessed 20 May 2016).

Vaizey, Ed (2011), Hansard HC Deb, 5 April 2011, c259WH, www.publications.parliament.uk/pa/cm201011/cmhansrd/cm110405/halltext/110405h0002.htm#11040557000591 (Accessed 20 May 2016).

Vaizey, Ed (2014) CMS 249920/asg 13562/13 COM (13) 634 Commission Communication: 'On the telecommunications single market', 16 May, available at http://europeanmemoranda.cabinetoffice.gov.uk/files/2014/05/13562-13_Min_Cor_16_May_2014_Vaizey-Cash.pdf (Accessed 20 May 2016).

Vallina-Rodriguez, Narseo with Srikanth Sundaresan, Christian Kreibich and Vern Paxson (2015) 'Header enrichment or ISP enrichment? Emerging privacy threats in mobile networks', paper presented at HotMiddlebox'15 (co-located with ACM SIGCOMM) 21 August, available at http://conferences.sigcomm.org/sigcomm/2015/pdf/papers/hotmiddlebox/p25.pdf (Accessed 20 May 2016).

van Eijk, N.A. N. M. (2011a) 'Net neutrality and audiovisual services', IRIS Plus, 2011–5, pp. 7–19, available at www.ivir.nl/publicaties/download/535 (Accessed 15 September 2016).

van Eijk, N. A. N. M. (2011b) 'About network neutrality 1.0, 2.0, 3.0 and 4.0', *Computers & Law Magazine*, 21:6, Amsterdam Law School Research Paper No. 2012–57, Institute for Information Law Research Paper No. 2012–15, available at http://papers.ssrn.com/sol3/papers.cfm?abstract_id=2038802 (Accessed 20 May 2016).

van Eijk, Nico (2014) 'The proof of the pudding is in the eating: net neutrality in practice, the Dutch Example', paper presented to the TPRC Conference 2014, 2 August, available at: http://dx.doi.org/10.2139/ssrn.2417933 (Accessed 20 May 2016).

Van Schewick, Barbara (2010) *Internet Architecture and Innovation*, Cambridge, MA: MIT Press.

Vatiero, Massimiliano (2015) 'Dominant market position and ordoliberalism', SSRN draft available at http://dx.doi.org/10.2139/ssrn.2585167 (Accessed 20 May 2016).

Verheyen, Sabine MEP (2015) Debates Tuesday, 27 October, 2. European single market for electronic communications (debate), available at www.europarl.europa.eu/sides/getDoc.do?pubRef=-//EP//TEXT+CRE+20151027+ITEM-002+DOC+XML+V0//EN&language=en&query=INTERV&detail=2-050-000 (Accessed 20 May 2016).

Veysey, Lawrence R. (1965) *The Emergence of the American University*. Chicago: University of Chicago Press.

Vilasau, Monica (2007) 'Traffic data retention v data protection: the new European Framework', *Computer and Telecommunications Law Review*, 13:2, pp. 52–59.

Viola, Robert (2015) 'What next for EU telecom rules?', 24 July, available at https://ec.europa.eu/digital-agenda/en/blog/what-next-eu-telecom-rules (Accessed 20 May 2016).

Wagner, Adam (2015) 'The European Court of Human Rights uncovered', UK Human Rights Blog, 14 August, available at http://ukhumanrightsblog.com/2015/08/14/the-european-court-of-human-rights-uncovered/ (Accessed 20 May 2016).

Waverman, Leonard and Francesc Trillas (2002) 'Corporate control and industry structure in global communications', *Telecommunications Policy: Special Issue*, 26:5–6, pp. 219–360.

Weiser, P. (2009) 'The future of internet regulation', *UC Davis Law Review*, 43, pp. 529–590.

Weiser, Phil (2004) 'Toward a next generation regulatory strategy', *Loyola University Law Journal*, 35, pp. 41–85.

Welinder, Yana and C. Schloeder (2014) 'Chilean regulator welcomes Wikipedia Zero', Wikimedia Blog, 22 September, available at http://blog.wikimedia.org/2014/09/22/chilean-regulator-welcomes-wikipedia-zero/ (Accessed 20 May 2016).

Wendell Holmes, Oliver (1881) *The Common Law*, ABA Classics, reprinted 2009 by American Bar Association.

Werbach, Kevin (2010) 'Off the hook', *Cornell Law Review*, 95:3, pp. 535–598.

Whish, Richard and David Bailey (2015) *Competition Law*, 8th Edition. Oxford: Oxford University Press.

Wien, Mathias (2015) *High Efficiency Video Coding: Coding Tools and Specification*. Heidelberg: Springer.

WIK (2015) 'Review of the Open Internet Codes', Study for Broadband Stakeholder Group (BSG), 16 November, available at www.broadbanduk.org/wp-content/uploads/2015/11/WIK-Review-of-the-Open-Internet-Codes-November-15.pdf (Accessed 20 May 2016).

Wikipedia (2015) Language demographics of Quebec, available at https://en.wikipedia.org/wiki/Language_demographics_of_Quebec#Numbers_of_native_speakers (Accessed 20 May 2016).

Williams, Christopher (2006), 'Ofcom to regulate switching ISP: Wants to be the MAC daddy', The Register 17 August, available at www.theregister.co.uk/2006/08/17/ofcom_mac_codes/ (Accessed 7 September 2016).

Williams, Christopher (2009a) 'UK.gov backs ISPs on charging content providers, throttling P2P: Carter kicks net neutrality corpse', The Register 29 January, available at www.theregister.co.uk/2009/01/29/carter_net_neut/ (Accessed 7 September 2016).

Williams, Christopher (2009b) 'EU threatens "formal action" against UK.gov on Phorm: Brussels increases pressure over secret trials', The Register 11 February, available at www.theregister.co.uk/2009/02/11/phorm_eu_action_threat (Accessed 5 September 2016).

Williams, Christopher (2009c) 'Virgin Media sticks with Phorm: There's more than one way to target ads', The Register 22 April, available at www.theregister.co.uk/2009/04/22/virgin_media_phorm_nma/ (Accessed 5 September 2016).

Wohlers, Marcio, Moacir Giansante, Antonio Carlos and Nathalia Fodich (2014) 'Shedding light on net neutrality: towards possible solutions for the Brazilian case', Conference Paper presented to International Telecommunications Society 20th Conference, Rio, 1 December, available at www.researchgate.net/publication/274310761_Shedding_light_on_net_neutrality_towards_possible_solutions_for_the_Brazilian_case (Accessed 20 May 2016).

World Bank (2015) World DataBank: Millennium Development Goals, available at http://databank.worldbank.org/data/reports.aspx?source=millennium-development-goals (Accessed 20 May 2016).

Wray, Richard (2009) 'Vodafone 360: mobile provider launches new applications service', *Observer*, 20 September, available at www.theguardian.com/business/2009/sep/20/vodafonegroup-telecoms (Accessed 20 May 2016).

Wu, T. (2003a) 'When code isn't law', *Virginia Law Review* [online], 89, pp. 103–170, available at http://papers.ssrn.com/sol3/papers.cfm?abstract_id=413201 (Accessed 20 May 2016).

Wu, T. (2003b) 'Network neutrality, broadband discrimination', *Journal on Telecommunications and High Technology Law*, 2, pp. 141–172.

Wu, Tim (2007) 'Wireless Carterfone', *International Journal of Communication*, 1, pp. 389–426, Columbia Public Law Research Paper No. 07-154, available at: http://ssrn.com/abstract=962027 (Accessed 20 May 2016).

Yeung, Karen (2012) 'Nudge as fudge', *Modern Law Review*, 75:1, pp. 122–148.

Yoo, Christopher S. (2006) 'Network neutrality and the economics of congestion', *Georgetown Law Journal*, 94, 1847–1908.

Zelnick, Bob (2013) *The Illusion of Net Neutrality: Political Alarmism, Regulatory Creep, and the Real Threat to Internet Freedom*. Stanford, CA: Hoover Institution Press.

Zimmerman, J. (2013) 'Neelie Kroes pushing telcos' agenda to end net neutrality', La Quadrature du Net, 30 August, available at www.laquadrature.net/en/neelie-kroes-pushing-telcos-agenda-to-end-net-neutrality (Accessed 20 May 2016).

Zittrain, J. (2008) *The Future of the Internet and How to Stop It*. New Haven, CT: Yale University Press.

Zuckerberg, M. (2015) 'Free Basics protects net neutrality: to connect a billion people, India must choose facts over fiction', *Times of India*, 28 December, available at http://blogs.timesofindia.indiatimes.com/toi-edit-page/free-basics-protects-net-neutrality/ (Accessed 20 May 2016).

Index